FUNDAMENTALS OF
SOUND AND VIBRATION
SECOND EDITION

FUNDAMENTALS OF SOUND AND VIBRATION
SECOND EDITION

Edited by
Frank Fahy
*(Formerly) Institute of Sound and Vibration Research
University of Southampton, UK*

David Thompson
*Institute of Sound and Vibration Research
University of Southampton, UK*

CRC Press
Taylor & Francis Group
Boca Raton London New York

CRC Press is an imprint of the
Taylor & Francis Group, an **informa** business

A SPON PRESS BOOK

CRC Press
Taylor & Francis Group
6000 Broken Sound Parkway NW, Suite 300
Boca Raton, FL 33487-2742

First issued in paperback 2019

© 2015 by Institute for Sound and Vibration Research
CRC Press is an imprint of Taylor & Francis Group, an Informa business

No claim to original U.S. Government works

ISBN-13: 978-0-415-56210-2 (hbk)
ISBN-13: 978-0-367-87487-2 (pbk)

This book contains information obtained from authentic and highly regarded sources. Reasonable efforts have been made to publish reliable data and information, but the author and publisher cannot assume responsibility for the validity of all materials or the consequences of their use. The authors and publishers have attempted to trace the copyright holders of all material reproduced in this publication and apologise to copyright holders if permission to publish in this form has not been obtained. If any copyright material has not been acknowledged please write and let us know so we may rectify in any future reprint.

Except as permitted under U.S. Copyright Law, no part of this book may be reprinted, reproduced, transmitted, or utilized in any form by any electronic, mechanical, or other means, now known or hereafter invented, including photocopying, microfilming, and recording, or in any information storage or retrieval system, without written permission from the publishers.

For permission to photocopy or use material electronically from this work, please access www.copyright.com (http://www.copyright.com/) or contact the Copyright Clearance Center, Inc. (CCC), 222 Rosewood Drive, Danvers, MA 01923, 978-750-8400. CCC is a not-for-profit organization that provides licenses and registration for a variety of users. For organizations that have been granted a photocopy license by the CCC, a separate system of payment has been arranged.

Trademark Notice: Product or corporate names may be trademarks or registered trademarks, and are used only for identification and explanation without intent to infringe.

Library of Congress Cataloging-in-Publication Data

Fundamentals of sound and vibration / edited by Frank Fahy and David Thompson. -- Second edition.
   pages cm
Includes bibliographical references and index.
ISBN 978-0-415-56210-2
   1. Acoustical engineering. 2. Sound-waves. 3. Vibration. 4. Sound. I. Fahy, Frank. II. Thompson, David (Acoustical engineer)

TA365.F86 2015
620.2--dc23   2014040864

Visit the Taylor & Francis Web site at
http://www.taylorandfrancis.com

and the CRC Press Web site at
http://www.crcpress.com

Contents

Preface xix
Acknowledgements xxi
Contributors xxiii

1 Introduction 1

FRANK FAHY

1.1 Sound and noise 1
1.2 Vibration 2
1.3 Signal processing and analysis 3
1.4 Experimental methods and equipment 3
1.5 Need for noise and vibration control technology 4
1.6 Challenges of noise and vibration control technology 6
1.7 Can software solve the problem? 7
1.8 An exciting field 8
Reference 8

2 Fundamentals of acoustics 9

DAVID THOMPSON AND PHILIP NELSON

2.1 Introduction to sound 9
    2.1.1 Sound pressure 10
    2.1.2 Harmonic motion and frequency 11
    2.1.3 Decibel scale 13
    2.1.4 Decibel arithmetic 15
    2.1.5 Complex exponential notation 16
2.2 One-dimensional wave equation 17
    2.2.1 Propagation of acoustic disturbances 17
    2.2.2 Fluctuating fluid properties 19
    2.2.3 Equation of mass conservation 20

- 2.2.4 Equation of momentum conservation 21
- 2.2.5 Relation between pressure and density 21
- 2.2.6 Wave equation 22
- 2.2.7 Solutions of the one-dimensional wave equation 23
- 2.2.8 Sound speed and specific acoustic impedance 26
- 2.2.9 Linearity and the superposition principle 27
- 2.2.10 Sound produced by a vibrating piston at the end of the tube 29
- 2.3 Acoustic energy density and intensity 30
  - 2.3.1 Energy density 30
  - 2.3.2 Acoustic intensity 31
  - 2.3.3 Power radiated by a piston source in a semi-infinite tube 33
- 2.4 The wave equation in three dimensions 35
  - 2.4.1 Derivation of the wave equation in three dimensions 35
  - 2.4.2 Spherically symmetric solutions of the three-dimensional wave equation 37
  - 2.4.3 Point sources of spherical radiation 40
  - 2.4.4 Acoustic power output from a pulsating sphere 41
  - 2.4.5 Two-dimensional solutions of the wave equation 43
- 2.5 Multipole sources 44
  - 2.5.1 Point dipole source 44
  - 2.5.2 Point quadrupole sources 48
  - 2.5.3 Sources near a rigid surface 50
  - 2.5.4 Rayleigh integral 53
- 2.6 Enclosed sound fields 55
  - 2.6.1 One-dimensional standing waves in a closed uniform tube 55
  - 2.6.2 One-dimensional standing waves in a uniform tube with an open end 58
  - 2.6.3 Enclosed sound fields in three dimensions at low frequencies 60
- 2.7 Room acoustics 63
  - 2.7.1 Enclosed sound fields in three dimensions at high frequencies 63
  - 2.7.2 One-sided intensity 66
  - 2.7.3 Power balance for an enclosure 68
  - 2.7.4 Direct and reverberant field in an enclosure 70
  - 2.7.5 Schroeder frequency 71
- 2.8 Further reading 72

2.9 Questions 72
References 74

3 Fundamentals of vibration 77
BRIAN MACE
- 3.1 Introduction 77
  - 3.1.1 Some terminology and definitions 77
  - 3.1.2 Degrees of freedom 78
  - 3.1.3 Methods of rigid body dynamics 79
    - 3.1.3.1 Newton's second law of motion 80
    - 3.1.3.2 Work–energy 80
    - 3.1.3.3 Impulse–momentum 81
  - 3.1.4 Simple harmonic motion 81
  - 3.1.5 Complex exponential notation 82
  - 3.1.6 Frequency-response functions 83
  - 3.1.7 Stiffness and flexibility 85
- 3.2 Free vibration of an SDOF system 88
  - 3.2.1 Undamped free vibration 88
  - 3.2.2 Energy: effective mass and stiffness 92
  - 3.2.3 Viscously damped SDOF system 95
  - 3.2.4 Estimation of SDOF-system parameters from measured free vibration 97
- 3.3 Time-harmonic forced vibration of an SDOF system 98
  - 3.3.1 Estimation of SDOF-system parameters from measured FRFs 103
  - 3.3.2 Vibration isolation 104
  - 3.3.3 Other applications 106
  - 3.3.4 Damping and loss factor 107
- 3.4 Response of an SDOF system to general excitation 108
  - 3.4.1 Transient excitation: impulse response 109
  - 3.4.2 Response to general excitation: convolution 110
  - 3.4.3 Periodic excitation 111
  - 3.4.4 Response to general excitation: Fourier transform and frequency response 112
  - 3.4.5 Response to random excitation 113
- 3.5 Multiple degrees of freedom systems 116
  - 3.5.1 Undamped free vibration of 2DOF systems 116
  - 3.5.2 Matrix notation: n arbitrary 119
  - 3.5.3 General free vibration: modal superposition 120
  - 3.5.4 Orthogonality 121

    3.5.5   *Modal decomposition: uncoupling
            the equations of motion 122*
3.6 *Forced response of MDOF systems 123*
    3.6.1   *Damped MDOF systems 124*
    3.6.2   *Response to harmonic forces: FRF 125*
    3.6.3   *Reciprocity 127*
    3.6.4   *Characteristics of FRFs 127*
    3.6.5   *Vibration absorber 129*
3.7 *Continuous systems 131*
    3.7.1   *Axial vibration of a rod 131*
    3.7.2   *Bending vibration of a beam 134*
    3.7.3   *Waves in structures 138*
3.8 *Concluding remarks 139*
3.9 *Questions 140*
References 144

4 Fundamentals of signal processing — 145
JOE HAMMOND AND PAUL WHITE
4.1 *Introduction 145*
    4.1.1   *Data classification, generation and processing 146*
            4.1.1.1 *Objectives of data analysis 147*
            4.1.1.2 *Methods of data analysis 147*
4.2 *Fourier analysis of continuous time signals 147*
    4.2.1   *Preliminary: delta function 147*
    4.2.2   *Fourier series and Fourier integral 149*
            4.2.2.1 *Random signals and nonstationarity 150*
4.3 *Some results in signal and system analysis 151*
    4.3.1   *Effect of delay 151*
            4.3.1.1 *Echoes 152*
    4.3.2   *Convolution 152*
    4.3.3   *Windowing 154*
    4.3.4   *Uncertainty principle 157*
    4.3.5   *Hilbert transform 158*
    4.3.6   *Instantaneous frequency and group delay 159*
4.4 *Effects of sampling 160*
    4.4.1   *Analogue-to-digital conversion 160*
    4.4.2   *Fourier transform of a sequence 160*
            4.4.2.1 *Impulse train modulation 161*
            4.4.2.2 *z-transform 162*

- 4.4.3 Aliasing *163*
- 4.4.4 Discrete Fourier transform *164*
- 4.4.5 Fast Fourier transform *167*
  - 4.4.5.1 Radix 2 FFT *167*
- 4.4.6 Practical issues associated with analogue-to-digital conversion *169*

4.5 Random processes *170*
- 4.5.1 Introduction *170*
- 4.5.2 Probability theory *170*
- 4.5.3 Random variables and probability distributions *171*
  - 4.5.3.1 Probability distributions for discrete random variables *171*
  - 4.5.3.2 Probability distribution function *171*
  - 4.5.3.3 Continuous distributions *172*
  - 4.5.3.4 Probability-density functions *172*
  - 4.5.3.5 Joint distributions *173*
- 4.5.4 Expectations of functions of a random variable *173*
  - 4.5.4.1 Moments of a random variable *173*
  - 4.5.4.2 Normal distribution (Gaussian distribution) *174*
  - 4.5.4.3 Higher moments *175*
  - 4.5.4.4 Bivariate processes *176*
- 4.5.5 Stochastic processes *176*
  - 4.5.5.1 Ensemble *177*
  - 4.5.5.2 Probability distributions associated with a stochastic process *177*
  - 4.5.5.3 Moments of a stochastic process *178*
  - 4.5.5.4 Stationarity *179*
  - 4.5.5.5 Covariance (correlation) functions *179*
  - 4.5.5.6 Cross-covariance (cross-correlation) function *180*
  - 4.5.5.7 Summary of the properties of covariance functions *181*
  - 4.5.5.8 Ergodic processes and time averages *181*
  - 4.5.5.9 Examples *182*
- 4.5.6 Spectra *185*
  - 4.5.6.1 Comments on the power-spectral density function *187*
  - 4.5.6.2 Examples of power spectra *187*

                4.5.6.3   Cross-spectral density function 188
                4.5.6.4   Properties and associated definitions 188
    4.6   Input–output relationships and system identification 189
        4.6.1   Single input–single output systems 189
        4.6.2   Coherence function 191
                4.6.2.1   Example 191
        4.6.3   System identification 192
                4.6.3.1   Frequency-response identification 194
                4.6.3.2   Effects of feedback 195
    4.7   Random processes and estimation 195
        4.7.1   Estimator errors and accuracy 195
                4.7.1.1   Bias error 196
                4.7.1.2   Variance 197
                4.7.1.3   Mean-square error 197
                4.7.1.4   Confidence intervals 197
        4.7.2   Estimators for stochastic processes 198
        4.7.3   Autocorrelation function 198
                4.7.3.1   Definition 198
                4.7.3.2   Statistical considerations 199
        4.7.4   Power-spectral density 199
                4.7.4.1   Fourier transform of the
                          autocorrelation function 199
                4.7.4.2   Statistical properties 200
                4.7.4.3   Smoothed spectral estimators 201
                4.7.4.4   Confidence intervals 202
                4.7.4.5   Segment averaging 203
        4.7.5   Cross-spectra, coherence and transfer functions 205
                4.7.5.1   Cross-spectra 205
                4.7.5.2   Coherence function 205
                4.7.5.3   Frequency-response functions 206
    4.8   Concluding remarks 206
    4.9   Questions 207
    References 209

5   **Noise control**   213
    DAVID THOMPSON
    5.1   Introduction 213
    5.2   Sound sources: categories and characterisation 214
        5.2.1   What is a sound source? 214
        5.2.2   Measurement quantities 215

            5.2.2.1   Amplitude 215
            5.2.2.2   Frequency spectra 215
            5.2.2.3   Weighted spectra 217
    5.2.3   Categories of sound source 217
    5.2.4   Acoustic efficiency 218
    5.2.5   Operational speed 220
            5.2.5.1   Aeroacoustic sources 220
            5.2.5.2   Mechanical sources 221
            5.2.5.3   Consequence of speed dependence 222
    5.2.6   Frequency spectra 223
    5.2.7   Directivity and geometric attenuation 225
            5.2.7.1   Dependence of sound pressure
                      on distance from source
                      (geometric spreading) 225
            5.2.7.2   Far-field directivity 226
5.3 Noise source quantification 227
    5.3.1   Purpose of source quantification 227
    5.3.2   Free-field sound power measurement methods 228
    5.3.3   Reverberant-field sound power
            measurement methods 229
    5.3.4   Measurement methods based on sound intensity 231
    5.3.5   Source substitution method 231
    5.3.6   Sound pressure standards 232
5.4 Principles of noise control 232
    5.4.1   Noise-path models 232
    5.4.2   Control at source 233
            5.4.2.1   General principles 234
            5.4.2.2   Impacts 234
    5.4.3   Control of airborne and structure-borne paths 236
5.5 Sound radiation from vibrating structures 237
    5.5.1   Radiation ratio 237
    5.5.2   Radiation from simple sources 239
    5.5.3   Rayleigh integral 239
    5.5.4   Bending waves and critical frequency 240
    5.5.5   Radiation from a plate in bending:
            engineering approach 243
            5.5.5.1   High frequencies 244
            5.5.5.2   Low frequencies 245
            5.5.5.3   Acoustic short-circuiting region 245
    5.5.6   Reducing sound radiation 247

5.6 Sound transmission through partitions 250
    5.6.1 Introduction 250
    5.6.2 Sound transmission under normal incidence 251
    5.6.3 Oblique and diffuse incidence 254
        5.6.3.1 Below the critical frequency 256
        5.6.3.2 Above the critical frequency 256
        5.6.3.3 Bounded partitions 257
    5.6.4 Double-leaf partitions 258
        5.6.4.1 Theoretical performance 259
        5.6.4.2 Practical considerations 261
    5.6.5 Sound transmission measurements 262
5.7 Noise control enclosures 265
5.8 Noise barriers 266
5.9 Sound absorption 269
    5.9.1 Local reaction 269
    5.9.2 Reflection and absorption at an impedance interface 270
    5.9.3 Parameters describing porous materials 272
    5.9.4 Equations for fluid in a porous material 273
    5.9.5 Absorption due to a layer of porous material 276
    5.9.6 Practical sound absorbers 278
        5.9.6.1 Porous layer in front of air cavity 278
        5.9.6.2 Porous layer covered by thin surface layer 279
        5.9.6.3 Helmholtz resonators 280
        5.9.6.4 Panel absorbers 280
        5.9.6.5 Micro-perforated panels 281
        5.9.6.6 Duct attenuators 282
    5.9.7 Measurements of absorption 283
        5.9.7.1 Impedance-tube method 283
        5.9.7.2 Reverberation room 285
5.10 Vibration control for noise reduction 286
    5.10.1 Impedance mismatch and vibration isolation 287
        5.10.1.1 Low-frequency model of vibration isolation 287
        5.10.1.2 Mobility and impedance 288
        5.10.1.3 Input power 289
        5.10.1.4 Impedance mismatch 290
        5.10.1.5 Force and velocity excitation 292

                5.10.1.6 High-frequency model for
                         vibration isolation 292
                5.10.1.7 Practical considerations 293
        5.10.2 Damping 295
                5.10.2.1 Effects of added damping 295
                5.10.2.2 Mean-square velocity of a plate 296
                5.10.2.3 Sound radiation from near
                         field around forcing point 297
                5.10.2.4 Methods of adding damping 297
                5.10.2.5 Material properties of
                         viscoelastic materials 298
                5.10.2.6 Unconstrained and constrained
                         damping layers 299
        5.10.3 Structural modification 299
                5.10.3.1 Added mass or stiffeners 299
                5.10.3.2 Effects of plate thickness and material 301
5.11 Further reading 302
5.12 Questions 303
References 305

6 Human response to sound   311
IAN FLINDELL
6.1 Introduction 311
        6.1.1 Objective standards 311
        6.1.2 Uncertainties 311
6.2 Auditory anatomy and function 312
        6.2.1 Research methods 312
        6.2.2 External ear 314
        6.2.3 Middle ear 315
        6.2.4 Inner ear 317
        6.2.5 Auditory nerves 321
        6.2.6 Overview 321
6.3 Auditory capabilities and acoustic metrics 322
        6.3.1 Auditory thresholds 322
        6.3.2 Just-noticeable differences 324
        6.3.3 Equal-loudness contours 326
        6.3.4 Time weightings and averaging 333
        6.3.5 Time-varying frequency spectra 340
        6.3.6 Spatial hearing 342

6.4 Range of noise effects on people 343
    6.4.1 Direct health effects 343
    6.4.2 Hearing damage risk 343
    6.4.3 Causal mechanisms for other adverse health effects 345
    6.4.4 Speech interference 346
    6.4.5 Activity disturbance and personal preference 350
    6.4.6 Sleep disturbance 352
    6.4.7 Annoyance 355
References 360

7 Human responses to vibration 363
MICHAEL J. GRIFFIN

7.1 Introduction 363
7.2 Measurement of vibration 364
7.3 Whole-body vibration 365
    7.3.1 Biodynamics 366
        7.3.1.1 Transmissibility of the human body 366
        7.3.1.2 Mechanical impedance of the human body 366
        7.3.1.3 Biodynamic models 367
    7.3.2 Vibration discomfort 369
        7.3.2.1 Effects of vibration magnitude 369
        7.3.2.2 Effects of vibration frequency and direction 370
        7.3.2.3 Effects of vibration duration 372
    7.3.3 Interference with activities 373
        7.3.3.1 Effects of vibration on vision 374
        7.3.3.2 Manual control 376
        7.3.3.3 Cognitive performance 379
    7.3.4 Health effects of whole-body vibration 379
        7.3.4.1 Evaluation of whole-body vibration according to BS 6841 (1987) and ISO 2631 (1997) 379
        7.3.4.2 Assessment of whole-body vibration according to BS 6841 (1987) and ISO 2631-1 (1997) 380
        7.3.4.3 EU Machinery Safety Directive 382
        7.3.4.4 EU Physical Agents (Vibration) Directive 383

        7.3.5   Disturbance in buildings 384
        7.3.6   Seating dynamics 384
  7.4   Motion sickness 387
        7.4.1   Causes of motion sickness 387
        7.4.2   Sickness caused by oscillatory motion 388
  7.5   Hand-transmitted vibration 389
        7.5.1   Sources of hand-transmitted vibration 390
        7.5.2   Health effects of hand-transmitted vibration 392
                7.5.2.1   Vascular disorders 392
                7.5.2.2   Neurological disorders 394
                7.5.2.3   Muscular effects 395
                7.5.2.4   Articular disorders 395
                7.5.2.5   Other effects 396
        7.5.3   Standards for evaluation of hand-transmitted vibration 396
                7.5.3.1   Vibration measurement 396
                7.5.3.2   Vibration evaluation according to ISO 5349-1 (2001) 396
                7.5.3.3   Vibration assessment according to ISO 5349-1 (2001) 398
                7.5.3.4   EU Machinery Safety Directive 399
                7.5.3.5   EU Physical Agents Directive (2002) 399
        7.5.4   Preventive measures 400
  References 402

8 Measurement of audio-frequency sound in air   411
  FRANK FAHY
  8.1   Introduction 411
  8.2   Condenser microphone 412
  8.3   Preamplifiers and transducer electronic data systems (TEDS) 415
  8.4   Calibration 416
  8.5   Specialised microphone applications 417
        8.5.1   MEMS microphones 417
        8.5.2   Probe microphone 417
        8.5.3   Artificial ear 418
        8.5.4   Directional microphones 419
        8.5.5   Microphone arrays 421
  8.6   Sound level meters and their use 426
        8.6.1   Sound level meter 426

   8.6.2 Sound level meter usage 430
   8.6.3 Computer-based sound level meters 431
 8.7 Sound intensity 431
   8.7.1 Introduction 431
   8.7.2 Sound intensity measurement probes 433
     8.7.2.1 Two-microphone sound intensity probe 433
     8.7.2.2 Microflown 436
 8.8 Applications of sound intensity measurement 438
   8.8.1 Determination of source sound power 438
   8.8.2 Determination of sound transmission through partitions 438
   8.8.3 Leak detection 439
   8.8.4 Measurement of sound absorption 439
 References 439

9 Vibration testing   443
TIM WATERS

 9.1 Introduction 443
 9.2 General test set-up 445
 9.3 Sensors 447
   9.3.1 Principle of force gauges 447
   9.3.2 Principle of accelerometers 448
   9.3.3 Accelerometer mounting considerations 450
   9.3.4 Signal conditioning 452
   9.3.5 Sensitivity and calibration 453
 9.4 Using an electrodynamic shaker 455
   9.4.1 Working principle 455
   9.4.2 Structure–shaker dynamics 456
     9.4.2.1 Force generation 457
     9.4.2.2 Force transmission 457
     9.4.2.3 Force measurement 460
     9.4.2.4 Supporting the structure 460
     9.4.2.5 Supporting the shaker 461
     9.4.2.6 Connecting the shaker to the structure 462
   9.4.3 Excitation signals 462
     9.4.3.1 Harmonic 463
     9.4.3.2 Periodic 464

                9.4.3.3   Transient 465
                9.4.3.4   Random 468
        9.4.4   Case study 469
9.5   Using an instrumented hammer 475
        9.5.1   Force spectrum 475
        9.5.2   Triggering 477
        9.5.3   Windowing 478
        9.5.4   Averaging 479
        9.5.5   Case study 480
References 482

Index   483

Preface

This introductory text covers the physical, mathematical and technical foundations of the principal elements of audio-frequency sound and vibration in clear and direct terms. Acoustics, vibration and the associated signal processing are presented in a single volume at a level suitable for students who have graduated with first degrees in engineering, physics or mathematics, or practising engineers having no prior formal training in the field.

The book is based on material taught in the first semester of the master's course in sound and vibration at the internationally acclaimed Institute of Sound and Vibration Research at the University of Southampton, United Kingdom, but written as a coherent textbook. It has been extensively revised and updated from its previous edition,* with a new introductory chapter and new chapters on the measurement of sound and of vibration. Other chapters include fundamentals of acoustics, fundamentals of vibration, signal processing, noise control, human response to sound and human response to vibration; many of these have been substantially revised. The more mathematical chapters include problems for self-study. In contrast to many introductory academic textbooks, extensive reference is made to recent technical publications for the benefit of those readers who wish to extend their knowledge and understanding of acoustic and vibration technology for professional purposes.

A multiauthor textbook inevitably lacks the uniformity of style of a single-author volume. However, we have attempted to maintain coherence and consistency of presentation by cross-referencing between chapters and by complementarity of material. The matter of mathematical notation is always problematic in such a book. Rather than present separate symbol lists for each chapter, as in the first edition, symbols are defined within the text of each chapter at the first appearance. It is felt to be unnecessary to indicate that quantities are mathematically complex by means of explicit notation, but rather to make such indication

---

* The first edition has the title *Fundamentals of Noise and Vibration*.

at appropriate points in the text. In this book, complex exponential representation of harmonic time variation uses the form $e^{i\omega t}$. Readers are asked to note this usage carefully, particularly with regard to impedances, admittances, mobilities and associated ratios of complex quantities.

Frank Fahy
David Thompson
Southampton

Acknowledgements

We are greatly indebted to Tony Moore of Taylor & Francis, who has remained remarkably patient and supportive in spite of the book's unintendedly lengthy gestation period.

We wish to thank John Shelton of AcSoft for allowing us to reproduce figures from the invited paper he presented to the ISVR 50th Anniversary Symposium and Keith Holland of the ISVR for supplying a source location image. The companies Brüel & Kjaer, Svantek, gfaitech, Microflown Technologies and G.R.A.S. Sound and Vibration generously granted permission to reproduce illustrations from their publications.

Finally, we wish to thank Daniel Lurcock, who kindly produced the cover design, and our various colleagues who, as contributing authors, have spent considerable time and effort in producing this volume.

Contributors

Frank Fahy retired in 1997 from his post of professor of engineering acoustics in the Institute of Sound and Vibration Research (ISVR) of the University of Southampton in the United Kingdom. He began his career in the late 1950s as an aeronautical engineer, designing and testing dynamic transonic wind tunnel models of aircraft such as Concorde. He joined the ISVR at its inception in 1963 as the British Aircraft Corporation research fellow working on interior noise in aircraft caused by turbulent boundary-layer flow over the fuselage. Professor Fahy played a major role in the development and standardisation of the means of measuring the magnitude and direction of energy flow in sound fields (sound intensity), which have greatly improved our ability to locate, distinguish and quantify individual sources of noise in the presence of other sources, such as in factories. He has researched many other areas of sound and vibration, including acoustic fatigue of gas-cooled nuclear-reactor ductwork, low-frequency noise transmission in buildings and ships, the modelling of sound fields in large factory spaces, violin acoustics, sound propagation in liquid-filled pipes, propeller noise in aircraft and optimisation of rocket-launcher nose fairings to protect satellite structures from launch noise. He has acted as a consultant on the acoustic design of over twenty theatres and multipurpose auditoria, on the problems of noise, vibration and fatigue in petrochemical plants, and to many industrial companies on diverse vibroacoustic problems.

For outstanding contributions to the field of acoustics, Professor Fahy has been awarded the Tyndall silver and Rayleigh gold medals of the Institute of Acoustics, of which he is an honorary fellow, the Helmholtz-Medaille of the German Acoustical Society (DEGA) for outstanding lifetime achievement, and the Médaille étrangère of the French Acoustical Society (SFA) for Anglo-French collaboration.

He is the author of five textbooks and monographs on sound and vibration and coeditor of six others.

Ian Flindell is widely known across Europe as an independent acoustics consultant and as a recently retired part-time teacher and researcher at the Institute of Sound and Vibration Research. He is a recognised expert in subjective acoustics specialising in the assessment and management of environmental noise, particularly around airports, and has contributed to the development of both objective and subjective standards and regulations and methods of assessment by United Kingdom standards committees and government departments, by the European Commission, and by the World Health Organization. He has acted as an expert witness at major airport planning inquiries (e.g. Heathrow Terminal 5, Stansted) over the past 30 years and conducted many minor assessments and research projects for airport authorities over the same period.

Michael J. Griffin is professor of human factors in the Institute of Sound and Vibration Research at the University of Southampton, where he has led research on human responses to vibration for the past 40 years. His research on whole-body vibration and hand-transmitted vibration has included studies of human performance, comfort, health, motion sickness, biodynamics and the dynamics of seats and gloves. He has authored more than 650 publications on human responses to vibration, including the *Handbook of Human Vibration*. He was awarded the Bartlett Medal of the Ergonomics Society in 1984, the R.W.B. Stephens Medal of the Institute of Acoustics in 2007 and the Taylor Award for advancing the understanding and prevention of the effects of hand-arm vibration in 2011.

Joseph (Joe) Hammond graduated in aeronautical engineering in 1966 at the University of Southampton. He completed his PhD in the Institute of Sound and Vibration Research (ISVR) in 1972 while a lecturer in the Mathematics Department at Portsmouth Polytechnic. He returned to Southampton in 1978 as a lecturer in the ISVR, and was later senior lecturer, professor, deputy director and then director of the ISVR from 1992 to 2001. In 2001, he became dean of the Faculty of Engineering and Applied Science, and in 2003 became dean of the Faculty of Engineering, Science and Mathematics. He retired in July 2007 and is an emeritus professor at Southampton.

Brian Mace is professor of mechatronics in the Department of Mechanical Engineering, University of Auckland. He graduated with a master of arts (with honours) in engineering science and subsequently DPhil (1977) from the University of Oxford. He was a research fellow at the Institute of Sound and Vibration Research (ISVR) (1977–1980), lecturer in the Department of Civil and Structural Engineering, University College, Cardiff, Wales (1980–1983), and then moved to the Department of Mechanical Engineering, University of Auckland. In 2000, he returned to the ISVR as a reader, and subsequently professor of structural dynamics, and moved again to the University of Auckland in 2011. Professor Mace's general research interests concern vibration, dynamics and control. Specific interests include structural dynamics, wave-based approaches at higher frequencies, energy methods, uncertainty modelling, hybrid-wave and finite-element modelling and applications, self-tuning adaptive-passive vibration control, active control and energy harvesting.

Philip Nelson is professor of acoustics at the University of Southampton, where he served as pro vice-chancellor for research and enterprise from 2005 to 2013, as director of the University's Institute of Sound and Vibration Research from 2001 to 2005 and as director of the Rolls-Royce University Technology Centre in Gas Turbine Noise from 1999 to 2001. He also served as president of the International Commission for Acoustics from 2004 to 2007 and is currently the chief executive of the United Kingdom Engineering and Physical Sciences Research Council. He has published two books, over 120 refereed journal publications and numerous other publications in the field of acoustics and related topics in the fields of signal processing, vibrations, fluid dynamics and control systems.

David Thompson is professor of railway noise and vibration at the Institute of Sound and Vibration Research (ISVR), University of Southampton, and head of the Dynamics Group. He graduated from Cambridge in mathematics and has a PhD from the ISVR. Prior to joining the ISVR in 1996, he worked at British Rail Research in Derby, United Kingdom and Nederlandse Organisatie voor Toegepast Natuurwetenschappelijk Onderzoek (TNO) in the Netherlands. His research interests cover most aspects of railway noise and vibration, and make use of a wide variety of analytical, numerical and experimental techniques for vibration and acoustics. He has published over 100 journal papers and his book on *Railway Noise and Vibration* was published by Elsevier in 2009. He teaches noise control and architectural and building acoustics.

Tim Waters is a senior lecturer in vibration engineering. Prior to joining the Institute of Sound and Vibration Research in 1998, he worked in the Aeroelasticity Group at British Aerospace (Airbus) Ltd. and completed his PhD in finite-element model validation at the University of Bristol. His research interests, on which he has authored or coauthored more than 70 journal and conference papers, include nonlinear vibration isolators, structural health monitoring and wave propagation in structures. He has also taught on over 40 short courses for industry.

Paul White has been professor of statistical signal processing in the Institute of Sound and Vibration Research (ISVR) at the University of Southampton since 2004. He studied for his PhD in signal processing at ISVR from

1985 and was appointed as a lecturer in 1988. In 2013, he was appointed the director of ISVR. Historically, Professor White's research areas have included signal processing in a variety of application fields, including sonar, condition monitoring and biomedicine. However, the primary research focus has always been in the area of underwater acoustics, and for many years now his interests have extended to include the study of underwater acoustics *per se*. This now includes the effect of sound on marine life and the use of signal-processing techniques to study the behaviour of marine mammals.

Chapter 1

Introduction

Frank Fahy

1.1 SOUND AND NOISE

Sound is a universal, ever-present and wonderful component of our environment. Human beings and many other animals use it for communication. It informs us about our environment: unlike our eyes, our ears can detect signals arriving from any direction and from hidden sources. Our auditory system provides us with a very sensitive warning system which, unlike our visual sense, is on guard even when we are asleep. Sound is the only form of wave that can travel over long distances under water and, as such, is vital to many forms of marine life, to the mapping and investigation of the seabed, to divers and fishermen and to the navies of the world. It informs physicians about the state of our cardiovascular and pulmonary systems and allows them to examine our inner structures non-invasively. It alerts operators to potential failure or malfunction of machinery and processes under their control. In the form of music, it has powerful aesthetic, psychological and emotive effects which have not yet been fully explained (Sachs 2007). Sound can propagate in both fluid (liquids and gases) and solid media. This book concentrates principally on sound in atmospheric air; but 'structure-borne' sound, which means the propagation of structural vibration in the audio-frequency range, features in Chapter 5 on Noise Control.

Sound has all the above and many other useful and desirable functions. But, in the form of noise, sound can cause undesirable effects on the physiological and psychological health of those who are overexposed to it; it can degrade the quality of communication, both verbal and musical; it can disturb sleep; and it can be a cause of social discord. The word *noise* has a wide range of meanings. The Oxford English Dictionary definitions of 'noise' that are relevant to the subject of this book are (i) a sound, especially a loud, harsh, disagreeable, disturbing, or intrusive sound; (ii) fluctuations or disturbances (usually random or irregular) which are not part of a signal, or which interfere with or obscure a signal. The first form of noise is properly described as *acoustic noise*; but, for the sake of conciseness, it will generally be referred to simply as *noise*. The context will clearly indicate where noise assumes its alternative meaning and relates to the disturbance of signals.

It is significant that the first definition recognises the fact that acoustic noise is both a physical and a perceptual phenomenon. The physical properties of noise are straightforward to determine using suitable measurement equipment. The subjective effects are far more complicated to quantify, describe and assess; but it is vital to do so because they have an influence on so many aspects of human well-being. The fundamentals of acoustics are introduced in Chapter 2, while human responses to sound are discussed in particular in Chapter 6.

1.2 VIBRATION

The distinction between 'sound' and 'vibration' is usually made on the basis of the means by which we detect their presence. Sound is commonly taken to be an audible phenomenon, essentially occurring in fluid media. Vibration (Chapter 3) is commonly taken to refer to the relatively low-frequency disturbances of solid structures, such as bridges or vehicles, which we can sense through 'whole-body' motion. However, vibrating surfaces constitute a large proportion of sources of noise over the entire audio-frequency range of 20 Hz to 20 kHz. Audible sound is transmitted through partitions such as building walls via vibration at amplitudes as small as a few nanometres. Vibration induced by intense sound generated at lift-off can cause damage to structures such as the multimillion-dollar payloads of rocket launchers. The response of the human body to excessive vibration can cause unpleasant consequences such as motion sickness; but, more seriously, it can damage body structures. The human body is a complex assemblage of solid and fluid components which alters its physical behaviour in response to external stimuli. Consequently, human vibration presents substantial difficulties in relation to the theoretical modelling of physical response, experimental design and testing, hazard assessment and prediction, and the setting of appropriate control targets (Chapter 7).

Although vibration is commonly perceived to be undesirable, it does have practical uses, although fewer than sound. For example, poured concrete is consolidated by the introduction of a vibrating probe. The application of vibrational massage has positive therapeutic effects on people suffering from a deterioration of their bones, tendons and muscles and it is also used by NASA to minimise the adverse effects of weightlessness on the bones of astronauts. Vibration is used to speed up chemical processes. Powdered materials can be transported by vibrating belts. Vibration is used in the sieving and selection of differently sized products of various forms. Dentists use ultrasonic vibration to remove tooth plaque. Many industrial processes such as chipping, drilling and surface cleaning also employ vibrating tools.

1.3 SIGNAL PROCESSING AND ANALYSIS

For many purposes it is necessary to quantify, analyse and characterise the time-varying physical quantities that constitute sound and vibration. These purposes include the evaluation of levels; the determination of frequency spectra and time histories; the comparison with limits imposed by regulations; the assessment of the likely effects on human beings and on inanimate structures and systems; diagnostic investigations designed to identify sources of undesirable levels or system malfunctions; the evaluation and improvement of the performances of systems that are designed to reduce acoustic noise and vibration; the evaluation of audio system performance; the comparison of the performances of competing products; and the safety and performance monitoring of vehicular transport systems.

Electrical signals that are analogues of the various time-varying physical quantities associated with sound and vibration, such as pressure and acceleration, are acquired and converted to digital forms, and are processed and analysed by specialised software running on computers. The forms of processing and outputs depend on the objectives of the signal analysis. The principles and methods of signal analysis and processing, which are central to the fields of acoustics and vibration, are introduced in Chapter 4.

There is also increasing interest in applying digital signal analysis to the investigation, interpretation and monitoring of time-varying quantities of interest to the medical and bioengineering sciences. These include blood flow, respiration, blood pressure, skeletal motion, eye motion, physical stresses and muscle operation.

1.4 EXPERIMENTAL METHODS AND EQUIPMENT

Transducers of various types are used to generate electrical signals that are analogues of the physical quantities under investigation. These include pressure microphones, particle velocity microphones, hydrophones, accelerometers, laser Doppler velocimeters, force gauges, impedance heads and strain gauges.

Experimental investigations of the acoustic and vibrational behaviours of systems are performed for the same purposes as those itemised in Section 1.3 and many others. The design of an experiment is a vital step in the achievement of valid, accurate and technically profitable results. Among the various scientific aspects that must be addressed are the following:

- What is the purpose of the experiment?
- What are the expected outcomes?

- What are the physical variables (more appropriately, non-dimensional parameters) that are most likely to influence the result(s)?
- Can each of these parameters be controlled sufficiently closely to guarantee that its influence on the results is clearly distinguished from that of other parameters?
- With what precision are the results required? There is little point in exceeding the minimum required.
- Are there uncontrolled parameters that might influence the results (e.g. temperature, humidity)?
- Is the system under investigation likely to behave linearly? If not, how will you check for nonlinear behaviour?
- Are the transducers sufficiently sensitive to give an adequate signal-to-noise ratio?
- Will the transducers be operating within their linear response range and within their intended frequency range?
- How will you check repeatability?
- What do you think are the most likely sources of errors or uncertainty in the results?

Sound and vibration measurement techniques are discussed in Chapters 8 and 9.

1.5 NEED FOR NOISE AND VIBRATION CONTROL TECHNOLOGY

As long as humankind builds and operates machinery, noise will always be with us. Mechanical efficiency may improve with increases in speed and reductions in volume and weight, but such changes will almost always lead to increased noise. The scope of the application of noise and vibration technology is very broad and the need for experts in the field is great.

Among the major motivations for the development and implementation of effective controls for noise are the following. Noise is a ubiquitous and noxious component of the environment that degrades the quality of life and the health of considerable fractions of the population worldwide, especially in densely populated areas. Exposure to excessive noise severely degrades the efficiency of communication. This can have potentially disastrous consequences where it affects people involved in collaborative activities such as the operation of industrial, marine and aeronautical systems. Its potentially damaging effect on hearing function constitutes a major hazard for those in industrial occupations. The inconsiderate imposition of noise on neighbours is a major cause of discord and conflict in multi-family dwellings. The noise from domestic, building services and construction equipment, and within vehicles, has a strong influence on market competitiveness. Excessive noise made in the oceans by military and commercial operations is known to cause distress, disorientation and even physiological damage to marine creatures.

Intense noise and associated high-frequency vibration, such as that generated by flow control valves, pose a threat to the structural integrity of many forms of industrial systems including petrochemical plants. The excessive response of extremely expensive payloads to noise that is generated by the engines of rocket launchers can fatally damage sensitive on-board components and destroy the mission. Aircraft structures are widely subject to acoustic fatigue damage, which has to be carefully monitored and controlled.

Vibration applied to the human body can have both physiological and psychological effects that are at best unpleasant and at worst hazardous to health. Improvements are needed in the designs of road, rail and marine vehicles, particularly in regard to seat design and construction. Hand–arm vibration that is caused by the use of vibrating handheld tools such as pneumatic jackhammers, petrol-driven chainsaws and electrical grinders constitutes a major industrial health hazard. Large-amplitude vibrational motion such as that experienced by the drivers of earth-moving equipment may cause serious long-term damage to skeleton and muscles. Vehicle motion is a common cause of travel sickness. Human responses to vibration are discussed in Chapter 7.

Ground-borne vibration that is generated by heavy road vehicles and trains (both over- and underground) can propagate over long distances and cause acoustic disturbance in dwellings, offices, libraries and auditoria for the performing arts. The theoretical modelling of this propagation is very challenging because of the inhomogeneous constitution of the ground and the complexity of the wave behaviour. The prediction of the disturbance that is likely to be caused in existing buildings by new activities and the identification of the sources of extant disturbances are both still subject to considerable uncertainty, which can be reduced by the development of improved models. Even more problematic is the identification of the vibrational source(s) of damage to buildings in cases of litigation.

Clearly, to achieve the technological aim of reducing noise and vibration and their undesirable effects, the practitioner requires a thorough understanding of the fundamentals of sound and vibration. Consequently, the book contains chapters that concentrate on the fundamental theoretical and analytical aspects of acoustics (Chapter 2) and vibration (Chapter 3). However, the practicability and effectiveness of control methods depend very much on the availability and durability of materials and the availability of space for the installation of control devices and materials, together with economic, safety and regulatory considerations. These aspects are too broad and complex to be given detailed coverage in this introductory volume, but suggestions for further relevant reading are provided.

1.6 CHALLENGES OF NOISE AND VIBRATION CONTROL TECHNOLOGY

A prerequisite for making proposals for noise control design and implementation is the setting of appropriate targets. Very often, those targets will be determined by legislation and regulations which will be expressed in terms of various metrics, such as sound level in dB(A) and L_{eq} (see Chapter 6). However, this is not by any means always the case, or even appropriate, and noise control engineers and acoustic consultants must apply their expertise to the task of deciding on desirable and technically achievable targets. There is also considerable scope for researchers to provide the basis for the development of improved procedures for setting and expressing targets.

The most effective means of controlling noise is to suppress it *at source* by eliminating or modifying the basic mechanism of sound generation. Noise sources are very diverse. They generate sound by different physical mechanisms, their radiated sound power depends on their physical and operational parameters in different ways, and their frequency spectra and temporal characteristics also vary considerably. However, apart from sources that are designed to be sound generators (such as loudspeakers) the radiated power of most sources is a very small fraction of the mechanical, fluid-mechanical, electrical or chemical power that they absorb or generate. This fraction is typically around one-millionth.

It is this very inefficiency of many acoustic sources that makes it extremely difficult to reduce their noise without making substantial changes to their design or operation. The theoretical modelling and analysis of source behaviour and the experimental identification and diagnosis of sources require a high degree of theoretical and experimental expertise. In order to achieve significant reductions by treating the sources, this expertise must be coupled with good engineering judgement. A broad appreciation is also required of the non-acoustic factors that constrain proposals for treatment; these include cost (supply, installation and maintenance), operational disruption, process efficiency, material limitations, weight, volume, fire resistance, robustness, chemical and mechanical stability and safety.

Noise can be controlled by means other than the modification of the source, including dynamic decoupling of the source from the propagation path. These are described in Chapter 5. Although the expertise that is necessary to achieve an effective reduction of noise by these means is somewhat less demanding than for control at source, it is by no means a trivial exercise, especially given the same constraints as those mentioned earlier. The prediction of the effectiveness of any noise and vibration control design requires a thorough understanding of the dynamic behaviour of the materials used in the control treatment. The complexities of sound propagation associated with atmospheric and topographical conditions also need to be considered. It is particularly important to be able to assess the reliability of various software programs for noise control prediction.

The control of the audio-frequency vibration of structures that radiate excessive noise is by no means straightforward. It necessitates a thorough grasp of the mechanics of vibration generation and propagation in structural elements such as panels, and of the influence of connections between elements. It also requires knowledge of the influence of structural modifications or applied materials on the structural vibration field. In addition, it requires an exhaustive understanding of the mechanics of sound radiation by vibrating surfaces and the influence of modifications to the vibration field on the sound radiation. However, this is not all that is demanded. It is common experience that theoretically desirable modifications to a structure often cannot be implemented for practical reasons such as lack of volume, unacceptable weight increase, excessive heat and operational requirements. This is where engineering expertise and ingenuity must come to the rescue.

1.7 CAN SOFTWARE SOLVE THE PROBLEM?

It may be thought that the recent rapid developments in software for noise and vibration control analysis and design have replaced the need for individual expertise. This is by no means the case. Software can help with the analysis of components of a problem, and very expensive packages can be used to model quite complicated systems, for example the sound radiation from a complete diesel engine. However, a major factor in attempting to apply noise control techniques to meet a specified target is the fundamental problem of uncertainty.

Considerable uncertainty is inherent in any computer-based analysis of physical noise and vibration control problems. A wide frequency range is involved in many such problems, which may mean that hundreds or even tens of thousands of acoustic and structural resonances must be taken into account. There are also often uncertainties associated with modelling the primary source of a problem or the propagation path of the noise, for example due to either atmospheric or topographical effects. Modelling structural vibration fields also involves considerable uncertainty because of the geometric and material complexity of the structures, nonlinearity, variations between nominally identical structures, variable joint dynamics and the difficulty of modelling damping. It is not at all clear how to combine the statistical estimates of uncertainty associated with individual components of a noise control calculation to estimate the overall uncertainty of any theoretical prediction. Finally, engineering expertise and knowledge have to be applied to select the most effective control strategy that is compatible with the conditions and constraints that apply.

Despite the advances in theoretical modelling and the associated software referred to earlier, there is still a need for good physical validation tests and personnel who are capable of designing, realising, operating, critically analysing and reporting the results of such tests.

1.8 AN EXCITING FIELD

We hope that this brief summary of the importance of sound and vibration has shown that this is an exciting field. Sound and vibration are major components of our environment that can have profound effects on our health and that of the engineering systems that we depend on. As mentioned earlier, they have a vast range of technological applications. The field of acoustics opens up many very worthwhile and satisfying careers, some of which will be of direct benefit to human health and others which will lead to improvements in the quality of life and comfort of many people.

REFERENCE

Sachs, O.W. (2007) *Musicophilia: Tales of Music and the Brain*, Alfred A. Knopf: New York.

Chapter 2

Fundamentals of acoustics*

David Thompson and Philip Nelson

2.1 INTRODUCTION TO SOUND

Acoustics is the science and study of sound. In this book, we concentrate on sound in air at audible frequencies. The field of acoustics extends to include many related topics such as environmental acoustics, musical acoustics, architectural acoustics, electro-acoustics, underwater acoustics and ultrasonics. This chapter introduces the physical processes involved in sound generation and propagation which form the basis for many of these related fields.

Sound is a wave phenomenon by which energy is transmitted through a medium via vibration of the medium and pressure (or stress) fluctuations within it. Before turning to the study of wave phenomena, however, it is helpful to consider the nature of sound signals arriving at a listener location, irrespective of the source of that sound. The acoustic quantity which is sensed by the human auditory system is the variation in air pressure, so we will concentrate on this quantity here.

A sound signal can be described in terms of its frequency components and associated amplitudes, as will be explained further in the following sections. We are familiar with the concepts from music, where the 'loudness' corresponds to the amplitude of the sound and the 'pitch' of a note corresponds to its frequency: at 'concert pitch', an A (the one above middle C) is tuned to a frequency of 440 Hz (cycles per second). Thus a piano string, for example, producing such a note will oscillate 440 times per second, although not necessarily with a sinusoidal motion. For audible sound in air, the range of frequencies and amplitudes that should be considered is defined to a large extent by the range of human hearing (see also Chapter 6). Thus the frequency range of the auditory system extends approximately from 20 to 20,000 Hz.

---

* Large parts of this chapter are based on material first published in Nelson, P.A. and Elliott, S.J., *Active Control of Sound*, chapter 1, Academic Press, London, 1992, copyright © Elsevier (1992), used by permission.

2.1.1 Sound pressure

Atmospheric air is subject to a hydrostatic pressure p_0 caused by the weight of the air above it. At sea level, this pressure has a nominal value of 1.013×10^5 Pa (Pascal or N m^{-2}), although it can vary typically by as much as ±5% due to meteorological conditions. The atmospheric pressure also decreases with increasing altitude.

Sound consists of very small, rapid fluctuations in pressure relative to the atmospheric pressure, as illustrated by Figure 2.1. This shows a random variation in pressure, varying by less than ±1 Pa, superimposed on the atmospheric pressure, for simplicity shown as 10^5 Pa. The acoustic pressure p is the fluctuating part of this signal, which, by definition, has a mean value of zero.

To quantify the magnitude of this pressure fluctuation, the mean value is clearly of no use. For a harmonic (single-frequency) signal, it is possible to define the amplitude, but for multi-frequency signals or random signals this is not well defined. It is, therefore, convenient to use the *mean-square* value of the fluctuation

$$\overline{p^2} = \lim_{T \to \infty} \frac{1}{T} \int_{t_1}^{t_1+T} p^2(t) dt \qquad (2.1)$$

in which t_1 is an arbitrary time. For random fluctuations with statistical properties that remain stationary over time, T should, in principle, extend to infinity. For this example, the squared pressure and its average are shown in Figure 2.2. Since this measure of magnitude has dimensions of pressure squared, we often work with the *root-mean-square* (r.m.s.) value associated with the fluctuation, which we will denote by p_{rms}, and which is given by the square root of the mean-square pressure, $p_{rms} = \sqrt{\overline{p^2}}$.

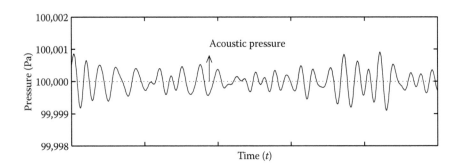

Figure 2.1 Illustration of pressure fluctuations relative to atmospheric pressure.

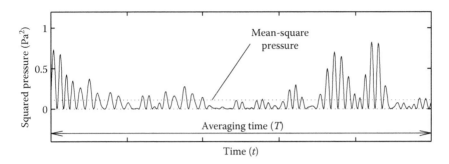

Figure 2.2 Squared acoustic pressure from Figure 2.1 and corresponding mean-square value.

In addition to pressure, an acoustic disturbance involves time-dependent variations in density and fluid motion. Their relationships will be considered in Section 2.2.

2.1.2 Harmonic motion and frequency

Acoustic pressure fluctuations are rapid: as we have seen, between 20 and 20,000 oscillations per second for audible sound. For a pressure fluctuation that repeats sinusoidally with a period T, the rapidity is described by the inverse of this period, known as its *frequency*. In acoustics, we often need to work with time-harmonic signals: signals consisting of sinusoidal variations (sine or cosine functions) at a single frequency, also known as pure tones. Although such sound signals may not occur often in practice, they are very useful. The equations of motion of linear acoustics can be readily solved for harmonic signals. Thus, harmonic signals are a basic building block for acoustic analysis. According to Fourier's theorem (see also Chapter 4), any arbitrary signal can be decomposed into an infinite number of sinusoidal components. The distributions of the amplitudes and phases of these sinusoids as functions of frequency are termed the amplitude and phase spectra. The human ear (see Chapter 6) responds to sound at different frequencies using different receptors, and thus performs its own frequency analysis. Each doubling of frequency is perceived as an equal interval, known as an octave.

Consider a time-harmonic variation of pressure shown in Figure 2.3a. This can be written as

$$p(t) = a\cos(\omega t + \varphi) \qquad (2.2)$$

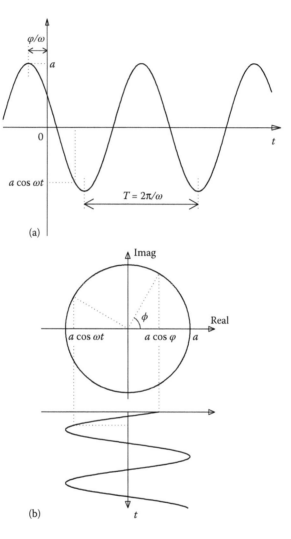

Figure 2.3 (a) Harmonic motion at circular frequency ω; (b) harmonic motion seen as a projection of circular motion.

where ω is called the *circular frequency* (in radians per second) and φ the *phase*. This signal can be seen to repeat with a period T

$$T = \frac{1}{f} = \frac{2\pi}{\omega} \qquad (2.3)$$

The inverse of T is the *frequency* f which is the number of cycles per second (or hertz, Hz). Both $\omega = 2\pi f$ and f are used for frequency throughout this book: ω is usually more convenient in mathematical derivations, whereas f is conventionally used for plotting spectra.

This harmonic motion can also be treated as a projection of a circular motion as shown in Figure 2.3b. The radius of the circle is the amplitude a, while the term $\theta = \omega t + \varphi$ is the angle traversed after time t. The phase φ is the angle (in radians) at time $t=0$, while ω can be seen to be the angular change per unit time (radians per second).

For a harmonic pressure variation, as given in Equation 2.2, the mean-square pressure can be found by integrating over a single period (or any number of periods)

$$\overline{p^2} = \frac{1}{T} \int_0^T a^2 \cos^2(\omega t + \varphi) \, dt = \frac{a^2}{2} \tag{2.4}$$

where use is made of the identity $\cos^2 \theta = \frac{1}{2}(1 + \cos 2\theta)$. The $\cos 2\theta$ term integrates to zero over the period, leaving the factor ½. Thus, for a harmonic signal, the r.m.s. value p_{rms} is $1/\sqrt{2}$ times the amplitude, although for a general signal this is not necessarily the case, as seen in Figure 2.2.

2.1.3 Decibel scale

The amplitude of acoustic pressure variations can vary typically from values as small as 10^{-5} Pa to greater than 10^2 Pa, but they remain small compared with the atmospheric pressure p_0. Due to the very large range of amplitudes that are possible, it is conventional to adopt a logarithmic measurement scale. This has the additional advantage that it mimics the human perception of loudness which exhibits an approximately logarithmic behaviour; each doubling in pressure amplitude corresponds approximately to an equal change in 'loudness' (see Chapter 6 for a discussion of loudness).

The decibel (dB) is defined in general as 10 times the logarithm to the base 10 of a ratio:

$$\text{Level} = 10 \log_{10} \left(\frac{\text{quantity}}{\text{reference value}} \right) \tag{2.5}$$

where the 'quantity' can be any relevant variable. The use of such a logarithmic scale means that a reference value has to be specified: 0 dB indicates that the quantity is equal to the reference value, while negative values are also possible, meaning that the quantity is smaller than the reference value.

Applying this to the measurement of sound pressure, the relevant decibel quantity is the sound pressure level, which, following Equation 2.5, is defined by

$$L_p = 10 \log_{10}\left(\frac{\overline{p^2}}{p_{\text{ref}}^2}\right) \quad (2.6)$$

where p_{ref} is the reference pressure, which usually takes the value of 20 μPa. This has been selected as the reference pressure used for sound pressure in air, since it corresponds to the r.m.s. pressure of a pure tone at 1 kHz which is just audible to the ear of the average young adult. This *threshold of hearing* at 1 kHz thus corresponds to a sound pressure level of 0 dB. The correct way to refer to these units is decibels *re* 20 μPa (or 2×10^{-5} Pa), but often the reference is omitted if the standard value is used. Using the properties of logarithms, an alternative definition is

$$L_p = 20 \log_{10}\left(\frac{p_{\text{rms}}}{p_{\text{ref}}}\right) \quad (2.7)$$

Some typical values of the approximate sound pressure level produced by various sources are shown in Table 2.1. Note that the word 'level' usually indicates the use of decibels, so that 'sound pressure' signifies p but 'sound-pressure level' means L_p. Decibels are also used for other quantities, such as sound power and sound intensity (see Section 2.3.3).

Table 2.1 Typical r.m.s. pressure fluctuations and the corresponding sound pressure levels

	p_{rms} (Pa)	L_p (dB re 2×10^{-5} Pa)
3 m from jet engine	200	140
Rock concert	20	120
2 m from pneumatic hammer	2	100
Vacuum cleaner	0.2	80
Conversational speech	0.02	60
Residential area at night	0.002	40
Rustling of leaves	0.0002	20
Threshold of hearing	0.00002	0

Source: Nelson, P.A. and Elliott, S.J., *Active Control of Sound*, chapter 1, Academic Press, London, 1992. With permission.

Note: The values shown are only rough approximations to the actual levels produced, which may vary considerably.

2.1.4 Decibel arithmetic

As has been seen, the use of a logarithmic scale is very convenient in reducing seven orders of magnitude to a linear range of 140 dB. However, it is not without its disadvantages, and it is important that the acoustician can correctly handle *decibel arithmetic*.

An important property of logarithms is that the operation of a multiplication is transformed into the addition of the logarithms:

$$\log_{10}(xy) = \log_{10} x + \log_{10} y \tag{2.8}$$

In acoustics, this means that operations such as the multiplication by a 'transfer function' (see e.g. Section 4.6.4) or by the gain of an amplifier can be expressed as the addition of its decibel equivalent.

It is more difficult, however, to deal with the logarithm of a sum. It is incorrect to think that two sound signals, each with a level of 80 dB, will have a combined level of 160 dB. To handle such cases, it is necessary to obtain the mean-square pressure from the sound pressure level using the inverse operation of Equation 2.6:

$$\overline{p^2} = p_{\text{ref}}^2 \times 10^{L_p/10} \tag{2.9}$$

If two sound signals can be considered to be independent ('uncorrelated', see Section 4.5.4.5), their mean-square values can be added to determine the combined mean-square pressure. Hence, combining uncorrelated signals with mean-square pressures $\overline{p_1^2}$ and $\overline{p_2^2}$, or sound pressure levels L_1 and L_2,

$$L_p = 10\log_{10}\left(\frac{\overline{p_1^2}}{p_{\text{ref}}^2} + \frac{\overline{p_2^2}}{p_{\text{ref}}^2}\right) = 10\log_{10}\left(10^{L_1/10} + 10^{L_2/10}\right) \tag{2.10}$$

Specifically, if $\overline{p_1^2} = \overline{p_2^2}$, the resulting sound pressure level will be $L_1 + 3$, that is, an increase of 3 dB over the level of a single signal (as $10 \log_{10} 2 = 3.01$).

From statistics, this relation holds for two *independent random* signals, since the variance of the sum of two independent random signals is equal to the sum of their variances (the variance of a zero-mean signal is equal to its mean-square value). It also holds for two harmonic signals with different frequencies. However, for harmonic signals with the same frequency, account has to be taken of their relative phase. We consider the particular case of two harmonic signals of circular frequency ω and equal amplitude a. As the time origin is arbitrary, it is convenient to set the phase of the first as

$\varphi_1 = 0$ and of the second as $\varphi_2 = \varphi$, their relative phase. Then, the combined sound pressure is given by

$$p(t) = a\cos(\omega t) + a\cos(\omega t + \varphi) \qquad (2.11)$$

After a little algebra, we find

$$p(t) = 2a\cos\left(\omega t + \frac{\varphi}{2}\right)\cos\left(\frac{\varphi}{2}\right) \qquad (2.12)$$

Clearly, if $\varphi = 0$, $p(t) = 2a\cos(\omega t)$ and the pressure amplitude is doubled. This corresponds to a 6 dB increase (20 $\log_{10} 2 = 6.02$), rather than the 3 dB found for uncorrelated signals. However, if $\varphi = \pi/2$, $p(t) = \sqrt{2}a\cos(\omega t + \pi/4)$ and the sound pressure level is increased by 3 dB, while if $\varphi = \pi$, $p(t) = 0$. Intriguingly, if $\varphi = 2\pi/3$, the sound pressure level is unchanged by the addition of the second signal.

2.1.5 Complex exponential notation

Returning to a time-harmonic variation of pressure, it is convenient to use complex exponential notation. At first sight, this may appear to be adding unnecessary complication, but it will turn out that the resulting mathematical derivations really are much simpler. From Euler's formula (a form of De Moivre's theorem),

$$e^{i\theta} = \cos\theta + i\sin\theta \qquad (2.13)$$

where $i = \sqrt{-1}$ is the unit imaginary number. Thus,

$$\cos\theta = \text{Re}(e^{i\theta}); \quad \sin\theta = \text{Im}(e^{i\theta}) \qquad (2.14)$$

in which Re(x) indicates the real part of x and Im(x) indicates the imaginary part of x. For the signal in Figure 2.3, it can be seen that the pressure in Equation 2.2 can be written as

$$p(t) = \text{Re}\left\{ae^{i(\omega t + \varphi)}\right\} = \text{Re}\left\{ae^{i\varphi}e^{i\omega t}\right\} \qquad (2.15)$$

or more generally as

$$p(t) = \text{Re}\left\{Ae^{i\omega t}\right\} \text{ with } A = ae^{i\varphi} \qquad (2.16)$$

In practice, in linear acoustics we can often drop the complex exponential term $e^{i\omega t}$ (treating it as implied) and work with the complex amplitude A, which includes both the magnitude a and the phase φ. It is important to remember, however, that the pressure is not the real part of A, but is the real part of the time-varying quantity in Equation 2.16.

2.2 ONE-DIMENSIONAL WAVE EQUATION

2.2.1 Propagation of acoustic disturbances

Having introduced pressure fluctuations at a point, we now turn to the propagation of sound waves. As stated in the introduction, sound is a wave phenomenon by which energy is transmitted through a medium via fluctuations in pressure and fluid vibration. To illustrate the physical process of the propagation of acoustic waves, Figure 2.4 shows a fluid (a gas or a liquid) enclosed in a semi-infinite tube with a sliding piston at one end. It is assumed that the tube is sufficiently narrow for all variations in fluid properties to depend only on the axial coordinate x (for a discussion of higher-order duct modes, see e.g. Fahy 2001). However, the influence of the walls is neglected. (In a narrow tube, viscous effects lead to energy loss in the vicinity of the walls but this is neglected here. Indeed, throughout this chapter, the fluid is assumed to be inviscid.) Such one-dimensional waves are called plane waves, as the acoustic fluctuations are uniform on any plane perpendicular to the axial direction.

We assume that the piston moves from rest with a constant velocity u. In reality, a finite velocity cannot be achieved instantaneously (Why?), but we will assume that the piston 'very rapidly' reaches the velocity u. Since the fluid inside the tube is compressible, the fluid immediately adjacent to the piston becomes squeezed and its pressure is increased because of the constraint applied by the inertia of the fluid. The compressibility of the fluid plays a central role in the propagation of acoustic disturbances. If, for example, the fluid in the tube were incompressible and did not deform at all, then all of the fluid in the tube would move at the same instant as the piston. However, if the fluid is compressible and has inertia, it takes a finite time for the disturbance caused by the motion of the piston to be transmitted to the fluid at the right-hand end of the tube. The speed at which this disturbance is transmitted is the speed of sound. The speed of sound is limited by the speed of molecular motion, which increases with temperature.

The physical process involved is illustrated in Figure 2.4. First, the fluid immediately in front of the piston becomes compressed. The compression is then transmitted down the tube. The leading edge of the disturbance, having an increased pressure over the ambient pressure, moves at the speed c_0, the speed of sound. As the compression moves down the tube, more and more of the fluid reaches the velocity u at which the piston is moving.

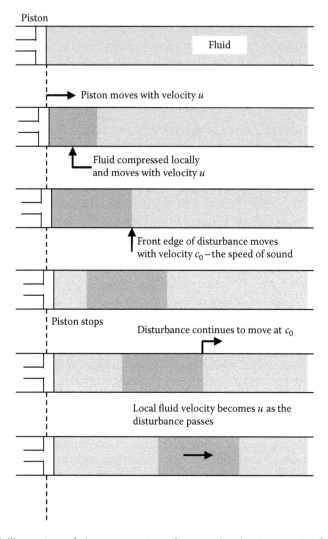

Figure 2.4 Illustration of the propagation of a sound pulse in a semi-infinite tube. The shaded area denotes a region of increased pressure in the tube. (Reprinted from Nelson, P.A. and Elliott, S.J., *Active Control of Sound*, chapter 1, Academic Press, London, 1992. With permission.)

If the piston is now suddenly brought to rest, then the fluid immediately in front of the piston will also come to rest; nevertheless, both the leading and trailing edges of the disturbance continue to propagate at the velocity c_0, and as the compression passes down the tube it imparts a velocity u to the fluid as it passes. This illustrates the important distinction between the velocity c_0 at which the compression propagates and the velocity u

which the fluid reaches as the compression passes through a given local region of fluid. The latter is known as the particle velocity, which corresponds to the passage of an acoustic wave. As we shall see later, the particle velocity generally has a magnitude which is very much less than the sound speed.

In general, a vibrating object, such as the piston shown here, will oscillate about a mean position with alternately positive and negative velocity, and this will correspond to both compression and rarefaction (reduction of pressure) in the fluid.

2.2.2 Fluctuating fluid properties

Having introduced the nature of acoustic propagation, we next introduce the physical properties of the acoustic medium and derive the wave equation. This is the fundamental equation describing the propagation of acoustic disturbances in a fluid such as air (or water). In this section, for simplicity, we restrict our consideration to a one-dimensional situation before turning in Section 2.4 to sound propagation in three dimensions.

Consider a tube of cross-sectional area S, as shown in Figure 2.5, containing air at ambient pressure p_0. As discussed in Section 2.1.1, the pressure consists of the sum of the ambient pressure and the acoustic pressure, so it can be written as $p_{tot} = p_0 + p$ where p is small compared with p_0. Similarly, as the fluid is compressed or rarefied, its density will vary from its nominal value ρ_0 and this can also be written as $\rho_{tot} = \rho_0 + \rho$, where ρ is the 'acoustic' fluctuation of the density, which is much smaller than ρ_0. Additionally, the fluid experiences motion within the tube which can be described by its particle velocity u in the x-direction. This represents the velocity of small elements of the fluid, although on a much larger scale than the molecular scale.

These three quantities, pressure, density and particle velocity, can all be used to describe acoustic fluctuations. In the following, three equations relating them will be obtained and then combined to derive the wave equation itself.

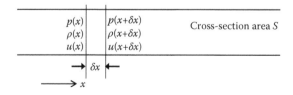

Figure 2.5 Tube of cross-sectional area S with a control volume of length δx. The acoustic variables are given by p, ρ, u at the position x.

2.2.3 Equation of mass conservation

Consider a fluid volume of infinitesimal length δx as shown in Figure 2.5. The mass of fluid within this volume is given by $\rho_{tot}\delta x S$. We are interested in the *rate* of increase of mass within the volume, which can be written as

$$\frac{\partial \rho_{tot}}{\partial t} \delta x S = \frac{\partial \rho}{\partial t} \delta x S \qquad (2.17)$$

as ρ_0 is independent of time. This must equal the flow of mass into the control volume. At the left-hand end, the rate of flow of mass into the volume is given by $\rho_{tot} u S$ evaluated at x, while at the right-hand end, the corresponding rate of flow of mass out of the volume is $\rho_{tot} u S$ evaluated at $x + \delta x$. Now, for any quantity h, to a first-order approximation

$$h(x + \delta x) \equiv h(x) + \frac{\partial h}{\partial x} \delta x \qquad (2.18)$$

Thus, the rate of mass inflow into the control volume is given by

$$\rho_{tot} u S - \left[\rho_{tot} u + \frac{\partial(\rho_{tot} u)}{\partial x} \delta x\right] S = -\frac{\partial(\rho_{tot} u)}{\partial x} \delta x S \qquad (2.19)$$

Expanding the derivative,

$$\frac{\partial(\rho_{tot} u)}{\partial x} = u \frac{\partial \rho_{tot}}{\partial x} + \rho_{tot} \frac{\partial u}{\partial x} \qquad (2.20)$$

Substituting $\rho_{tot} = \rho_0 + \rho$ and noting that $\partial \rho_0 / \partial x = 0$

$$\frac{\partial(\rho_{tot} u)}{\partial x} = u \frac{\partial \rho}{\partial x} + (\rho_0 + \rho) \frac{\partial u}{\partial x} \qquad (2.21)$$

However, for small amplitudes, the terms $u(\partial \rho/\partial x)$ and $\rho(\partial u/\partial x)$, which are the product of two small quantities, can be neglected, leaving only $\rho_0(\partial u/\partial x)$ on the right-hand side. Substituting this into Equation 2.19 and equating it with Equation 2.17 yields

$$\frac{\partial \rho}{\partial t} + \rho_0 \frac{\partial u}{\partial x} = 0 \qquad (2.22)$$

which is known as the *linearised equation of mass conservation* because it only contains terms which are linear functions of the fluctuating quantities.

2.2.4 Equation of momentum conservation

The second equation required can be found by applying the principle of momentum conservation to a fluid element; in this case, we will apply the principle to an element that moves with the fluid. The net force acting on the fluid element in the positive x-direction is given by the difference in p_{tot} at the two sides of the element multiplied by the area S,

$$p_{tot}S - \left(p_{tot} + \frac{\partial p_{tot}}{\partial x}\delta x\right)S = -\frac{\partial p_{tot}}{\partial x}\delta x\, S = -\frac{\partial p}{\partial x}\delta x\, S \qquad (2.23)$$

The momentum conservation principle states that this net applied force must be balanced by the acceleration of the fluid element multiplied by its mass, $m = \rho_{tot}\delta x S \approx \rho_0 \delta x S$ (i.e. a form of Newton's second law). The acceleration of the moving fluid element is given by

$$\frac{Du}{Dt} = u\frac{\partial u}{\partial x} + \frac{\partial u}{\partial t} \qquad (2.24)$$

where D/Dt is the total convective derivative. However, the first term, which expresses the effect of convection of the fluid, is the product of small quantities and can be neglected, so that the mass times the acceleration of the fluid element is given by $\rho_0 \delta x S(\partial u/\partial t)$. Equating this to the net force on the fluid element, Equation 2.23, results in

$$\rho_0 \frac{\partial u}{\partial t} + \frac{\partial p}{\partial x} = 0 \qquad (2.25)$$

This relationship is known as the *linearised equation of momentum conservation* or *Euler's equation*; again, the terms in the equation are only linearly dependent on the small fluctuating quantities.

2.2.5 Relation between pressure and density

A third equation is needed, in this case relating the pressure and density. For the most part, acoustic compressions are rapid and do not involve an appreciable heat transfer from the element of fluid that becomes compressed to the surrounding fluid. Under these circumstances, where the compression can be considered *isentropic* (i.e. constant entropy; see e.g. Pierce 1981, chapter 1 for a full discussion) we can assume that the density change in the fluid is determined solely by the pressure change in the fluid and is not dependent on the temperature change. This is despite the fact that the temperature of the fluid will fluctuate slightly during the course of

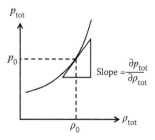

Figure 2.6 General form of the relationship between total instantaneous pressure and density.

the compression and rarefaction. In the particular case of a perfect gas (to which air is a good approximation) undergoing such an *adiabatic* compression (one in which no heat escapes from the compressed fluid), the relationship between the total pressure and total density is given by

$$p_{tot} = p_0 \left(\frac{\rho_{tot}}{\rho_0} \right)^{\gamma} \qquad (2.26)$$

where γ is the ratio of the specific heats of the gas, which for air has the value 1.4 (see e.g. Kinsler et al. 2000, appendix 9).

This nonlinear relation is shown schematically in Figure 2.6. The relation between the acoustic pressure and density can be found from the gradient of Equation 2.26 evaluated at density ρ_0. This can be written as

$$\frac{p}{\rho} = \left.\frac{\partial p_{tot}}{\partial \rho_{tot}}\right|_{\rho_{tot}=\rho_0} = \gamma p_0 \left.\frac{(\rho_{tot})^{\gamma-1}}{(\rho_0)^{\gamma}}\right|_{\rho_{tot}=\rho_0} = \frac{\gamma p_0}{\rho_0} \qquad (2.27)$$

2.2.6 Wave equation

If we now differentiate Equation 2.22 with respect to t and Equation 2.25 with respect to x, then eliminating the common term $\rho_0 \partial^2 u/\partial x \partial t$ gives

$$\frac{\partial^2 p}{\partial x^2} - \frac{\partial^2 \rho}{\partial t^2} = 0 \qquad (2.28)$$

The acoustic pressure and density fluctuations are related by Equation 2.27. Substituting this into Equation 2.28 yields the one-dimensional wave equation in terms of acoustic pressure fluctuations:

$$\frac{\partial^2 p}{\partial x^2} - \frac{1}{c_0^2}\frac{\partial^2 p}{\partial t^2} = 0 \tag{2.29}$$

where c_0 is given by

$$c_0^2 = \frac{\gamma p_0}{\rho_0} \tag{2.30}$$

As will be seen in the next section, c_0 is the speed at which acoustic disturbances travel in the medium. Note, from Equation 2.27, that

$$p = c_0^2 \rho \tag{2.31}$$

in all linear acoustic fields, so that the density also satisfies the wave equation (as c_0 is a constant for a given medium), as does the particle velocity.

It also follows from the ideal gas equation, $p_{tot} = \rho_{tot} R T_a$, where R is the specific gas constant and T_a is the absolute temperature (in Kelvin), that

$$c_0^2 = \gamma R T_a \tag{2.32}$$

Equation 2.29 relates the way in which the acoustic pressure fluctuations behave with respect to space (as a function of the coordinate distance x) and with respect to time t. In the derivation of this equation, the process of linearisation was a central assumption. We will now proceed to find solutions that satisfy the wave equation.

2.2.7 Solutions of the one-dimensional wave equation

A general form of solution to the one-dimensional wave equation was first given by d'Alembert (see Pierce 1981):

$$p(x,t) = f(t - x/c_0) + g(t + x/c_0) \tag{2.33}$$

where f and g are arbitrary functions. The fact that these are solutions can be readily demonstrated by differentiating the functions f and g. Thus, for f:

$$\frac{\partial^2}{\partial t^2}\{f(t - x/c_0)\} = f''(t - x/c_0) \tag{2.34}$$

where the dash indicates the derivative of the function, and

$$\frac{\partial^2}{\partial x^2}\{f(t-x/c_0)\} = \frac{1}{c_0^2}f''(t-x/c_0) \quad (2.35)$$

which indeed satisfy Equation 2.29. The same applies to g. These two functions also have a well-defined physical interpretation. The first function $f(t - x/c_0)$ describes a pressure disturbance which is travelling in the positive x-direction. Conversely, the function $g(t + x/c_0)$ describes a fluctuation which travels in the negative x-direction. For waves travelling in the positive x-direction, Equation 2.33 implies that, if a disturbance $f(t)$ is generated at $x=0$, this is reproduced at a position a distance x along the tube at a time x/c_0 later. The form of the disturbance is preserved exactly as it travels at a constant speed along the tube. Thus c_0 can be seen to be the speed at which the disturbance is propagated through the medium, as described in Figure 2.4. The same applies to fluctuations $g(t)$ travelling in the negative x-direction.

Of particular interest are functions f (and g) that are harmonic in time. Using complex notation (see Section 2.1.5), for a time-dependence $e^{i\omega t}$, the corresponding function f is given by

$$f(t-x/c_0) = \text{Re}\{Ae^{i\omega(t-x/c_0)}\} = \text{Re}\{Ae^{i(\omega t-kx)}\} \quad (2.36)$$

or, writing $A = ae^{i\varphi}$:

$$f(t-x/c_0) = a\cos(\omega t - kx + \varphi) \quad (2.37)$$

The quantity $k = \omega/c_0$ is known as the *wavenumber*, and can be seen to correspond to the phase change per unit distance, having units of radians per metre. It can be thought of as a spatial frequency.

Figure 2.7 shows such a harmonic disturbance varying in time and space. The period of oscillation T was given in Equation 2.3. Similarly, the spatial variation can be seen to have a wavelength λ, the distance between two consecutive wavefronts (surfaces of equal phase), given by

$$\lambda = \frac{2\pi}{k} \quad (2.38)$$

Substituting for k, this leads to

$$\lambda = \frac{c_0}{f} = c_0 T \quad (2.39)$$

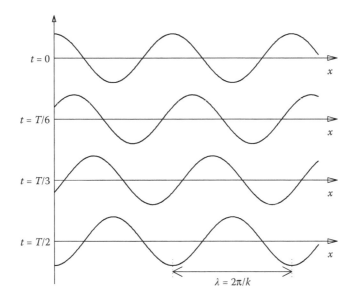

Figure 2.7 Propagation of a harmonic disturbance as a one-dimensional wave.

It is also useful in dealing with acoustical problems at a single frequency to define a spatially dependent complex pressure, which here we will denote simply by $p(x)$. The actual pressure fluctuation can be recovered from

$$p(x,t) = \text{Re}\left\{p(x)e^{i\omega t}\right\} \tag{2.40}$$

In the particular case of positive-going harmonic plane waves introduced in Equation 2.36, then the complex pressure is given by $p(x) = Ae^{-ikx}$, while for negative-going waves $p(x) = Be^{ikx}$. Substitution of Equation 2.40 into the wave equation gives

$$\left(\frac{\partial^2}{\partial x^2} - \frac{1}{c_0^2}\frac{\partial^2}{\partial t^2}\right)\text{Re}\left\{p(x)e^{i\omega t}\right\} = 0 \tag{2.41}$$

The operation of taking the real part can be taken outside the differential operators, allowing us to write this equation as

$$\text{Re}\left\{\left[\frac{d^2 p(x)}{dx^2} + \frac{\omega^2}{c_0^2}p(x)\right]e^{i\omega t}\right\} = 0 \tag{2.42}$$

where the double differentiation with respect to time (of the term $e^{i\omega t}$) yields a factor $-\omega^2$ in the second term. This equation must be satisfied for all values of time t including the cases $e^{i\omega t} = 1$ or $e^{i\omega t} = i$. To satisfy both these conditions, we require that both the real part and imaginary part of the term in square brackets are zero, that is

$$\frac{d^2 p(x)}{dx^2} + k^2 p(x) = 0 \tag{2.43}$$

where $k = \omega/c_0$. This equation is known as the one-dimensional Helmholtz equation, which must be satisfied by the complex pressure amplitude $p(x)$. It is easy to verify that $p(x) = Ae^{-ikx}$ and $p(x) = Be^{-ikx}$ are solutions of this equation. Note that the Helmholtz equation is an ordinary differential equation, whereas the wave equation is a partial differential equation; this simplification is an advantage of assuming harmonic fluctuations.

2.2.8 Sound speed and specific acoustic impedance

The relationship given in Equation 2.32 defines the dependence of the sound speed on the temperature of a gas. In air, at atmospheric pressure, it can also be written as

$$c_0 = 331\sqrt{1 + \frac{T_c}{273}} \approx 331 + 0.6 T_c \tag{2.44}$$

where T is the temperature in Celsius. At 20°C, this gives a wavespeed of $c_0 = 343$ m s^{-1}. More thorough discussions of the dependence of the sound speed on ambient fluid properties are given by, for example, Pierce (1981, chapter 1) and Lighthill (1978, chapter 1).*

For an arbitrary forward-travelling wave, f, in Equation 2.33, the linearised equation of momentum conservation, Equation 2.25, becomes

$$\rho_0 \frac{\partial u}{\partial t} = \frac{1}{c_0} f'(t - x/c_0) \tag{2.45}$$

---

* The absolute temperature of a gas is proportional to the average kinetic energy of the molecules, which is proportional to the squares of their speeds. The molecules rush around in all directions with equal probability. The average speed of molecules in any one direction is proportional to the square root of the average squared molecular speed. The acoustic disturbances are passed on by mutual collision from molecule to molecule. Hence, the speed of sound in any individual direction is proportional to the square root of the absolute temperature.

from which it is found that $\partial p/\partial t = \rho_0 c_0 \partial u/\partial t$, and hence

$$p = \rho_0 c_0 u \qquad (2.46)$$

Similarly, for a negative-travelling wave, $p = -\rho_0 c_0 u$. Thus, in any one-dimensional propagating wave, the pressure and particle velocity in the direction of propagation are in phase and related by the quantity $\rho_0 c_0$, which is known as the *characteristic specific acoustic impedance* of the medium. For a detailed discussion of the different types of acoustic impedance, see Fahy (2001). In air at atmospheric pressure and a temperature of 20°C, $\rho_0 = 1.21$ kg m^{-3} and $\rho_0 c_0$ takes a value of 415 kg m^{-2} s^{-1} (also called Rayl). In fresh water at 20°C, $c_0 = 1480$ m s^{-1}, $\rho_0 = 1000$ kg m^{-3} and the impedance $\rho_0 c_0$ is equal to 1.48×10^6 kg m^{-2} s^{-1}.

2.2.9 Linearity and the superposition principle

In our discussion of acoustic wave motion, we have emphasised that the acoustic variables are small perturbations, that is to say, p and ρ are very small compared with their ambient values p_0 and ρ_0, and u is also very small, in particular compared with the speed of sound c_0. We have seen that, as a direct consequence of this, the variables p, ρ and u are, to a very good approximation, all linearly related to one another. We have also seen that we can neglect the products of small fluctuating quantities in deriving a wave equation which governs the form of acoustic pressure fluctuations in one dimension. This equation is, therefore, linear in the pressure fluctuation $p(x,t)$ and does not contain any nonlinear terms such as $p^2(x,t)$. In addition, the (partial) differential operators in the wave equation are also 'linear' operators. That is to say, given a pressure fluctuation $p_1(x,t)$ and another pressure fluctuation $p_2(x,t)$, then

$$\frac{\partial^2}{\partial t^2}\left[p_1(x,t) + p_2(x,t)\right] = \frac{\partial^2}{\partial t^2} p_1(x,t) + \frac{\partial^2}{\partial t^2} p_2(x,t) \qquad (2.47)$$

and

$$\frac{\partial^2}{\partial x^2}\left[p_1(x,t) + p_2(x,t)\right] = \frac{\partial^2}{\partial x^2} p_1(x,t) + \frac{\partial^2}{\partial x^2} p_2(x,t) \qquad (2.48)$$

As a direct consequence of the wave equation both being linear in the fluctuation $p(x,t)$ and containing only linear differential operators, we can show that two sound-pressure fluctuations, each of which satisfies the wave equation, can also be simply added together to produce another pressure fluctuation which also satisfies the wave equation. Thus, if we have

$$\left(\frac{\partial^2}{\partial x^2} - \frac{1}{c_0^2}\frac{\partial^2}{\partial t^2}\right)p_1(x,t) = 0 \text{ and } \left(\frac{\partial^2}{\partial x^2} - \frac{1}{c_0^2}\frac{\partial^2}{\partial t^2}\right)p_2(x,t) = 0 \quad (2.49\text{a,b})$$

then it also follows that

$$\left(\frac{\partial^2}{\partial x^2} - \frac{1}{c_0^2}\frac{\partial^2}{\partial t^2}\right)[p_1(x,t) + p_2(x,t)] = 0 \quad (2.50)$$

This is known as the principle of superposition. That is, two sound fields which have pressure fluctuations that vary both as a function of the spatial coordinate x and as a function of time t can simply be added at each position x and time t to yield the net pressure fluctuation produced:

$$p(x,t) = p_1(x,t) + p_2(x,t) \quad (2.51)$$

In practical terms, this means that one pressure fluctuation will not become distorted by the presence of another; if the radio is on it will not *distort* the sound of another person's voice in the same room; the two sound-pressure fluctuations simply add up at each instant. We have already used this concept in Equation 2.11.

There are circumstances where the neglect of the nonlinear terms in the basic equations used to derive the wave equation will become significant and sound propagation is no longer governed by the linear wave equation derived here. This occurs, in particular, when dealing with pressure fluctuations of very large amplitude and with propagation over very large distances, such as thunder or explosions. Typically, nonlinear effects will produce significant distortion of an acoustic waveform in air having a sound pressure level of 120 dB when the wave propagates over a distance of a few kilometres (see Hall 1987, chapter 9; Pierce 1981, chapter 11; Morse and Ingard 1968, chapter 14). However, for most circumstances of practical interest, the sound propagation can safely be assumed to be linear and the principle of superposition can be applied.

2.2.10 Sound produced by a vibrating piston at the end of the tube

Consider a vibrating piston located at the left-hand end of a semi-infinite tube, as shown in Figure 2.8. The piston is assumed to vibrate uniformly with velocity $v(t)$. In turn, as discussed in Section 2.2.1, this causes the fluid adjacent to the piston to vibrate. Applying continuity boundary conditions at $x=0$, the fluid particle velocity $u(0,t)$ must perfectly match the velocity of the piston.

Assuming that this is the only source of sound in the tube, and that there are no reflections from further along the tube, only positive-going waves are generated. For a general positive-going wave, $p(x,t) = f(t-x/c_0)$, the particle velocity at position x is given by $u = p/\rho_0 c_0$ (see Equation 2.46). Hence, at $x=0$,

$$v(t) = u(0,t) = \frac{1}{\rho_0 c_0} f(t) \qquad (2.52)$$

Thus, the acoustic pressure in the tube is given by

$$p(x,t) = \rho_0 c_0 v(t - x/c_0) \qquad (2.53)$$

For the particular case of a harmonically vibrating piston with frequency ω and amplitude v_0, $v(t) = v_0 e^{i\omega t}$, and the complex pressure amplitude $p(x)$ at a position x along the tube is given by

$$p(x) = \rho_0 c_0 v_0 e^{-ikx} \qquad (2.54)$$

where $k = \omega/c_0$ is the wavenumber, as before, and the corresponding complex particle velocity is $u(x) = v_0 e^{-ikx}$.

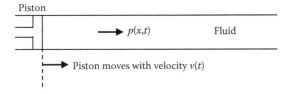

Figure 2.8 Excitation of plane waves in a semi-infinite tube by a vibrating piston.

2.3 ACOUSTIC ENERGY DENSITY AND INTENSITY

2.3.1 Energy density

As stated in the introduction, in a sound wave, energy is transmitted through the medium via vibration and pressure fluctuations. Sound energy has two components, the kinetic energy of fluid vibration and the potential energy due to fluid compression and rarefaction, which both vary with time.

Considering again the control volume shown in Figure 2.5, the instantaneous kinetic energy of the element can be written as one-half of its mass times the square of its speed

$$E_k(t) = \frac{1}{2}\rho_0 V_0 u^2(t) \tag{2.55}$$

where the volume $V_0 = S\delta x$.

An expression for the potential energy is a little more difficult to derive. Suppose that the element of fluid is compressed from its initial volume V to a volume $V - \delta V$. The work done on the fluid element to achieve this is given by $p\delta V$. As the fluid is compressed, its density increases from $(\rho_0 + \rho)$ to $(\rho_0 + \rho + \delta\rho)$. By mass conservation this gives

$$(\rho_0 + \rho + \delta\rho)(V - \delta V) = (\rho_0 + \rho)V \tag{2.56}$$

For small changes in volume and density of the fluid, products of small quantities can be neglected, so this can be written as $\rho_0 \delta V = \delta\rho V_0$. Since the compression is adiabatic, using $p = c_0^2 \rho$, or $\delta p = c_0^2 \delta\rho$, we have $\delta V = \left(V_0 / \rho_0 c_0^2\right) \delta p$. The potential energy is given by the integral of the work done, $p\delta V$, in increasing the acoustic pressure from 0 to p. The integral is given by

$$E_p(t) = \frac{V_0}{\rho_0 c_0^2} \int_0^p p(t)\,dp = \frac{1}{2} \frac{V_0}{\rho_0 c_0^2} p^2(t) \tag{2.57}$$

This is the expression which defines the potential energy of the element resulting from the increase or decrease in acoustic pressure p. The total instantaneous energy density of the fluid, that is, the energy per unit volume, $e(t) = E(t)/V_0$, is given by summing the potential and kinetic energies and dividing by the volume V_0

$$e(t) = e_k(t) + e_p(t) = \frac{1}{2}\rho_0\left[u^2(t) + \frac{p^2(t)}{(\rho_0 c_0)^2}\right] \tag{2.58}$$

This expression applies at any instant in time, and since both u and p are time dependent, the energy density also fluctuates as the pressure and velocity rise and fall due to the passage of acoustic disturbances. Note that this expression for the acoustic energy density holds for all linear acoustic fields (see e.g. Pierce 1981, chapter 1). However, for the special case of plane travelling wave disturbances, $p(t)$ and $u(t)$ are simply related by $u(t) = p(t)/\rho_0 c_0$ and, therefore, the instantaneous energy density associated with the passage of a plane travelling wave is given simply by

$$e(t) = \frac{p^2(t)}{\rho_0 c_0^2} \tag{2.59}$$

It is also useful to determine the time average of these fluctuations in energy density. From the definition of the mean-square value,

$$e = \frac{\overline{p^2}}{\rho_0 c_0^2} = \frac{p_{rms}^2}{\rho_0 c_0^2} \tag{2.60}$$

In the case of a harmonic plane wave with a complex pressure amplitude that can be written as $p(x)$, making use of Equation 2.4,

$$e = \frac{|p(x)|^2}{2\rho_0 c_0^2} \tag{2.61}$$

The time-averaged energy density due to the passage of the plane wave is therefore related simply to the mean-square amplitude of the pressure fluctuations.

2.3.2 Acoustic intensity

It is also evident that, as the waves propagate through the fluid, there is a transmission of energy from one part of the fluid to another; for example, traffic noise can make windows vibrate. The instantaneous rate at which work is done by an element of fluid on an adjacent element, per unit area, is $p(t)u(t)$, the energy flux density. The *time-averaged* rate at which energy is transferred per unit area is called the acoustic intensity. This is given by

$$I = \frac{1}{T}\int_0^T p(t)u(t)\,dt \tag{2.62}$$

and has units of watts per metre squared (W m^{-2}). In the case of plane travelling waves, then $u(t)$ and $p(t)$ are simply related by $u(t)=p(t)/\rho_0 c_0$ and, undertaking the time average in the same way as in the case of the energy density, this produces an expression for the intensity:

$$I = \frac{\overline{p^2}}{\rho_0 c_0} \quad \text{(for plane travelling waves)} \tag{2.63}$$

For harmonic plane travelling waves, this is related to the amplitude of the pressure fluctuations by

$$I = \frac{|p(x)|^2}{2\rho_0 c_0} \tag{2.64}$$

Comparing these with Equations 2.60 and 2.61, it can be seen that, for plane travelling waves, the time-averaged intensity and energy density are related by $I = c_0 e$.

More generally, the pressure and particle velocity are rarely simply related as in the case of a plane travelling wave, and Equations 2.63 and 2.64 are not then applicable. For a harmonic fluctuation associated with complex pressure $p(x)$ and complex particle velocity $u(x)$ at a given frequency, it is useful to derive an expression for the time-averaged intensity in terms of the complex numbers $p(x)$ and $u(x)$. When undertaking time averages, we have to be careful to choose only the real part of these two quantities. If $p(x)$ is written as the complex number $p_R + ip_I$, we can write the real part of the pressure fluctuation as

$$\operatorname{Re}\{p(x)e^{i\omega t}\} = \operatorname{Re}\{(p_R + ip_I)(\cos\omega t + i\sin\omega t)\} = p_R \cos\omega t - p_I \sin\omega t \tag{2.65}$$

Similarly, the real part of the velocity fluctuation is given by

$$\operatorname{Re}\{u(x)e^{i\omega t}\} = u_R \cos\omega t - u_I \sin\omega t \tag{2.66}$$

Therefore, the time-averaged acoustic intensity for a general harmonic fluctuation in pressure and velocity is given by

$$\frac{1}{T}\int_0^T p(t)u(t)\,dt = \frac{1}{T}\int_0^T (p_R \cos\omega t - p_I \sin\omega t)(u_R \cos\omega t - u_I \sin\omega t)\,dt \quad (2.67)$$

Note that the average of $\cos^2(\omega t)$ and $\sin^2(\omega t)$ over T is ½ in each case, whereas the products $\cos(\omega t)\sin(\omega t)$ will integrate to zero. The result of this integration is therefore given by

$$I = \frac{1}{2}(p_R u_R + p_I u_I) \quad (2.68)$$

This can be shown to be equal to

$$I = \frac{1}{2}\text{Re}\{p^*(x)u(x)\} = \frac{1}{2}\text{Re}\{p(x)u^*(x)\} \quad (2.69)$$

where $p^*(x)$ and $u^*(x)$ are the complex conjugates of $p(x)$ and $u(x)$ given by $p^*(x) = p_R - ip_I$ and $u^*(x) = u_R - iu_I$. This can be used to evaluate the time-averaged acoustic intensity associated with harmonic fluctuations in pressure and velocity.

2.3.3 Power radiated by a piston source in a semi-infinite tube

Considering the case of a piston vibrating in the end of a tube, shown in Figure 2.8, the pressure was given in terms of the velocity fluctuations $v(t)$ by Equation 2.52. Similarly to Equation 2.62, the instantaneous rate at which work is done by the piston on an adjacent element of fluid per unit area is $p(t)v(t)$. Thus, the time-averaged rate at which energy is transferred from the piston to the fluid per unit area is given by

$$\frac{W}{S} = \frac{1}{T}\int_0^T p(t)v(t)\,dt = \frac{1}{T}\int_0^T \rho_0 c_0 (v(t))^2\,dt = \rho_0 c_0 v_{\text{rms}}^2 \quad (2.70)$$

W is the power radiated by the piston and has units of watts (W). Considering the case of a harmonically vibrating piston with velocity amplitude v_0, this can be written as

$$W = \rho_0 c_0 S \frac{|v_0|^2}{2} \quad (2.71)$$

Similarly, it is clear from Equation 2.64 that the time-averaged intensity at any point in the tube is given by

$$I = \frac{|p(x)|^2}{2\rho_0 c_0} = \rho_0 c_0 \frac{|v_0|^2}{2} \tag{2.72}$$

as the exponential term e^{-ikx} has unit magnitude for all x.

As in the case of the mean-square sound pressures encountered in practice, the range of acoustic power outputs is similarly vast. For example, the power output of a whispered voice is as small as 10^{-9} W, whereas the power output of a jet engine can be as large as 10^4 W. To compress this range, a logarithmic measurement scale is defined, as was the case with sound pressure. We therefore define the sound power level in decibels by

$$L_W = 10 \log_{10} \frac{W}{W_{ref}} \tag{2.73}$$

where the reference sound power W_{ref} is defined as 10^{-12} W. Some typical sound power levels are shown in Table 2.2. Similarly, the sound intensity can also be expressed as a decibel quantity: the sound intensity level is

$$L_I = 10 \log_{10} \frac{I}{I_{ref}} \tag{2.74}$$

where the reference sound intensity I_{ref} is defined as 10^{-12} W m^{-2}.

For a plane propagating wave, the mean-square pressure and intensity are related by Equation 2.63. Consequently, the sound intensity level and sound pressure level are related by

$$L_I = 10 \log_{10} \left(\frac{\overline{p^2}}{\rho_0 c_0 I_{ref}} \right) = 10 \log_{10} \left(\frac{\overline{p^2}}{p_{ref}^2} \right) + 10 \log_{10} \left(\frac{p_{ref}^2}{\rho_0 c_0 I_{ref}} \right) \tag{2.75}$$

Table 2.2 Typical sound power outputs

	Power (W)	L_W (dB re 10^{-12} W)
Jet aircraft	10,000	160
Chainsaw	1	120
Normal voice	0.00001	70
Whisper	0.000000001	30

Source: Nelson, P.A. and Elliott, S.J., *Active Control of Sound*, chapter 1, Academic Press, London, 1992. With permission.

In the final form, the first term is L_p and the second term is approximately zero as a consequence of the choice of reference values I_{ref} and p_{ref}. Therefore $L_I \approx L_p$. Note, however, that this is not the case for a general sound field but is approximately true in a free field at large distances from the source.

2.4 THE WAVE EQUATION IN THREE DIMENSIONS

2.4.1 Derivation of the wave equation in three dimensions

Turning to acoustic fluctuations in a three-dimensional space, where wave propagation is not just constrained to one dimension as analysed in Section 2.2, we again use the principles of mass and momentum conservation to relate the acoustic variables pressure, density and particle velocity of the fluid. The steps are the same as previously, but we now use vector notation to account for the three-dimensional geometry.

For the three-dimensional situation depicted in Figure 2.9, any position in space can be defined by the vector **x** which has components $x_1\mathbf{i}$, $x_2\mathbf{j}$ and $x_3\mathbf{k}$, where **i**, **j** and **k** are the unit vectors in the x_1, x_2 and x_3 directions, respectively. The velocity of the fluid at any given position can be denoted by the vector **u**, with components $u_1\mathbf{i}$, $u_2\mathbf{j}$ and $u_3\mathbf{k}$ in the three coordinate directions. Consideration of the balance between the net inflow of mass into a small three-dimensional element and the net increase of mass of the element leads to a similar mass conservation equation to that derived in Section 2.2.3. Again, to the accuracy of linear approximation, the equation that results for the mass balance in a small element is given by

$$\frac{\partial \rho}{\partial t} + \rho_0 \left(\frac{\partial u_1}{\partial x_1} + \frac{\partial u_2}{\partial x_2} + \frac{\partial u_3}{\partial x_3} \right) = 0 \qquad (2.76)$$

This can be written in vector notation as

$$\frac{\partial \rho}{\partial t} + \rho_0 \nabla \cdot \mathbf{u} = 0 \qquad (2.77)$$

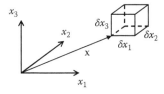

Figure 2.9 Small control volume used to derive the three-dimensional linearised continuity equation.

where $\nabla.$ is the divergence operator, defined by $\nabla = (\mathbf{i}\partial/\partial x_1 + \mathbf{j}\partial/\partial x_2 + \mathbf{k}\partial/\partial x_3)$. This equation expresses essentially the same relationship between the rate of increase of density and the rate of inflow of mass that was derived for the one-dimensional case considered in more detail in Section 2.2.

Similarly, by applying the principle of momentum conservation to a fluid element, we can relate the components of acceleration of a given element of fluid in three coordinate directions to the pressure gradients in those three directions. There are three independent momentum equations associated with the three coordinate directions, which can again be written to a linear approximation as

$$\rho_0 \frac{\partial u_1}{\partial t} + \frac{\partial p}{\partial x_1} = 0, \quad \rho_0 \frac{\partial u_2}{\partial t} + \frac{\partial p}{\partial x_2} = 0 \text{ and } \rho_0 \frac{\partial u_3}{\partial t} + \frac{\partial p}{\partial x_3} = 0 \qquad (2.78\text{a,b,c})$$

These can also be written as a single vector equation

$$\rho_0 \frac{\partial \mathbf{u}}{\partial t} + \nabla p = 0 \qquad (2.79)$$

where ∇p is the gradient of p which has the three components, $\mathbf{i}(\partial p/\partial x_1)$, $\mathbf{j}(\partial p/\partial x_2)$ and $\mathbf{k}(\partial p/\partial x_3)$. This is the three-dimensional linearised equation of momentum conservation. Following the same steps as used in arriving at Equation 2.28, if we now take the divergence of Equation 2.79 (i.e. operate on this equation with the divergence operator $\nabla.$) and if we also differentiate Equation 2.77 with respect to time, we can eliminate $\nabla.(\partial \mathbf{u}/\partial t)$ to give

$$\nabla^2 p - \frac{\partial^2 \rho}{\partial t^2} = 0 \qquad (2.80)$$

where, in rectangular Cartesian coordinates, $\nabla^2 p$ is given by

$$\nabla^2 p = \nabla.\nabla p = \left(\frac{\partial^2}{\partial x_1^2} + \frac{\partial^2}{\partial x_2^2} + \frac{\partial^2}{\partial x_3^2} \right) p \qquad (2.81)$$

Since the acoustic pressure and density are linearly related by $p = c_0^2 \rho$, we have

$$\nabla^2 p - \frac{1}{c_0^2} \frac{\partial^2 p}{\partial t^2} = 0 \qquad (2.82)$$

This is the wave equation which relates the space- and time-dependence of the pressure in a three-dimensional sound field. This wave equation must be satisfied by acoustic disturbances propagating in the medium. However, to find solutions to this equation, we must specify in more detail the form that the sound field takes. In particular, as in Section 2.2.10, we have to apply boundary conditions to find the form of the sound field.

Finally, for harmonic time-dependence, by using an argument that is identical to that used in Section 2.2.7, we can define a spatially dependent complex pressure amplitude, $p(\mathbf{x})$, that must satisfy

$$\nabla^2 p(\mathbf{x}) + k^2 p(\mathbf{x}) = 0 \tag{2.83}$$

which is the three-dimensional Helmholtz equation. We should also note that the principle of linear superposition also applies to solutions of both the wave equation (Equation 2.82) and the Helmholtz equation (Equation 2.83). This can be justified by the argument presented in Section 2.2.9.

2.4.2 Spherically symmetric solutions of the three-dimensional wave equation

There are many forms of solution of the three-dimensional wave equation, including, of course, plane waves, which may propagate in each of the coordinate directions (or any other direction). Other forms of solution can be found by rewriting the wave equation in, for example, spherical or cylindrical polar coordinates. The full solution in spherical coordinates consists of terms that describe the dependence on the radial, azimuthal and polar-angle directions. A particularly interesting form of solution can be derived by assuming that the propagation taking place is spherically symmetric, that is, dependent only on the radial coordinate r. This is the type of wave motion that would be generated by a sphere, the surface of which pulsates uniformly and generates sound waves which travel radially outwards into free space (i.e. a 'free field'). If the sound field is assumed to be spherically symmetric then the spatial operator on the pressure in the three-dimensional wave equation can be written as

$$\nabla^2 p = \frac{1}{r^2}\frac{\partial}{\partial r}\left(r^2 \frac{\partial p}{\partial r}\right) \tag{2.84}$$

and the wave equation for spherically symmetric wave propagation is

$$\frac{1}{r^2}\frac{\partial}{\partial r}\left(r^2 \frac{\partial p}{\partial r}\right) - \frac{1}{c_0^2}\frac{\partial^2 p}{\partial t^2} = 0 \qquad (2.85)$$

After some algebra, by using the laws of the derivatives of the product of two functions, it can be shown that this equation is identical to

$$\frac{\partial^2}{\partial r^2}(rp) - \frac{1}{c_0^2}\frac{\partial^2 (rp)}{\partial t^2} = 0 \qquad (2.86)$$

This has a form which is equivalent to that of the one-dimensional wave equation discussed in Section 2.2, although, in this case, it is not just the pressure p which is the variable considered, but the product of r and p. We know that the solution of this equation can be written as

$$rp(r,t) = f\left(t - \frac{r}{c_0}\right) + g\left(t + \frac{r}{c_0}\right) \qquad (2.87)$$

where f and g are arbitrary functions. Hence, the solution for the pressure can be written in the form

$$p(r,t) = \frac{f(t - r/c_0)}{r} + \frac{g(t + r/c_0)}{r} \qquad (2.88)$$

Similarly to the one-dimensional case, the first term in this equation represents pressure fluctuations which travel outwards from the origin of the spherical coordinates, while the second term represents pressure fluctuations which travel inwards towards the origin. Note that the pressure amplitude falls as the inverse of the radial distance in the sound field, unlike the plane waves considered in Section 2.2.7. For an unbounded medium, the *Sommerfeld radiation condition* requires

$$\lim_{r \to \infty}\left\{r\left(\frac{\partial p(r,t)}{\partial r} + \frac{1}{c_0}\frac{\partial p(r,t)}{\partial t}\right)\right\} = 0 \qquad (2.89)$$

which is satisfied only by the first term of Equation 2.88. Waves of the form of the function g can therefore be neglected in a free field. Again, it is useful to consider a harmonic function of $(t - r/c_0)$ of the type described by

$$p(r,t) = \text{Re}\left\{\frac{Ae^{i\omega(t-r/c_0)}}{r}\right\} = \text{Re}\left\{\frac{Ae^{i(\omega t - kr)}}{r}\right\} \qquad (2.90)$$

where A is again an arbitrary complex number that specifies both the amplitude and phase of the pressure fluctuation. We can, recover the particle velocity associated with this pressure fluctuation by applying the momentum conservation equation. For the spherically symmetric case, where u_r denotes the radial component of the particle velocity, the momentum conservation equation (Equation 2.79) reduces to

$$\rho_0 \frac{\partial u_r}{\partial t} + \frac{\partial p}{\partial r} = 0 \qquad (2.91)$$

We can define a radially dependent complex pressure amplitude $p(r)$ as

$$p(r) = \frac{A e^{-ikr}}{r} \qquad (2.92)$$

and again we recover the actual pressure fluctuation from $p(r,t) = \text{Re}\{p(r)e^{i\omega t}\}$. We can, similarly, define a complex particle velocity which has a radial component $u_r(r)$ where the actual radial particle velocity fluctuation is given by $u_r(r,t) = \text{Re}\{u_r(r)e^{i\omega t}\}$. In the same way that we used the one-dimensional equation of conservation of momentum in Section 2.2.8, it can be shown that Equation 2.91 can be used to relate the complex pressure and particle velocity. In this case, the equation reduces to

$$i\omega \rho_0 u_r(r) + \frac{\partial p(r)}{\partial r} = 0 \qquad (2.93)$$

Differentiation of Equation 2.92 with respect to r and substitution into Equation 2.93 results in the following expression for the complex radial particle velocity

$$u_r(r) = \frac{A}{i\omega \rho_0} \left[\frac{ik}{r} + \frac{1}{r^2} \right] e^{-ikr} \qquad (2.94)$$

Note that this has both real and imaginary parts, and therefore, there is one component of the particle velocity which is in phase with the acoustic pressure and another which is in quadrature (i.e. 90° out of phase). As the radial distance r becomes increasingly large, the second term in the brackets of Equation 2.94 tends to zero, and the relationship between the pressure and particle velocity tends to that which applies in a plane one-dimensional travelling wave, $u_r(r) = p(r)/\rho_0 c_0$. This makes sense physically, since, at very large distances from the source of the waves, the wavefronts (regions in phase with one another) will locally appear planar. However, when r becomes very small, the velocity will tend to become 90° out of phase with the pressure. This can be demonstrated more clearly by writing an expression for the specific acoustic impedance z as a function of the radial distance r. This is given by

$$z(r) = \frac{p(r)}{u_r(r)} = \rho_0 c_0 \left[\frac{ikr}{1+ikr} \right] \quad (2.95)$$

Thus, as kr becomes large compared with unity, the specific acoustic impedance tends to $\rho_0 c_0$, whereas when kr is very small, the specific acoustic impedance becomes imaginary. Note that the product kr is equivalent to the ratio $2\pi r/\lambda$ and, therefore, it is the ratio of the radial distance to the acoustic wavelength which is important in determining the specific acoustic impedance in spherical wave propagation.

2.4.3 Point sources of spherical radiation

The source of the spherical radiation can be assumed to be a pulsating sphere of radius a, as shown in Figure 2.10. If this has a harmonic velocity of amplitude v_0, we can now apply boundary conditions at the surface of the sphere such that the particle velocity described by Equation 2.94 must exactly match the surface velocity of the pulsating sphere. The expression for the complex pressure at $r=a$ can be written as

$$p(a) = \frac{Ae^{-ika}}{a} = \rho_0 c_0 \left[\frac{ika}{1+ika} \right] v_0 \quad (2.96)$$

From this, we can deduce the relationship between A and v_0, and it follows that the complex pressure as a function of the radial distance r from the centre of the sphere can be written as

$$p(r) = \rho_0 c_0 a \left[\frac{ika}{1+ika} \right] v_0 \frac{e^{-ik(r-a)}}{r} \quad (2.97)$$

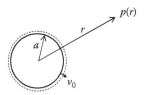

Figure 2.10 Pulsating sphere of radius a.

We define the *volume velocity*, q, as the product of the surface area of the sphere and its radial velocity. Thus, in this case, the volume velocity is a harmonic quantity with amplitude given by

$$q = 4\pi a^2 v_0 \tag{2.98}$$

Writing the pressure at a radial distance r in terms of q, we can eliminate v_0:

$$p(r) = \rho_0 c_0 \left[\frac{ik}{1+ika}\right] \frac{q e^{-ik(r-a)}}{4\pi r} \tag{2.99}$$

In the limit as $a \to 0$, the pulsating sphere can be considered to be a point source or *point monopole* source. If q is finite, $v_0 \to \infty$ as $a \to 0$; q is known as the source strength. In the limit $ka \to 0$,

$$p(r) = i\omega\rho_0 q \left(\frac{e^{-ikr}}{4\pi r}\right) \tag{2.100}$$

Note that the sound-pressure amplitude for a point source decreases in inverse proportion to the radial distance r. Equivalently, we can state that the sound pressure level decreases by 6 dB per doubling of distance from a point source. The term in brackets in Equation 2.100 is known as the free-field Green's function.

2.4.4 Acoustic power output from a pulsating sphere

We now derive the acoustic power radiated by the pulsating sphere. Consider the time-averaged acoustic intensity associated with spherical wave propagation. In a general three-dimensional field, the time-averaged intensity is a vector quantity given by

$$\mathbf{I} = \frac{1}{T}\int_0^T p(t)\mathbf{u}(t)dt \tag{2.101}$$

which depends on the magnitude, phase and direction of the particle velocity vector $\mathbf{u}$. For the spherically symmetric propagation described in Section 2.4.2, the particle velocity is directed radially and we can evaluate the intensity vector in the direction of the radial coordinate r. For harmonic fluctuations, this can be expressed in terms of the complex pressure and radial particle velocity as (see Equation 2.69)

$$I_r = \frac{1}{2}\text{Re}\{p^*(r)u_r(r)\} \tag{2.102}$$

Using this relation, Equation 2.95 shows that

$$I_r = \frac{|p(r)|^2}{2\rho_0 c_0}\text{Re}\left\{\frac{1+ikr}{ikr}\right\} = \frac{|p(r)|^2}{2\rho_0 c_0} \tag{2.103}$$

Recall that the time-averaged intensity, Equation 2.62, represents the average rate of energy flow per unit area of cross-section normal to the direction of propagation. Therefore, this shows that the intensity depends only on the modulus squared of the complex pressure but, of course, this value decreases as the square of the radial distance from the source, as is evident from Equation 2.92. The total power output of a source is given by integrating the intensity over a surface which totally encloses the source. The time-averaged sound power output is defined by

$$W = \int_S \mathbf{I}.\mathbf{n}\,dS \tag{2.104}$$

where S is a closed surface and $\mathbf{n}$ is the unit normal vector pointing outwards from S. In the case of the spherically symmetric propagation described in Section 2.4.2, the expression of the sound power output can be written as $W = I_r(4\pi r^2)$, and is therefore related to the pressure amplitude of the wave $|p| = |A|/r$ by

$$W = \frac{|A|^2}{2\rho_0 c_0 r^2}4\pi r^2 = \frac{2\pi|A|^2}{\rho_0 c_0} \tag{2.105}$$

This is clearly independent of r and, as one would expect under steady conditions of harmonic motion, there is a constant time-averaged power flowing from the spherically radiating source. Substituting for A from Equation 2.96, we find

$$W = 2\pi a^2 \rho_0 c_0 |v_0|^2 \frac{(ka)^2}{1+(ka)^2} \tag{2.106}$$

As the product ka becomes very large, this tends to the expression for a power output of a plane vibrating surface of area $4\pi a^2$ radiating plane acoustic waves; see Equation 2.71. However, as the product ka becomes very small, the expression for the power output can be approximated by

$$W = 2\pi a^2 \rho_0 c_0 |v_0|^2 (ka)^2 = \frac{\rho_0 c_0 k^2 |q|^2}{8\pi}, \quad ka \ll 1 \qquad (2.107)$$

The product ka is simply the ratio $2\pi a/\lambda$ and, therefore, the efficiency of radiation of the pulsating sphere depends on the ratio of the circumference of the sphere to the wavelength of sound at the frequency of excitation. In general, for vibrating bodies which are large compared with the wavelength, sound can be very effectively radiated by the vibration of the surface, but bodies which are very small compared with the wavelength are very inefficient radiators of acoustic power.

2.4.5 Two-dimensional solutions of the wave equation

In some situations, such as pipework, power cables or dense lines of traffic, it is interesting to find solutions to the wave equation that are effectively two-dimensional. For example, a very long cylindrical body vibrating uniformly can radiate sound in a way that is independent of the axial direction. For the particular case where the sound field is axisymmetric and independent of the axial direction, the wave equation reduces to

$$\frac{\partial^2 p}{\partial r^2} + \frac{1}{r}\frac{\partial p}{\partial r} - \frac{1}{c_0^2}\frac{\partial^2 p}{\partial t^2} = 0 \qquad (2.108)$$

For harmonic solutions at frequency ω, solutions take the form of a Bessel function $J_0(kr)$ and Neumann function $Y_0(kr)$ of order 0 (see e.g. Spiegel 1974). These can also be combined into Hankel functions (in a similar way to that in which a combination of cosine and sine functions forms a complex exponential; see Equation 2.13), to give

$$p(r) = A H_0^{(2)}(kr) + B H_0^{(1)}(kr) \qquad (2.109)$$

where:
  A and B are complex amplitudes
  $H_0^{(2)}(kr) = J_0(kr) - iY_0(kr)$ corresponds to an outgoing wave
  $H_0^{(1)}(kr) = J_0(kr) + iY_0(kr)$ corresponds to an incoming wave

In a free field, applying the Sommerfeld radiation condition (Equation 2.89) means that only the outgoing wave can exist. At large distances (i.e. when kr is large), this has the approximate form

$$p(r) = AH_0^{(2)}(kr) \approx A\left(\frac{1}{\pi kr}\right)^{1/2}(1+i)e^{-ikr} \tag{2.110}$$

Thus, the sound-pressure amplitude from a line source reduces in proportion to $r^{-1/2}$, or the sound pressure level decreases by 3 dB per doubling of distance. This is important, for example, in the design of traffic noise barriers, because a dense line of cars radiates sound in a similar manner.

2.5 MULTIPOLE SOURCES

2.5.1 Point dipole source

The monopole described in Section 2.4.3 is a fundamental simple source. Combinations of such simple sources, called multipole sources, are also fundamental building blocks for acoustic analysis. The first such multipole we consider is the point dipole. This results from two monopole sources of equal and opposite strength placed a distance d apart, which is very small compared with the acoustic wavelength. The point dipole results in the limit when the product of the strength of the component sources and their separation distance is held constant as the separation distance is allowed to vanish. There is then no net monopole source strength (volume velocity), but a net force is applied to the fluid. With reference to Figure 2.11, the complex pressure produced by this arrangement of two equal and opposite sources can be written, from Equation 2.100, as

$$p(\mathbf{x}) = i\omega\rho_0 q\left[\frac{e^{-ikr_1}}{4\pi r_1} - \frac{e^{-ikr_2}}{4\pi r_2}\right] \tag{2.111}$$

If we assume that the vector $\mathbf{x}$ (corresponding to the coordinates x_1, x_2, x_3) specifies the position at which we wish to calculate the sound field and that the vector $\mathbf{y}$ (corresponding to the coordinates y_1, y_2, y_3) specifies the position of the source, then for the particular arrangement illustrated in Figure 2.11, we can write the radial distances r_1 and r_2 from the two component monopole sources as

$$r_1 = \left[(x_1 - (y_1 + \varepsilon_1))^2 + (x_2 - y_2)^2 + (x_3 - y_3)^2\right]^{1/2} \tag{2.112}$$

Fundamentals of acoustics 45

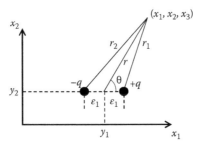

Figure 2.11 Geometrical arrangement of two sources of equal and opposite strength separated by a distance $d = 2\varepsilon_1$, centred at (y_1, y_2, y_3). The x_3 axis is suppressed for clarity.

and

$$r_2 = \left[(x_1 - (y_1 - \varepsilon_1))^2 + (x_2 - y_2)^2 + (x_3 - y_3)^2 \right]^{1/2} \quad (2.113)$$

where $\varepsilon_1 = d/2$ is a distance which is small compared with the acoustic wavelength. The terms in the square brackets of Equation 2.111 can now be regarded as functions of the type $f(y_1 \pm \varepsilon_1)$, such that we can use a Taylor series expansion of the form

$$f(y_1 + \varepsilon_1) = f(y_1) + \varepsilon_1 \frac{\partial}{\partial y_1} f(y_1) + \frac{1}{2!} \varepsilon_1^2 \frac{\partial^2}{\partial y_1^2} f(y_1) + \ldots \quad (2.114)$$

where ε_1 is small. Thus we can write

$$\frac{e^{-ikr_1}}{4\pi r_1} = \frac{e^{-ikr}}{4\pi r} + \varepsilon_1 \frac{\partial}{\partial y_1}\left[\frac{e^{-ikr}}{4\pi r}\right] + \frac{1}{2!}\varepsilon_1^2 \frac{\partial^2}{\partial y_1^2}\left[\frac{e^{-ikr}}{4\pi r}\right] + \ldots \quad (2.115)$$

and

$$\frac{e^{-ikr_2}}{4\pi r_2} = \frac{e^{-ikr}}{4\pi r} - \varepsilon_1 \frac{\partial}{\partial y_1}\left[\frac{e^{-ikr}}{4\pi r}\right] + \frac{1}{2!}\varepsilon_1^2 \frac{\partial^2}{\partial y_1^2}\left[\frac{e^{-ikr}}{4\pi r}\right] + \ldots \quad (2.116)$$

to express the terms of Equation 2.111 as functions of the distance r from the centre of the dipole pair. In the limit as $\varepsilon_1 \to 0$, we need only consider the leading-order terms in the series, and the expression for the radiated pressure reduces to

$$p(r) = i\omega\rho_0(q2\varepsilon_1)\frac{\partial}{\partial y_1}\left[\frac{e^{-ikr}}{4\pi r}\right] \quad (2.117)$$

We can substitute $d = 2\varepsilon_1$ and $\partial/\partial y_1 = (\partial r/\partial y_1)\,\partial/\partial r$. From Equation 2.112, $\partial r/\partial y_1 = (y_1 - x_1)/r = -\cos\theta$ (see Figure 2.11). In addition,

$$\frac{\partial}{\partial r}\left[\frac{e^{-ikr}}{4\pi r}\right] = \frac{-ik\,e^{-ikr}}{4\pi r}\left(1 - \frac{i}{kr}\right) \quad (2.118)$$

and, therefore, the radiated pressure can be written as

$$p(r) = -\frac{\omega\rho_0 qd\,e^{-ikr}}{4\pi r}(k\cos\theta)\left(1 + \frac{1}{ikr}\right) \quad (2.119)$$

This is the sound field produced by a *point dipole* source. The product (qd) is known as the 'strength' of the dipole. Note that in the far field, at a distance from the source that is many wavelengths, such that $kr \to \infty$, this expression reduces to $p = p_M(ikd\cos\theta)$, where p_M represents the far-field pressure radiated by a point monopole source of strength q (see Equation 2.100). Thus, the far-field pressure radiated differs from that of a monopole in two respects. First, it is reduced in amplitude by a factor kd, and second, it has a directional dependence specified by the term $\cos\theta$. This dependence of $|p|$ on θ is shown in Figure 2.12. In addition, close to the source, the term $1/ikr$ becomes important and characterises the more intense near field of the dipole source. This near field has large velocity fluctuations associated with it as fluid oscillates to and fro between the two component sources of the dipole. We can calculate the form of the velocity fluctuation by evaluating the radial component of particle velocity u_r. We know from the momentum conservation equation that $\rho_0\,\partial u_r/\partial t = -\partial p/\partial r$, and therefore for harmonic fluctuations (for which $\partial u_r/\partial t = i\omega u_r$), $u_r = (i/\omega\rho_0)\,\partial p/\partial r$. It follows from differentiation of Equation 2.119 that the radial component of the particle velocity due to the dipole can be written as

$$u_r(r,\theta) = \frac{-q\,e^{-ikr}}{4\pi r}(k^2 d\cos\theta)\left[1 - \frac{2i}{kr} - \frac{2}{(kr)^2}\right] \quad (2.120)$$

Similarly, we can calculate the tangential component of particle velocity u_θ, which is given by (see Morse and Ingard 1968):

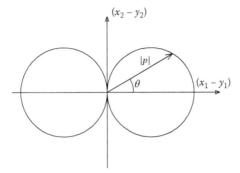

Figure 2.12 Directivity of a dipole showing the angular variation of the far-field pressure amplitude. The x_3 axis is suppressed for clarity. The directivity is axisymmetric about the x_1–y_1 axis.

$$u_\theta(r,\theta) = \frac{-q\,e^{-ikr}}{4\pi r}\left(k^2 d \sin\theta\right)\left[\frac{i}{kr} + \frac{1}{(kr)^2}\right] \qquad (2.121)$$

It can be shown that an identical sound field is produced by the application of a force in the x_1 direction to the medium. The origin of this force is associated with the rate of change of momentum of the fluid flowing between the two component monopoles (see e.g. Nelson and Elliott 1992, chapter 5). Moreover, an identical solution is obtained for the case of an oscillating sphere (see Pierce 1981). This can be shown by matching the normal velocity on the sphere, $v_0\cos\theta$, to the radial particle velocity at $r=a$, from which an equivalent dipole strength qd can be obtained. The sound power radiated by a dipole is smaller than the corresponding monopole by a factor proportional to $(kd)^2$, as a result of the partial cancellation of the sound field from the two constituent monopoles. It is, therefore, important to connect audio loudspeakers with the correct polarity, as otherwise they will form a dipole at low frequencies and the bass frequencies will be lost.

It is also important to note that, in the three-dimensional case, the sound field produced is a function of the orientation of the axis on which the two component monopole sources lie. It is, therefore, possible to derive a more general form of these results. If **d** is the vector distance between the two monopoles, the expression for the pressure field can be written in the form

$$p = i\omega\rho_0 q\mathbf{d}\cdot\nabla_y\left(\frac{e^{-ikr}}{4\pi r}\right) \qquad (2.122)$$

The product $q\mathbf{d}$ is now the vector dipole strength, which can be directly related to the force vector applied to the fluid.

2.5.2 Point quadrupole sources

The next order of multipole is a quadrupole, consisting of a pair of dipoles. There is now no net force, but a net stress is applied to the fluid. Consider the source arrangement depicted in Figure 2.13a, where a source of strength $-2q$ is located at the position (y_1, y_2, y_3) and two further sources of strength $+q$ at positions $(y_1+\varepsilon_1, y_2, y_3)$ and $(y_1-\varepsilon_1, y_2, y_3)$. This can also be considered as two dipoles located back-to-back. The pressure field radiated is

$$p = i\omega\rho_0 q \left[\frac{e^{-ikr_1}}{4\pi r_1} - \frac{2e^{-ikr}}{4\pi r} + \frac{e^{-ikr_2}}{4\pi r_2} \right] \tag{2.123}$$

where r, r_1 and r_2 are the distances indicated in Figure 2.13a. If we use the series expansions given by Equations 2.115 and 2.116, then the terms of order $O(\varepsilon_1)$ add to zero and only the terms of order $O(\varepsilon_1^2)$ remain. The expression for the pressure field that results is thus given by

$$p = i\omega\rho_0 q d^2 \frac{\partial^2}{\partial y_1^2} \left[\frac{e^{-ikr}}{4\pi r} \right] \tag{2.124}$$

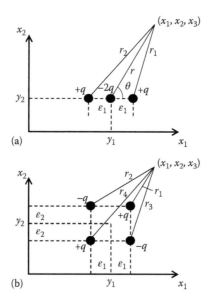

Figure 2.13 Coordinate systems used for the analysis of (a) a longitudinal quadrupole and (b) a lateral quadrupole, centred at (y_1, y_2, y_3). The x_3 axis is suppressed for clarity.

where we have now written $d=\varepsilon_1$ for the distance between the component sources. This is the sound field produced by a *longitudinal quadrupole source*. The product qd^2 is known as the strength of the quadrupole. An explicit expression for the radiated sound field can be derived by evaluating the second derivative with respect to y_1 in Equation 2.124. After some algebraic effort, the expression that results is given by

$$p = -\frac{i\omega\rho_0 q e^{-ikr}}{4\pi r}(kd)^2\left[\cos^2\theta\left\{1+\frac{3}{ikr}+\frac{3}{(ikr)^2}\right\}-\frac{1}{ikr}-\frac{1}{(ikr)^2}\right] \quad (2.125)$$

Note the form of this expression in the far field. Thus, when $kr \to \infty$, the expression takes the form $p = p_M(kd\cos\theta)^2$, where p_M is again the pressure radiated by a point monopole source of strength q. Thus, the amplitude of the far-field pressure is now a factor $(kd)^2$ smaller than that of a monopole, and the radiated power is smaller than that of a monopole by a factor proportional to $(kd)^4$. The directivity of the longitudinal quadrupole source is dependent on $\cos^2\theta$ and is shown in Figure 2.14. In addition, as kr becomes small, terms that are $O(1/(kr)^2)$ dominate the pressure in the near field of the quadrupole.

Another form of quadrupole field is that produced by the source arrangement depicted in Figure 2.13b, in which the two constituent dipoles are

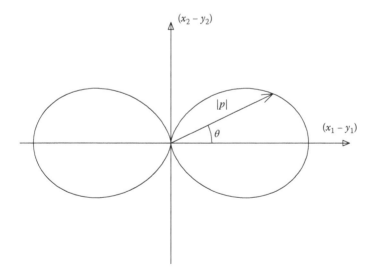

Figure 2.14 Directivity of a longitudinal quadrupole showing the angular variation of the far-field pressure amplitude. The x_3 axis is suppressed for clarity. The directivity is axisymmetric about the $x_1 - y_1$ axis.

arranged side by side. This type of arrangement is known as a lateral quadrupole. It has the pressure field given by

$$p = i\omega\rho_0 q \left[\frac{e^{-ikr_1}}{4\pi r_1} - \frac{e^{-ikr_2}}{4\pi r_2} - \frac{e^{-ikr_3}}{4\pi r_3} + \frac{e^{-ikr_4}}{4\pi r_4} \right] \quad (2.126)$$

If, as illustrated in Figure 2.13b, we regard the centre of the quadrupole source array to have coordinates (y_1, y_2, y_3), then using similar arguments, the expression for the pressure field reduces to

$$p = i\omega\rho_0 q d^2 \frac{\partial^2}{\partial y_1 \partial y_2} \left[\frac{e^{-ikr}}{4\pi r} \right] \quad (2.127)$$

where the distances between the sources in this case have been assumed to be such that $d = 2\varepsilon_1 = 2\varepsilon_2$. Evaluation of the partial derivatives with respect to y_1 and y_2 shows, again after some effort, that the pressure field can be written as

$$p = -\frac{i\omega\rho_0 q e^{-ikr}}{4\pi r}(kd)^2 \left[\cos\theta \sin\theta \left\{ 1 + \frac{3}{ikr} + \frac{3}{(ikr)^2} \right\} \right] \quad (2.128)$$

where θ is again the angle of the vector from the centre of the quadrupole **y** to the receiver point **x** relative to the x_1 axis. Note, again, that the far-field pressure due to this source array is a factor $(kd)^2$ smaller than that due to a monopole source of strength q, making the quadrupole much less efficient in radiating sound. The directivity of the far-field radiation from a lateral quadrupole of this type, that is, $\cos\theta \sin\theta = \frac{1}{2} \sin 2\theta$, is shown in Figure 2.15.

In addition to the particular quadrupole arrangements considered here, one can also conceive of a longitudinal quadrupole which has its axis in an arbitrary orientation, and similarly for lateral quadrupoles. Quadrupole sources are particularly important in aeroacoustics; Lighthill (1952) explained that the noise produced by free turbulence in a jet can be explained in terms of equivalent quadrupoles corresponding to fluid shear stresses.

2.5.3 Sources near a rigid surface

We next consider an infinite, flat, rigid surface located in the plane $x_3 = 0$. If a monopole sound source, with strength q, is located a distance h above the surface, the sound at a receiver position will consist of the direct sound and a contribution reflected in the rigid surface. The requirement that the

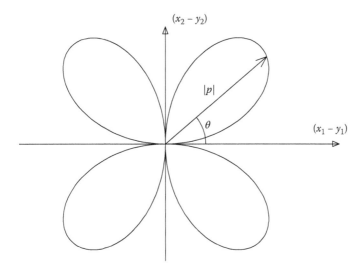

Figure 2.15 Directivity of a lateral quadrupole showing the angular variation of the far-field pressure amplitude. The x_3 axis is suppressed for clarity. The directivity is axisymmetric about the $x_1 - y_1$ axis.

normal particle velocity, u_3, is zero on the surface is satisfied by a solution of the form

$$p(\mathbf{x}) = i\omega\rho_0 q \left[\frac{e^{-ikr}}{4\pi r} + \frac{e^{-ikr'}}{4\pi r'} \right] \qquad (2.129)$$

where $r = |\mathbf{x} - \mathbf{y}|$ is the distance from the source location $\mathbf{y}$ to the receiver $\mathbf{x}$ and $r' = |\mathbf{x} + \mathbf{y}|$ is the distance from its image reflected in the plane to the receiver, as shown in Figure 2.16. The term in brackets is also known as the half-space Green's function. The first term in the brackets can be associated with the direct propagation from the source to the receiver and the second term with propagation from its image source.

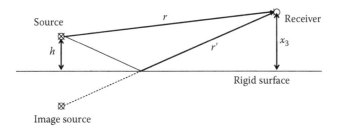

Figure 2.16 Source and its image source in a rigid reflecting plane.

At low frequencies, if the distance from the source to the ground is small compared with the wavelength, the two contributions are approximately equal and the sound-pressure amplitude is doubled by the presence of the rigid surface, that is, an increase of 6 dB. However, as frequency increases, the phase difference between the direct and reflected sound becomes significant, even if the amplitude difference between them remains small. The phase difference is $k(r - r')$, which is proportional to the frequency ω. At a certain frequency, this phase difference will be equal to π and the combined sound pressure will have a minimum. At twice this frequency, the phase difference will be equal to 2π and a maximum will occur. An example of this behaviour is shown in Figure 2.17. At high frequencies, there are many peaks and dips but, if an average is taken over a large enough frequency band, such as a one-third octave band (see Chapter 5), the combined sound pressure tends to a limit of $\sqrt{2}$ times the free-field value, that is, an increase of 3 dB.

As the source is brought closer to the surface, the source and its image coalesce and the pressure at the receiver is given by

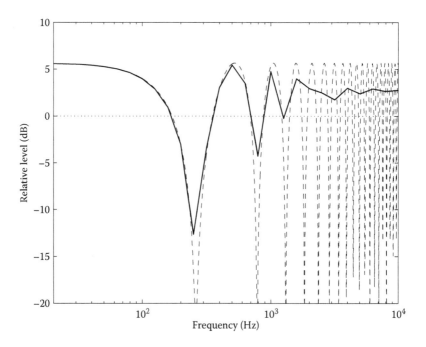

Figure 2.17 Relative sound pressure level due to a source located above a rigid reflecting plane, calculated at narrow frequency spacing (- - -); averaged into one-third octave bands (—).

$$p(\mathbf{x}) = i\omega\rho_0 q \frac{e^{-ikr}}{2\pi r} \tag{2.130}$$

This forms the basis of the Rayleigh integral, which can be used for predicting the sound radiation from a vibrating plane surface (Rayleigh 1896), described in the next section.

2.5.4 Rayleigh integral

We now consider a finite plane surface that is vibrating with normal velocity $v_n(\mathbf{y})$. Suppose that this vibrating surface is located in the plane $x_3=0$ and that the remainder of this plane is a rigid coplanar baffle that satisfies $u_3=0$. While this is not representative of all vibrating objects, it is a convenient arrangement, since an analytical solution to the wave equation can be obtained, and the radiation behaviour of such a 'baffled' surface contains many of the features of real sources.

A surface element of area dS and normal complex velocity amplitude v_n can be replaced by a point source of strength $q = v_n \mathrm{d}S$ located at the surface. The sound field produced by this source is given by Equation 2.130. The total complex pressure at the position $\mathbf{x}$ is found by integrating over all surface elements dS:

$$p(\mathbf{x}) = \int_{\mathbf{y}\in S} \frac{i\omega\rho_0 v_n(\mathbf{y}) e^{-ikr}}{2\pi r} \mathrm{d}S \tag{2.131}$$

where $r = |\mathbf{x} - \mathbf{y}|$. The integral outside the area S is also included, but since $v_n = 0$ it has no contribution to the pressure.

To illustrate this, we consider the example of a circular piston of radius a, set in an infinite baffle, as shown in Figure 2.18, in which the velocity $v_n e^{i\omega t}$ is equal over the whole surface. By symmetry, the solution is the same at any azimuthal angle, so we select $x_2 = 0$. Substituting the polar coordinates (η, φ) for the point $\mathbf{y}$ on the piston and (R, θ) for the field point $\mathbf{x}$,

$$y_1 = \eta\cos\varphi, \; y_2 = \eta\sin\varphi, \; x_1 = R\cos\theta, \; x_2 = 0, \; x_3 = R\sin\theta \tag{2.132}$$

the distance between the surface element $\mathbf{y}$ and the field point $\mathbf{x}$ is given (after some algebra) by

$$r^2 = (x_1 - y_1)^2 + (x_2 - y_2)^2 + (x_3 - y_3)^2 = R^2 - 2R\eta\sin\theta\cos\varphi + \eta^2$$
$$\tag{2.133}$$

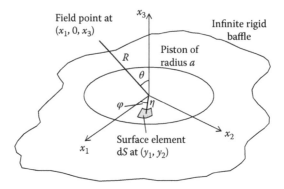

Figure 2.18 Circular piston set in a rigid coplanar baffle.

In the far field, where $R \gg a$, we can make an approximation for r, giving

$$r = R\left(1 - \frac{2\eta}{R}\sin\theta\cos\varphi + \frac{\eta^2}{R^2}\right)^{1/2} \approx R - \eta\sin\theta\cos\varphi \quad (2.134)$$

We can also make the approximation $r \approx R$ in the denominator of Equation 2.131, but not in the phase term. This allows the Rayleigh integral for the far field to be written as

$$p(\mathbf{x}) = \frac{i\omega\rho_0 v_n e^{-ikr}}{2\pi r} \int_{\mathbf{y}\in S} e^{ik\eta\sin\theta\cos\varphi}\,dS \quad (2.135)$$

The surface integral can be evaluated by separating the terms dependent on η and φ, with $dS = \eta\,d\eta\,d\varphi$. Thus,

$$\int_S e^{ik\eta\sin\theta\cos\varphi}\,dS = \int_0^a \left\{\int_0^{2\pi} e^{ik\eta\sin\theta\cos\varphi}\,d\varphi\right\}\eta\,d\eta \quad (2.136)$$

We can use the identity

$$J_0(z) = \int_0^{2\pi} e^{iz\cos\varphi}\,d\varphi \quad (2.137)$$

where J_0 is the Bessel function of order zero, so that the integral in Equation 2.136 reduces to

$$\int_S e^{ik\eta\sin\theta\cos\varphi}\,dS = 2\pi\int_0^a \eta J_0(k\eta\sin\theta)\,d\eta \tag{2.138}$$

This, in turn, can be evaluated using the identity

$$zJ_1(z) = \int_0^z yJ_0(y)\,dy \tag{2.139}$$

to give

$$\int_S e^{ik\eta\sin\theta\cos\varphi}\,dS = 2\pi a\frac{J_1(ka\sin\theta)}{k\sin\theta} \tag{2.140}$$

Thus, the far-field pressure is given by

$$p(\mathbf{x}) = \frac{i\omega\rho_0 a^2 v_0}{2r}e^{-ikr}\frac{2J_1(ka\sin\theta)}{ka\sin\theta} \tag{2.141}$$

In the limit $z \to 0$, $2J_1(z)/z \to 1$. Thus, as $ka \to 0$, the radiated field is identical to that of a simple source of strength, $q = \pi a^2 v_0$, when mounted on an infinite baffle. As ka increases, this omnidirectional behaviour gives way to a more directional beam pattern with a series of nodes occurring at the zeroes of $2J_1(z)/z$. Examples are shown in Figure 2.19.

The Rayleigh integral can also be used numerically to determine the sound radiation from plane vibrating sources of arbitrary geometry and vibration distribution, under the assumption that they are located in a coplanar infinite baffle. However, it is now more common to use numerical techniques, such as the boundary element method, to determine the sound radiation from vibrating structures (see e.g. Petyt and Jones 2004 and Wu 2000).

2.6 ENCLOSED SOUND FIELDS

2.6.1 One-dimensional standing waves in a closed uniform tube

Up to this point, we have considered only propagating acoustic disturbances. In practice, sound fields in enclosed spaces are very important, and the

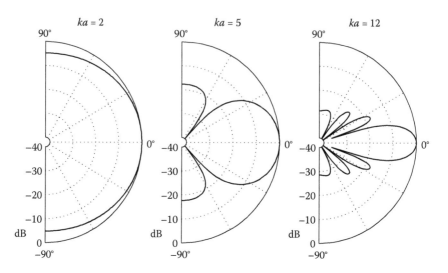

Figure 2.19 Sound pressure level distribution obtained from a circular piston set in a rigid baffle for different values of *ka*, shown relative to the value on the axis.

influence of reflections must be taken into account. We consider first a one-dimensional case, in which the fluid in a uniform tube of length L is terminated by rigid boundaries at both the left-hand end and the right-hand end. For harmonic fluctuations, one solution which will satisfy the wave equation is given by the superposition of the complex pressures associated with a positive-going and a negative-going plane travelling wave. Thus, we can write

$$p(x) = A e^{-ikx} + B e^{ikx} \tag{2.142}$$

where A and B are arbitrary complex numbers which represent the amplitude and relative phases of waves travelling in the positive and negative x-directions respectively. Use of the linearised momentum equation (Equation 2.25) enables the complex particle velocities associated with each of these component travelling plane waves to be deduced from the corresponding pressure gradient, as described in Section 2.2.8. This yields an expression for the complex particle velocity, which is given by

$$u(x) = \frac{A e^{-ikx}}{\rho_0 c_0} - \frac{B e^{ikx}}{\rho_0 c_0} \tag{2.143}$$

We can now apply the boundary conditions. We know that the particle velocity fluctuation at the left-hand end of the tube, $x = 0$, is zero. Thus, we can use the boundary condition $u(0)=0$ in Equation 2.143 to give $A=B$. Hence, from Equation 2.142, the pressure can be written as

$$p(x) = 2A\cos(kx) \tag{2.144}$$

and from Equation 2.143, the particle velocity is given by

$$u(x) = -\frac{2iA}{\rho_0 c_0}\sin(kx) \tag{2.145}$$

The two plane travelling waves in Equation 2.142, therefore, interfere to produce a standing wave with pressure and particle velocity fluctuations that are functions of distance x along the tube. Similarly, since we know that there must be zero particle velocity at the right-hand end of the tube, $x=L$, then the condition $u(L)=0$ in Equation 2.145 results in the relationship

$$A\sin(kL) = 0 \tag{2.146}$$

Equation 2.146 can only be satisfied for $A \neq 0$ at frequencies where $kL=n\pi$, with n an integer; that is

$$\omega_n = \frac{n\pi c_0}{L}, \, n = 1, 2, 3,\ldots \tag{2.147}$$

This corresponds to the relationship $L=n\lambda/2$; that is, it occurs at frequencies where an integer number of half wavelengths can fit exactly into the length of the tube. Frequencies at which this occurs are known as the natural frequencies of the fluid in the tube. The *natural frequencies* are those at which free vibration can occur in the absence of excitation. The corresponding pressure amplitudes can be written as

$$p(x) = 2A\cos\left(\frac{n\pi x}{L}\right) \tag{2.148}$$

These particular cosinusoidal functions of the pressure amplitude along the tube represent the pressure *mode shapes* corresponding to the natural frequencies of the fluid in the tube, and each natural frequency specified by a value of the integer n is associated with a different mode shape. The first few mode shapes are shown in Figure 2.20. The pressure has a maximum amplitude at each end of the tube as well as elsewhere for $n>1$. The integer

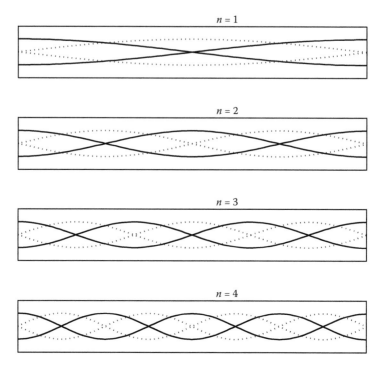

Figure 2.20 Pressure (—) and axial particle velocity (·····) mode shapes of a pipe closed at both ends.

n corresponds to the number of zeroes (nodal points) in the pressure distribution along the length L. In contrast, the particle velocity has a mode shape which is zero at each end of the tube and has its maxima at the nodal points in the pressure field. Note, also, from Equations 2.144 and 2.145, that the particle velocity is in quadrature (90° out of phase) with the pressure. Thus, from Equation 2.69, the time-averaged intensity is everywhere zero (although the instantaneous energy flux density may be nonzero).

2.6.2 One-dimensional standing waves in a uniform tube with an open end

Another interesting case is a finite tube with an open end at $x=L$. Instead of $u(L)=0$, the boundary condition at the open end becomes approximately $p(L)=0$. However, the open end of the tube radiates sound into the surrounding fluid, so that the corresponding acoustic impedance $z(L)$ is nonzero. Consequently, it is found that the mode shape is such that the pressure would be zero at a location slightly beyond the end of the tube, $x=L'=L+0.6a$, where $0.6a$ is known as the end correction, in this case for an unflanged circular tube (see Kinsler et al. 2000); for a flanged end it is $L' = 8a/3\pi = 0.85a$. Thus, the boundary condition at the open end is

$$p(L') = 2A\cos(kL') = 0 \qquad (2.149)$$

Hence, the natural frequencies of the tube which is open at one end are given by $kL' = (2n - 1)\pi/2$, or

$$\omega_n = \frac{(2n-1)\pi c_0}{2L'}, \text{ n} = 1, 2, 3, \ldots \qquad (2.150)$$

with $L' = L + 0.6a$. The corresponding pressure and velocity mode shapes are shown in Figure 2.21. Note: the radiation of sound from the open end introduces damping into the modes of the pipe; this limits the effective range of wind instruments which radiate effectively from the open end at high frequencies.

It may be noted that the natural frequencies of both the open and closed uniform pipe occur in a regular sequence. This is important in producing

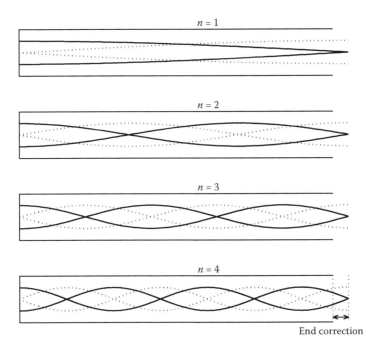

Figure 2.21 Pressure (—) and axial particle velocity (·····) mode shapes of an open-closed pipe.

pleasant musical sounds in woodwind instruments and organ pipes. The closed pipe (and a pipe that is open at both ends) has a full sequence of 'harmonics' or 'overtones' of the fundamental frequency, whereas the open-closed pipe has only odd 'harmonics'. The appropriate case to use in modelling a particular wind instrument depends on the nature of the excitation which may be considered as either an open or a closed end. For more details, see Fletcher and Rossing (1998).

2.6.3 Enclosed sound fields in three dimensions at low frequencies

We now extend the foregoing principles to the case of sound fields in three-dimensional enclosures by considering the solution to the three-dimensional wave equation when the fluid is enclosed by a rigid-walled rectangular box, as shown in Figure 2.22. In this case, for harmonic pressure fluctuations of a particular frequency in the enclosure, the complex pressure amplitude must satisfy the Helmholtz equation (Equation 2.83), which for Cartesian coordinates can be written as

$$\left(\frac{\partial^2}{\partial x_1^2}+\frac{\partial^2}{\partial x_2^2}+\frac{\partial^2}{\partial x_3^2}\right)p(\mathbf{x})+k^2 p(\mathbf{x})=0 \qquad (2.151)$$

To solve this equation, we use a technique known as separation of variables. We assume that the pressure fluctuation can be expressed as separate functions of x_1, x_2 and x_3, such that

$$p(\mathbf{x})=f_1(x_1)f_2(x_2)f_3(x_3) \qquad (2.152)$$

Substituting this into Equation 2.151 and dividing by $p(\mathbf{x})$ gives

$$\frac{1}{f_1(x_1)}\frac{\partial^2 f_1(x_1)}{\partial x_1^2}+\frac{1}{f_2(x_2)}\frac{\partial^2 f_2(x_2)}{\partial x_2^2}+\frac{1}{f_3(x_3)}\frac{\partial^2 f_3(x_3)}{\partial x_3^2}+k^2=0 \qquad (2.153)$$

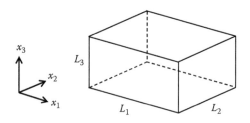

Figure 2.22 Rigid-walled rectangular enclosure.

Each of these terms depends on only one of the three independent variables and, hence, must be equal to a constant,

$$\frac{1}{f_1(x_1)}\frac{d^2 f_1(x_1)}{dx_1^2} = k_1^2; \quad \frac{1}{f_2(x_2)}\frac{d^2 f_2(x_2)}{dx_2^2} = k_2^2; \quad \frac{1}{f_3(x_3)}\frac{d^2 f_3(x_3)}{dx_3^2} = k_3^2$$

(2.154a,b,c)

where the constants satisfy

$$k_1^2 + k_2^2 + k_3^2 = k^2 \qquad (2.155)$$

For plane propagating wave solutions in three dimensions, the terms k_1, k_2, k_3 are components of a *wavenumber vector* **k**, with magnitude $k = \omega/c_0$ and direction corresponding to the propagation direction of the wave.

The function $f_1(x_1)$ must also satisfy the rigid-wall boundary conditions of zero normal particle velocity at $x_1 = 0$ and $x_1 = L_1$. From Section 2.6.1, it is clear that solutions of Equation 2.154a that satisfy the boundary conditions are $f_1(x_1) = \cos k_1 x_1$, where $k_1 = n_1 \pi / L_1$ and n_1 is an integer. Similar solutions are found for the other two differential equations. The modes of the enclosure occur for frequencies such that Equation 2.155 is satisfied, that is, similarly to Equation 2.147,

$$\frac{\omega_n}{c_0} = \sqrt{\left(\frac{n_1 \pi}{L_1}\right)^2 + \left(\frac{n_2 \pi}{L_2}\right)^2 + \left(\frac{n_3 \pi}{L_3}\right)^2} \qquad (2.156)$$

for integer values n_1, n_2 and n_3 and where the integer n denotes the trio of integers (n_1, n_2, n_3). The corresponding pressure mode shapes are given by

$$p_n(\mathbf{x}) = \cos\frac{n_1 \pi x_1}{L_1} \cos\frac{n_2 \pi x_2}{L_2} \cos\frac{n_3 \pi x_3}{L_3} \qquad (2.157)$$

Note the similarity of this solution to that of the problem considered in Section 2.6.1 for fluid in a one-dimensional closed tube. At the natural frequencies of the closed tube, an integer number of half wavelengths defines a given mode shape associated with the pressure fluctuations in the tube. In this three-dimensional case, exactly similar pressure fluctuations are generated, but now for a given natural frequency, there is not only a spatial dependence on the x_1 coordinate but also on the x_2 and x_3 coordinates. Modes which have only one nonzero index, that is, of the form $(n_1, 0, 0)$, $(0, n_2, 0)$ or $(0, 0, n_3)$, are known as axial modes and have a mode shape

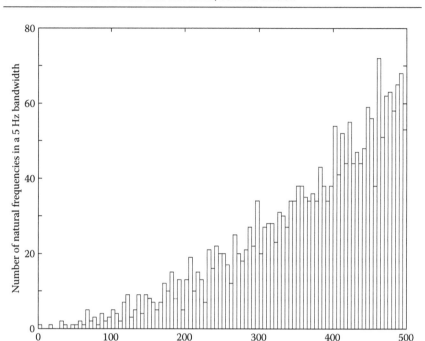

Figure 2.23 Number of natural frequencies in each 5 Hz bandwidth up to 500 Hz for a rectangular enclosure of dimensions 10 m×5 m×3 m. (Reprinted from Nelson, P.A. and Elliott, S.J., *Active Control of Sound*, chapter 1, Academic Press, London, 1992. With permission.)

similar to the one-dimensional tube. Modes with two nonzero indices are known as tangential modes and those with three nonzero indices are called oblique modes.

The solution given in Equation 2.157 is applicable *only* to unforced motion at these particular natural frequencies. For the perfectly rigid-walled enclosure considered here with *no source*, a sound field can exist only at the particular values of ω_n specified by Equation 2.156. The corresponding mode shapes associated with rigid-wall boundary conditions can be used to construct approximate solutions for the pressure field in an enclosure with walls having a high but finite impedance (see e.g. Nelson and Elliott 1992, chapter 10).

The practical usefulness of this solution is further limited by the fact that, in many enclosures at the frequencies of interest, there are potentially very many modal contributions. For example, Figure 2.23 shows, for a typical room, the number of modes having natural frequencies in a given frequency bandwidth. In addition, in the presence of a source, each mode can be *excited* over a range of frequencies, not just at its natural frequency. Thus, at high excitation frequencies, a large number of enclosure modes may be excited at any particular frequency. Computation of the sound field by using this modal model then becomes extremely laborious and we adopt the approach described in the next section, where the sound field is assumed to consist of an infinitely large number of plane travelling wave contributions. This is not inconsistent with the modal model presented here, since each mode may be considered to consist of a superposition of a number of plane travelling wave contributions, just as we saw in Section 2.6.1 that a one-dimensional standing wave can be considered to consist of two plane propagating waves (see e.g. Kinsler et al. 2000, chapter 9 for a full discussion).

2.7 ROOM ACOUSTICS

2.7.1 Enclosed sound fields in three dimensions at high frequencies

As discussed in the last section, to deal with enclosed sound fields at high frequencies we can adopt a model in which we assume that the sound field at a point consists of a superposition of the contributions from a large number of uncorrelated plane travelling waves. The plane waves are assumed to be arriving at the point from all possible propagation directions. One can visualise this by assuming that the point in the sound field is at the centre of a sphere the surface of which is divided into a very large number of segments of equal area, as shown in Figure 2.24. The line passing through the centre of one of these segments and the centre of the sphere defines the propagation direction associated with a particular plane wave. The sound field consists of an infinite number of such plane waves, each associated with a different propagation direction. These propagation directions are defined by assuming

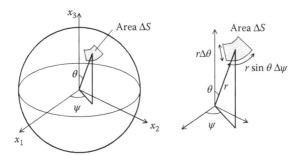

Figure 2.24 Plane-wave construction of a diffuse-sound field model. The surface of the sphere of radius r surrounding a point is divided into N tiny segments of equal area. Each area element has an associated plane wave propagating in a direction defined by the normal to the surface area element. Thus the segment of area ΔS contains $(\Delta S N/4\pi r^2)$ tiny elements.

that all the segments of equal area on the surface of the sphere become infinitesimally small. Thus, the sound field at a single frequency has a complex pressure at a point **x** that is given by

$$p(\mathbf{x}) = \lim_{N \to \infty} \frac{1}{\sqrt{N}} \sum_{n=1}^{N} p_n(\mathbf{x}) \qquad (2.158)$$

where $p_n(\mathbf{x}) = |p_n| e^{i\varphi_n(\mathbf{x})}$ is the complex pressure associated with the nth plane wave, having modulus $|p_n|$ and phase $\varphi_n(\mathbf{x})$. The phase changes as a function of position **x** as a result of the propagation of the waves through space and at a given location can be considered as a random variable. The factor $1/\sqrt{N}$ is included to ensure that the limit remains finite as the number N of contributing waves becomes infinite.

We can compute the squared modulus of the complex pressure at the position **x** by taking the product of $p(\mathbf{x})$ and its complex conjugate. Thus,

$$|p(\mathbf{x})|^2 = \lim_{N \to \infty} \frac{1}{N} \left(\sum_{n=1}^{N} p_n(\mathbf{x}) \right) \left(\sum_{m=1}^{N} p_m(\mathbf{x}) \right)^* \qquad (2.159)$$

where the additional index m has been introduced to denote the mth plane wave in the conjugate term. Now, note that the product of terms for which $n \neq m$ takes the form

$$p_n(\mathbf{x}) p_m^*(\mathbf{x}) = |p_n| |p_m| e^{i(\varphi_n(\mathbf{x}) - \varphi_m(\mathbf{x}))} \qquad (2.160)$$

which can also be written as

$$p_n(\mathbf{x}) p_m^*(\mathbf{x}) = |p_n| |p_m| \left[\cos(\varphi_n(\mathbf{x}) - \varphi_m(\mathbf{x})) + i \sin(\varphi_n(\mathbf{x}) - \varphi_m(\mathbf{x})) \right] \qquad (2.161)$$

If we now move through a volume of space which has dimensions that are large compared with the acoustic wavelength at the frequency considered (but still much smaller than the dimensions of the enclosure), the terms in the square brackets of Equation 2.161 will have an average value of zero, since positive and negative values of the sine and cosine terms are equally likely to occur. This is a direct result of the assumption that φ_n and φ_m are unrelated (the waves are uncorrelated) and that the two wave propagation directions are different. Thus, if we define a local spatial average of the squared pressure modulus, where we compute the average value of $|p(\mathbf{x})|^2$

by taking samples of the squared pressure field over a volume of space with dimensions larger than an acoustic wavelength, then all the terms for which $n \neq m$ will average to zero and only terms for which $n=m$ will contribute. The space-averaged modulus-squared pressure thus becomes

$$\langle |p(\mathbf{x})|^2 \rangle = \lim_{N \to \infty} \frac{1}{N} \sum_{n=1}^{N} |p_n|^2 \tag{2.162}$$

where we have used $\langle \ \rangle$ to denote the operation of space averaging.

We can use very similar arguments to this in deducing expressions for the space-averaged energy density and intensity in the sound field. First, we can use the expression derived in Section 2.3.1 for the energy density of a three-dimensional sound field, provided that we interpret $u(t)$ as the instantaneous modulus of the particle-velocity vector at a position in the sound field. For harmonic excitation, we can use an argument identical to that presented in Section 2.3.1 to compute an expression for the time-averaged energy density. This is given by

$$e(\mathbf{x}) = \frac{1}{4} \rho_0 \left[|u(\mathbf{x})|^2 + \frac{|p(\mathbf{x})|^2}{(\rho_0 c_0)^2} \right] \tag{2.163}$$

where $p(\mathbf{x})$ is the complex pressure amplitude and $u(\mathbf{x})$ is the complex particle velocity in the direction of propagation. If we now compute a local spatial average of this time-averaged energy density, we will again find that the contribution of terms arising from dissimilar propagation directions will average to zero. It follows that

$$\langle e(\mathbf{x}) \rangle = \lim_{N \to \infty} \frac{1}{N} \sum_{n=1}^{N} \frac{1}{4} \rho_0 \left[|u_n|^2 + \frac{|p_n|^2}{(\rho_0 c_0)^2} \right] \tag{2.164}$$

Since we know that for the nth plane wave, $|u_n| = |p_n|/\rho_0 c_0$, this leads to

$$\langle e(\mathbf{x}) \rangle = \lim_{N \to \infty} \frac{1}{N} \sum_{n=1}^{N} \left[\frac{|p_n|^2}{2\rho_0 c_0^2} \right] = \frac{\langle |p(\mathbf{x})|^2 \rangle}{2\rho_0 c_0^2} \tag{2.165}$$

Finally, we will assume that all these directions of propagation contribute equally to the space-averaged energy density, that is, all the p_n are equal. A sound field comprising plane waves of random phases and coming from all possible directions which also satisfies this criterion is said to be *diffuse* (see e.g. Pierce 1981, chapter 6). The space average can often also be approximated by a frequency average.

2.7.2 One-sided intensity

Theoretically, in a diffuse field, the net time-averaged sound intensity, found by adding the contributions from all directions, will be zero. (Note: this is not the case in a real room excited by a source, as energy flows from the source to the absorbent surfaces; however, it is a reasonable assumption for almost rigid walls.) However, it is useful to determine the so-called one-sided intensity, that is, the intensity due to the waves falling on one side of a hypothetical plane surface in the sound field. This is visualised in Figure 2.25. We can use the diffuse field assumption to derive an expression for this one-sided intensity, which is the intensity produced by the waves whose directions are specified by a hemisphere centred on a point in space. The time-averaged intensity in a direction normal to this surface can be written as

$$I(\mathbf{x}) = \frac{1}{2}\operatorname{Re}\{p^*(\mathbf{x})u(\mathbf{x})\} \tag{2.166}$$

where $u(\mathbf{x})$ is the component of the complex particle velocity in the direction normal to the surface. For a single propagating wave n, this velocity component can be written as

$$u_n(\mathbf{x}) = \frac{p_n(\mathbf{x})}{\rho_0 c_0}\cos\theta_n = \frac{p(\mathbf{x})}{\rho_0 c_0}\cos\theta_n \tag{2.167}$$

where θ_n defines the angle of the propagation direction of the nth plane wave from the normal to the hypothetical surface. The final form in Equation 2.167 applies in a diffuse field, where the wave amplitudes p_n are

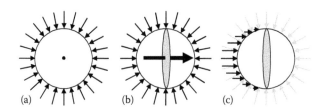

Figure 2.25 Visualisation of the sound intensity at a point in a diffuse field: (a) intensity vectors of plane waves arriving at a point from different directions; (b) the one-sided intensity is that which passes through a surface from one side; (c) components of the intensity vectors of plane waves that contribute to the one-sided intensity.

assumed to be equal for all the directions of propagation considered. The velocity component in Equation 2.166 can, therefore, be written as

$$u(\mathbf{x}) = \frac{p(\mathbf{x})}{\rho_0 c_0} \lim_{N \to \infty} \frac{1}{\sqrt{N}} \sum_{n=1}^{N/2} \cos \theta_n \qquad (2.168)$$

where the summation extends only over half of the wave components, that is, those incident from one side.

If we now evaluate the local space-average value of $I(\mathbf{x})$, again assuming that the cross-terms associated with different propagation directions average to zero, the expression that results is given by

$$\langle I(\mathbf{x}) \rangle_{1-s} = \frac{\langle |p(\mathbf{x})|^2 \rangle}{2\rho_0 c_0} \lim_{N \to \infty} \frac{1}{N} \sum_{n=1}^{N/2} \cos \theta_n \qquad (2.169)$$

We can evaluate the summation over $\cos \theta_n$ by reducing the summation to an integral. Figure 2.24 shows an element of surface area of a sphere surrounding a point in space. The elemental area is defined by the angles $\Delta \theta$ and $\Delta \psi$. Since the number of plane waves passing through a unit area of the sphere is given by $N/4\pi r^2$, where r is the radius of the sphere, the number of plane waves passing through the element of surface area ΔS is given by

$$\frac{N}{4\pi r^2} \Delta S = \left(\frac{N}{4\pi r^2} \right) r^2 \sin \theta \, \Delta \theta \Delta \psi \qquad (2.170)$$

We therefore write the summation over all directions of propagation as a summation over all the elements of area. Thus,

$$\lim_{N \to \infty} \frac{1}{N} \sum_{n=1}^{N/2} \cos \theta_n = \lim_{\Delta \theta, \Delta \psi \to 0} \frac{1}{N} \sum_{\Delta \theta} \sum_{\Delta \psi} \frac{N}{4\pi} \cos \theta \sin \theta \, \Delta \theta \Delta \psi \qquad (2.171)$$

The summations over $\Delta \psi$ and $\Delta \theta$ can thus be reduced to integrals and

$$\lim_{N \to \infty} \frac{1}{N} \sum_{n=1}^{N/2} \cos \theta_n = \frac{1}{4\pi} \int_0^{2\pi} \int_0^{\pi/2} \cos \theta \sin \theta \, d\theta \, d\psi = \frac{1}{4} \qquad (2.172)$$

Therefore, from Equation 2.169, the expression for the 'one-sided' space-averaged, time-averaged intensity is given by

$$\langle I(\mathbf{x})\rangle_{1-s} = \frac{\langle |p(\mathbf{x})|^2 \rangle}{8\rho_0 c_0} = \frac{c_0}{4}\langle e(\mathbf{x})\rangle \tag{2.173}$$

where we have used Equation 2.165 to relate the intensity to the space- and time-averaged energy density. Compared with the intensity in a plane travelling wave (Equation 2.64), the expression for the one-sided intensity contains a factor of ¼. This corresponds to the ratio between the areas of a circle of radius a (πa^2) and a sphere of the same radius ($4\pi a^2$); as indicated in Figure 2.25, the total pressure corresponds to the intensity acting over the surface of the sphere, whereas the one-sided intensity corresponds to the component incident on one side of the circular disc.

2.7.3 Power balance for an enclosure

We can use this result for the one-sided intensity to write a power-balance equation for the whole enclosure (of volume V, say). The central assumption that we make is that the difference between the rate of acoustic energy input (input power) to the enclosure and the rate of energy lost by the absorption of sound at the enclosure walls must be balanced by the rate of change of the total room acoustic energy. The rate of energy lost as a result of the incidence of the diffuse field on a given plane area S_m of the wall of the enclosure can be written as $\langle I(\mathbf{x})\rangle_{1-s} S_m \alpha_m$. The term α_m is the *absorption coefficient* which defines the ratio of the intensity absorbed by the surface to the intensity incident on the surface. Note that here, we are assuming that the normal component of intensity incident on an enclosure surface is equal to the 'one-sided' intensity in the diffuse-field model. The total power absorbed by all the surfaces of the enclosure can thus be written as

$$W_{abs} = \langle I(\mathbf{x})\rangle_{1-s}[S_1\alpha_1 + S_2\alpha_2 + \cdots + S_M\alpha_M] \tag{2.174}$$

where we add all the powers absorbed by the M different surfaces comprising the walls of the enclosure. We can define an *average absorption coefficient* by

$$\bar{\alpha} = \frac{S_1\alpha_1 + S_2\alpha_2 + \cdots + S_M\alpha_M}{S} \tag{2.175}$$

where S is the total surface area of the enclosure walls. The power-balance equation for the enclosure can then be written as

$$\frac{d}{dt}\left[V\langle e(\mathbf{x})\rangle\right] = W_{in} - \langle I(\mathbf{x})\rangle_{1-s} S\bar{a} \tag{2.176}$$

where W_{in} is the power input to the diffuse field and V is the volume of the enclosure. Under steady-state conditions, when the power input to the diffuse field is balanced by the power lost by absorption, the derivative term is zero. This leads directly to a relationship between the space-averaged squared pressure and the power input to the diffuse field. This is given by

$$\langle |p(\mathbf{x})|^2 \rangle = \frac{8\rho_0 c_0}{S\bar{a}} W_{in} \tag{2.177}$$

This expression is used in the measurement of sound power radiated by a source into a reverberation chamber (ISO 2010). It should be noted that the sound field in an enclosure in which the surfaces have a nonzero absorption coefficient cannot be truly diffuse, because power has to flow from the source into the surfaces and, therefore, the time-averaged intensity cannot be zero. However, the diffuse field is a good approximation if the surface sound absorption coefficient is much less than unity.

The transient solution to the power-balance equation is also of significance. If the power input to the enclosure is suddenly switched off, the use of Equation 2.173 and the power-balance equation (Equation 2.176) shows that the space-averaged squared pressure will satisfy

$$\frac{dE}{dt} = -\frac{S\bar{a}c_0}{4V}E \tag{2.178}$$

where, for convenience, we have written $E = \langle |p(\mathbf{x})|^2 \rangle$. The solution of this first-order differential equation is an exponential decay of the form

$$E(t) = E_0\, e^{-[S\bar{a}c_0/4V]t} \tag{2.179}$$

where E_0 is the initial value of the space-averaged squared pressure. If we evaluate the time $t = T_{60}$ taken for the squared pressure to decay to 10^{-6} times its original value (i.e. for the sound pressure level to decay by 60 dB), this is given by

$$\frac{S\bar{a}c_0 T_{60}}{4V} = \ln(10^6) \tag{2.180}$$

from which it follows that

$$T_{60} = \frac{0.161V}{S\bar{a}} \qquad (2.181)$$

where the constant $0.161 = 24\ln 10/c_0$ holds for air at 20°C. Terms may also be added to $S\bar{a}$ to represent air absorption, which can become significant at high frequencies in large rooms. T_{60} is known as the *reverberation time* of the enclosure and Equation 2.181 is known as the *Sabine equation* after W. C. Sabine, who first established this relationship empirically (Sabine 1898). This equation is useful in characterising the acoustical properties of enclosed sound fields. In fact, direct measurements of the reverberation time of an enclosure can be made to quantify its total room absorption $S\bar{a}$. This is one technique used for measuring the effective (or equivalent) diffuse field absorption coefficient (see Chapter 5). The reverberation time of an empty reverberant room is first measured and its absorption calculated. A sample of material (typically 10 m² in area) is then laid in the room and the reverberation time measured again. The difference in the room absorption obtained then enables a calculation to be made of the diffuse field absorption coefficient of the material (ISO 2003).

A good discussion of the limitations of the Sabine equation is presented by Pierce (1981). For rooms with large absorption, as $\bar{a} \to 1$ the Sabine equation tends to a finite value of T_{60}, whereas it should tend to zero. Eyring (1930) proposed a refinement of the Sabine model which results in

$$T_{60} = \frac{-0.161V}{S\ln(1-\bar{a})} \qquad (2.182)$$

according to which, T_{60} does indeed tend to zero for large absorption. Although this provides an improvement, it still relies on an equal distribution of absorption and a room geometry which does not have a large aspect ratio.

2.7.4 Direct and reverberant field in an enclosure

For a source located in a reverberant enclosure, a direct field will exist, which is radiated by the source prior to being absorbed or reflected at the boundaries. From Section 2.4.4, the mean-square pressure in the direct field can be given by

$$p_{D,\text{rms}}^2 = \frac{\rho_0 c_0 Q_\theta}{4\pi r^2} W_{\text{in}} \qquad (2.183)$$

where we have introduced Q_θ, which is the directivity factor quantifying the ratio of the mean-square pressure produced in a particular direction to that produced by an omnidirectionally radiating source of the same power output.

The power input to the reverberant field is that proportion resulting from the first reflection, that is, $(1-\bar{a})W_{in}$. Consequently, the reverberant sound pressure is given by

$$p_{R,rms}^2 = \frac{4\rho_0 c_0 (1-\bar{a})}{S\bar{a}} W_{in} \qquad (2.184)$$

The factor $S\bar{a}/(1-\bar{a})$ is known as the room constant. The direct field reduces with distance from the source, whereas the ideal reverberant field is uniform throughout the room (although this is never the case in practice, especially at low frequencies).

2.7.5 Schroeder frequency

Finally, returning to the modal description of a room, it has been seen in Figure 2.23 that the number of modes which have their natural frequency in a given frequency band becomes larger as the frequency increases. The total number of modes with natural frequency less than a given frequency ω can be given asymptotically by

$$N(\omega) = \frac{1}{6}\frac{V}{\pi^2}\left(\frac{\omega}{c_0}\right)^3 + \frac{1}{16}\frac{S}{\pi}\left(\frac{\omega}{c_0}\right)^2 + \frac{1}{16}\frac{L}{\pi}\left(\frac{\omega}{c_0}\right) + \frac{1}{8} \qquad (2.185)$$

where:
    $V = L_x L_y L_z$ is the room volume
    $S = 2(L_x L_y + L_x L_z + L_y L_z)$ is the surface area of the walls
    $L = 4(L_x + L_y + L_z)$ is the total length of all the edges of the room

At high enough frequencies, the first term in Equation 2.185 dominates. The average modal density, the number of modes in a unit frequency spacing, is given by

$$\frac{dN}{d\omega} \approx \frac{1}{2}\frac{V}{\pi^2}\left(\frac{\omega}{c_0}\right)^2 = \frac{1}{(\Delta\omega)_{mode}} \qquad (2.186)$$

where $(\Delta\omega)_{mode}$ is the average frequency spacing between modes. Due to the influence of absorption, the resonance peak in the frequency response associated with each mode has a certain bandwidth (see Equation 3.67) which can be written as

$$\Delta\omega = \frac{2.2}{2\pi T_{60}} \qquad (2.187)$$

This is known as the half-power bandwidth, as the squared amplitude of the modal response reduces by a factor of 2 (3 dB) from the resonance ω_n to $\omega_n \pm \Delta\omega$. Schroeder (1962) indicated that the sound field in a room can be considered to be *diffuse*, providing that an average of three modes have their natural frequencies within the half-power bandwidth of each mode. Combining Equations 2.186 and 2.187, this gives a limiting frequency (in Hz) of

$$f_S \approx 2000 \left(\frac{T_{60}}{V} \right)^{1/2} \tag{2.188}$$

which is known as the Schroeder frequency (see also Kuttruff 2009).

2.8 FURTHER READING

There are many books on acoustics that serve as an introduction to the subject. Particularly useful basic texts written primarily for undergraduate students are those by Fahy (2001), Hall (1987) and Kinsler et al. (2000). An extended version of the present treatment is given in Nelson and Elliott (1992). More advanced treatments of the subject are presented, for example, by Pierce (1981), Morse and Ingard (1968) and Skudrzyk (1971). Various applications of acoustics are discussed in Fahy and Walker (2004). The *Dictionary of Acoustics* (Morfey 2001) is also an invaluable reference.

2.9 QUESTIONS

1. Evaluate the mean-square pressures and the corresponding sound-pressure levels of the continuous sound-pressure signals shown below:

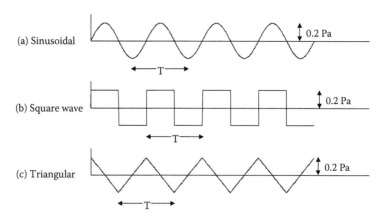

2. What is the r.m.s. pressure amplitude of a sound signal with a sound-pressure level of (i) 80 dB, (ii) 94 dB, (iii) 0 dB (re 2×10^{-5} Pa)?
3. Three independent sound sources produce sound pressure levels at a receiver position of 80, 82 and 85 dB re 2×10^{-5} Pa respectively. Calculate the total sound pressure level when all three operate simultaneously. What would be the maximum possible sound pressure level if the three sources were coherent?
4. A sliding piston at the end of a semi-infinite tube moves with a maximum velocity of 0.1 m s^{-1} for a short duration. Given that the gas in the tube is air, with a density $\rho_0=1.21$ kg m^{-3}, an ambient pressure $p_0=10^5$ N m^{-2} and a ratio of specific heats $\gamma=1.4$, calculate the maximum pressure and density changes produced by the motion of the piston.
5. A harmonic plane wave having a pressure amplitude of 5×10^{-3} Pa travels in air. Calculate the amplitude of the acoustic particle-velocity fluctuation and the value of the time-averaged acoustic intensity in the direction of propagation of the wave. If a wave with the same pressure amplitude travels in water (having an ambient density ρ_0 of 1000 kg m^{-3} and a sound speed c_0 of 1480 m s^{-1}), what is the corresponding amplitude of the acoustic particle velocity and the value of the time-averaged acoustic intensity?
6. Two travelling harmonic plane waves with different amplitudes propagate in a direction parallel to the x-axis, one in the positive x-direction and the other in the negative x-direction. The waves produce pressure fluctuations that are in phase at $x=0$. Derive an expression for the modulus and phase of the net pressure fluctuation produced by the interference of the waves as a function of x and as a function of the ratio of the amplitudes of the positive- and negative-going waves. Illustrate your results graphically with a sketch of the dependence on x of the modulus and phase of the net pressure fluctuation for the case where the ratio of amplitudes is 2.
7. Two uncorrelated spherically radiating acoustic sources (in air) are respectively 15 m and 5 m from a receiver point at which the sound-pressure level is 80 dB re 2×10^{-5} Pa. Given that the sound power level of the more distant source is 110 dB re 10^{-12} W, calculate the sound power level of the nearer source.
8. A point monopole source radiating single-frequency sound is placed 1 m above a rigid reflecting surface. The sound pressure is measured at a distance of 10 m from the source at the same height as the source above the plane. Calculate the complex ratio of this pressure to that produced at the same position in the absence of the reflecting plane and plot its magnitude as a function of frequency.
9. (a) A rectangular air-filled enclosure has dimensions 5 m×7 m×3 m. All the surfaces are covered with plaster having an absorption

coefficient of 0.07 at 500 Hz, except the floor (which measures 5 m×7 m), which is covered with carpet having an absorption coefficient of 0.2 at 500 Hz. Calculate the average absorption coefficient at 500 Hz and hence estimate the reverberation time.
(b) A material having an unknown absorption coefficient is used to cover one of the walls (measuring 5 m × 3 m) of the enclosure, as a result of which the reverberation time in the enclosure is reduced to three-quarters of its original value. Estimate the absorption coefficient of the material.
(c) Explain why this estimate is likely to be unreliable.
10. Derive an expression relating the sound power level of a spherically radiating source to the sound pressure level (in dB re 2×10^{-5} Pa) at a radial distance r from the source. Calculate the sound pressure level at 10 m from such a source of which the sound power is 5 W.

REFERENCES

Eyring, C.F. (1930) Reverberation time in 'dead' rooms, *Journal of the Acoustical Society of America*, 1, 217–241.
Fahy, F. (2001) *Foundations of Engineering Acoustics*, Academic Press: London.
Fahy, F. and Walker, J. (eds) (2004) *Advanced Applications of Acoustics, Noise and Vibration*, Spon Press: London.
Fletcher, N.H. and Rossing, T.D. (1998) *The Physics of Musical Instruments*, 2nd edn, Springer: New York.
Hall, D.E. (1987) *Basic Acoustics*, John Wiley and Sons: New York.
ISO (International Organization for Standardization) (2003) *ISO 354:2003: Acoustics: Measurement of Sound Absorption in a Reverberation Room*, International Organization for Standardization, Geneva.
ISO (International Organization for Standardization) (2010) *ISO 3741:2010: Acoustics: Determination of Sound Power Levels of Noise Sources Using Sound Pressure: Precision Methods for Reverberation Rooms*, International Organization for Standardization, Geneva.
Kinsler, L.E., Frey, A.R., Coppens, A.B., and Sanders, J.V. (2000) *Fundamentals of Acoustics*, 4th edn, John Wiley: New York.
Kuttruff, H. (2009) *Room Acoustics*, 5th edn, Spon Press: London.
Lighthill, M.J. (1952) On sound generated aerodynamically, I. General theory. *Proceedings of the Royal Society of London*, **211A**, 564–587.
Lighthill, M.J. (1978) *Waves in Fluids*, Cambridge University Press: Cambridge.
Morfey, C.L. (2001) *The Dictionary of Acoustics*, Academic Press: London.
Morse, P.M. and Ingard, K.U. (1968) *Theoretical Acoustics*, McGraw-Hill: New York.
Nelson, P.A. and Elliott, S.J. (1992) *Active Control of Sound*, Academic Press: London.

Petyt, M. and Jones, C.J.C. (2004) Numerical methods in acoustics, in Fahy, F. and Walker, J. (eds), *Advanced Applications of Acoustics, Noise and Vibration*, chapter 2, Spon Press: London.

Pierce, A.D. (1981) *Acoustics: An Introduction to Its Physical Properties and Applications*. McGraw-Hill: New York, reprinted by Acoustical Society of America: Woodbury NY, 1989.

Rayleigh, L. (1896) *The Theory of Sound*, 2nd edn, reprinted by Dover: New York, 1945.

Sabine, W.C. (1898) Architectural acoustics. *Engineering Records*, 38, 520–522.

Schroeder, M.R. (1962) Frequency correlation functions of frequency responses in rooms, *Journal of the Acoustical Society of America*, 34, 1819–1823.

Spiegel, M.R. (1974) *Theory and Problems of Fourier Analysis*, McGraw-Hill: New York.

Skudrzyk, E. (1971) *The Foundations of Acoustics*, Springer: New York.

Wu, T.W. (2000) *Boundary Element Acoustics: Fundamentals and Computer Codes*, WIT Press: Southampton.

Chapter 3

Fundamentals of vibration

Brian Mace

3.1 INTRODUCTION

Vibration of engineering structures is commonplace: for example, cars vibrate due to forces produced by the engine, or by road roughness; aircraft vibrate due to aerodynamic loads and turbulent boundary layers; and buildings vibrate due to forces produced by wind loading and earthquakes. We need to be able to model, analyse and design structures so that their vibrational behaviour is acceptable, and this requires us to understand the fundamentals behind the physics of vibration.

Vibration is usually, but not always, a small-amplitude oscillatory motion about a static equilibrium position. Its effects are often unwanted: large displacements and stresses may occur, especially due to resonance, and these may lead to fatigue, breakage, wear or improper operation; structural vibration produces sound which radiates into the surrounding air (indeed, sound is also a form of vibration); vibration can produce physical discomfort, motion sickness or physiological effects such as vibration white finger; instabilities can occur that lead to growing vibrations that often end in failure, such as flutter and galloping. However, not all vibration is undesirable: music is one example.

This chapter concerns the fundamental aspects of vibration. Emphasis will be placed on relatively simple systems (with one or two degrees of freedom – see next section) and the fundamentals of their physical behaviour. In practice, vibration analysis often involves a finite element (FE) model of which the number of elements can be very large. Although this might be complicated, the underlying physics is the same – natural frequencies, resonance, modes and so on – and hence, understanding of the fundamentals is essential.

3.1.1 Some terminology and definitions

The basic properties of a vibrating system are inertia (mass) and stiffness, which store kinetic and potential energy respectively, damping, which dissipates energy, and external force, which provides energy. A state of *free vibration* is one in which no external forces act, while a *forced vibration* is one where external forces operate. The external forces might be *harmonic* (i.e. vary sinusoidally with time), *transient* (existing only for a finite time), *periodic* (the forces repeat at regular intervals) or *random* (unpredictable, such as produced by a rough road or by turbulence). Systems may be theoretically modelled as being *damped* or *undamped*, although all passive physical systems are damped to some extent.

We will only consider *linear* vibrations. For a linear system the principle of *superposition* holds. In practice, linearity is an idealisation, since all real systems are to some extent nonlinear, but very often they are only very weakly nonlinear, so that their behaviour can be predicted accurately enough by a linear model.

A *discrete* system is one that is made up of a finite number of rigid masses connected by massless springs, while a *continuous* system is one for which there is a continuous distribution of mass and stiffness throughout the system – under loads, their deformed shapes are continuous in space, like a cantilever beam in bending.

3.1.2 Degrees of freedom

The number of *degrees of freedom (DOFs)* of a system is defined as the number of *independent* coordinates necessary to describe its motion. These coordinates can be any response variable that we choose: the displacement of some point, rotation, relative displacement and so on (in Sections 3.5 through 3.7 we will use *modal coordinates* to describe the motion of a system). A single DOF (SDOF) system has just one DOF, while a multiple DOF (MDOF) system has more than one. Engineers form models of structures to predict their vibration. The number of DOFs depends on how complicated the system is, what modelling simplifications and assumptions we make, but most importantly how we choose to model the system and what results we want from our model.

As an example, consider the vibration modelling of a car. Figure 3.1a shows an SDOF quarter-car model comprising the mass m_1 of one-quarter of the car body supported by a spring of stiffness k_1, which represents the suspension at one wheel. The damper, having viscous damping coefficient c_1, represents the shock absorber. The road provides the vibration input and we assume that the mass vibrates vertically with a displacement $x_1(t)$.

This model would be suitable for the first stages of the design of the suspension system (i.e. what values should we choose for k and c?). A slightly more sophisticated model is shown in Figure 3.1b. This has two DOFs, the second representing the unsprung mass associated with the axle, wheel hub and so forth; the stiffness being that of the tyre. This model might be used for more refined suspension design.

Fundamentals of vibration 79

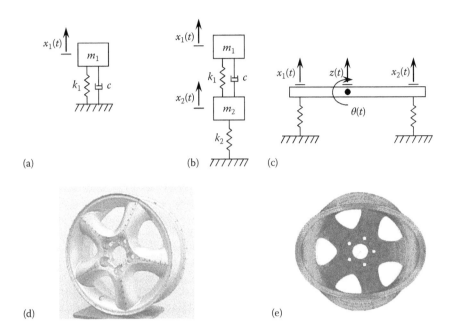

Figure 3.1 Models for the vibration of a car: (a) SDOF ¼ car model; (b) 2DOF ¼ car model; (c) 2DOF model including pitch; (d) wheel rim and (e) FE model, mode 1.

A third model includes pitching motion (Figure 3.1c) and gives information regarding vehicle handling and how the suspension affects it. Note that four coordinates are marked – x_1, x_2, z and θ – but they are not independent. We can describe the motion by choosing any two (e.g. (x_1, x_2) or (z, θ)) and there are thus two DOFs. Finally, suppose we are interested in vibration transmission from a rough road, through the tyre and wheel rim and hence to the car body. This is affected by the design of the wheel rim (Figure 3.1d). In particular, it is important to make sure there are no rim resonances in certain frequency bands. We might model this using an FE model with many thousands of DOFs. In summary, we might use different models of the same system with different numbers of DOFs depending on how we choose the model and what we want it to predict.

3.1.3 Methods of rigid body dynamics

There are various methods available to find the equations of motion of the system and some of these are summarised in the following subsections. It is assumed that the reader is familiar with these methods, which are described in more detail in undergraduate texts on engineering mechanics (e.g. Meriam and Kraige 2003, or introductory texts on vibration such as Rao 2010 and Meirovitch 2000).

3.1.3.1 Newton's second law of motion

Newton's second law of motion relates the resultant external force *vector* **F** acting on a body of mass m to the acceleration **a** of its centre of mass by

$$\mathbf{F} = m\mathbf{a} \tag{3.1}$$

while for motion in the plane

$$M = I\alpha \tag{3.2}$$

where:
- M is the resultant moment about the centre of mass, or a fixed point
- I is the second moment of mass (also known as the moment of inertia or mass moment of inertia) about that point
- α is the angular acceleration

When applying Newton's laws we should draw free-body diagrams and may need to consider constraints (i.e. fixed relations between the motions of various points in the system).

3.1.3.2 Work–energy

The kinetic energy of a rigid body moving in a plane is

$$T = \tfrac{1}{2}mv^2 + \tfrac{1}{2}I\omega^2 \tag{3.3}$$

where:
- v is the magnitude of the velocity of the centre of mass
- ω is the magnitude of the angular velocity

Here, I is the second moment of mass about the centre of mass. Potential energy arises partly from gravity (mgh, h being the height of the mass above some reference plane) and partly from the strain energy stored in elastic deformation, such as a spring (see Section 3.1.7).

The work done by an external force vector **F** when its point of application undergoes a displacement vector **x** equals

$$U = \int \mathbf{F}.d\mathbf{x} \tag{3.4}$$

The total work done (by external and friction forces, etc.) as a body moves equals the change in its total energy.

3.1.3.3 Impulse–momentum

The impulse of a force F(t) is defined as

$$I = \int_{t_1}^{t_2} \mathbf{F}(t)\,dt \qquad (3.5)$$

The impulse is related to the change in momentum mv of the mass *m* by

$$I = \int_{t_1}^{t_2} m\mathbf{a}(t)\,dt = m\mathbf{v}(t_2) - m\mathbf{v}(t_1) \qquad (3.6)$$

assuming the mass is constant.

3.1.4 Simple harmonic motion

When analysing vibration we will frequently come across differential equations of the form

$$\ddot{x} + \omega_n^2 x = 0 \qquad (3.7)$$

where $\ddot{x} = d^2x/dt^2$ and ω_n is a constant. Here $x(t)$ might represent the displacement of a vibrating system. The solution can be written as any one of

$$x(t) = A\sin\omega_n t + B\cos\omega_n t$$
$$x(t) = C\sin(\omega_n t + \phi)$$
$$x(t) = C\cos(\omega_n t + \theta) \qquad (3.8)$$

This is *simple harmonic motion* (SHM) (Figure 3.2). The variable $x(t)$ is *time harmonic*: it varies sinusoidally with time at a frequency ω_n radians per second or $f_n = \omega_n/2\pi$ Hz (cycles per second). The *period* of the motion $\tau = 1/f_n$ represents the duration of each cycle of vibration. The constant $C = \sqrt{A^2 + B^2}$ is the *magnitude*, while ϕ and θ represent the *phase* of the motion with respect to some reference time (when *t* is chosen to be zero).

Note from Equation 3.8 that the time-average displacement $\overline{x(t)} = 0$, that is to say, the displacement is an oscillatory motion about $x = 0$. The mean-square displacement is

$$\overline{x^2(t)} = \frac{1}{\tau}\int_{t}^{t+\tau} x^2(t)\,dt = \frac{1}{2}C^2 \qquad (3.9)$$

and hence the root-mean-square (r.m.s.) displacement is $x_{\text{rms}} = C/\sqrt{2}$. This r.m.s. value represents the 'typical size' of the vibration.

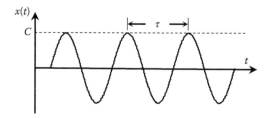

Figure 3.2 Time-harmonic motion x(t).

For SHM, the displacement, velocity and acceleration may be expressed by

$$x(t) = C\sin(\omega_n t + \phi)$$
$$\dot{x}(t) = \omega_n C \cos(\omega_n t + \phi) = \omega_n C \sin(\omega_n t + \phi + \pi/2)$$
$$\ddot{x}(t) = -\omega_n^2 C \sin(\omega_n t + \phi) = \omega_n^2 C \sin(\omega_n t + \phi + \pi) \tag{3.10}$$

(where we note that $\cos\alpha = \sin(\alpha + \pi/2)$). Thus the velocity *leads* the displacement by $\pi/2$ radians, or 90°, while the velocity *lags* the acceleration by 90° (Figure 3.3). The displacement and acceleration are 180° out of phase. Phase angles, leads and lags thus represent the relative phase between different time-harmonic motions of the same frequency.

3.1.5 Complex exponential notation

Complex exponential notation (CEN) is used widely in frequency-response methods (see also Section 2.1.5). We introduce complex numbers, but the advantage over the explicit use of sine and cosine terms is that the mathematics is much easier. Using CEN, a time-harmonic quantity is written as

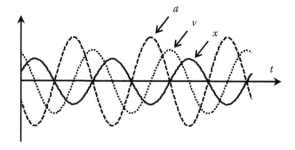

Figure 3.3 Displacement, velocity and acceleration for SHM.

$$x(t) = Xe^{i\omega t} \tag{3.11}$$

where $X = |X|\exp(i\theta)$ is the amplitude and is generally complex. Here $|X|$ is the magnitude and θ is a phase angle. In the physical world we only see or measure the real part of $x(t)$, that is,

$$x(t) = \text{Re}\{Xe^{i\omega t}\} \tag{3.12}$$

Noting that $\exp(i\omega t) = \cos \omega t + i \sin \omega t$, it follows that

$$x(t) = \text{Re}\{|X|e^{i(\omega t + \theta)}\} = |X|\cos(\omega t + \theta) \tag{3.13}$$

This is, of course, SHM as defined in Equation 3.8. For simplicity, we will write $x(t) = X \exp(i\omega t)$ as in Equation 3.11, but we should bear in mind that to recover a physically meaningful solution from an analysis based on CEN, we must retain only the real part: implicitly we only 'see' the real part, that is, Equation 3.13, in the real, physical world.

The advantages of CEN are that time derivatives follow simply, for example,

$$\dot{x}(t) = i\omega Xe^{i\omega t} = i\omega x(t)$$
$$\ddot{x}(t) = -\omega^2 Xe^{i\omega t} = -\omega^2 x(t) \tag{3.14}$$

Thus, in a differential equation, the time derivative d/dt is replaced by the quantity $(i\omega)$, turning our differential equation into an algebraic equation in $(i\omega)$. In Section 3.3, the benefits of this will become clear.

3.1.6 Frequency-response functions

We will often be concerned with time-harmonic behaviour; for example, when a time-harmonic force $f(t) = F \exp(i\omega t)$ acts on a system to produce a response. In the *steady state*, when *transients* have decayed to zero (Section 3.3) the displacement of our vibrating system $x(t) = X \exp(i\omega t)$ is also time harmonic. In practice, more general excitations can be regarded as the superposition of a number of harmonics, and the response is also a superposition of harmonics: see Section 3.4. The *frequency-response function* (FRF) of a system is defined as the ratio of the response (i.e. the output) of the system to a time-harmonic excitation (input) in the steady state, when both are time harmonic. (Note: the response of a nonlinear system

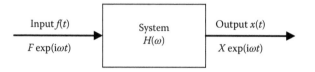

Figure 3.4 Frequency response of a system.

to a harmonic excitation contains frequencies other than the excitation frequency). Thus for the system in Figure 3.4, the input and output are

$$f(t) = Fe^{i\omega t}; \quad x(t) = Xe^{i\omega t} \tag{3.15}$$

and the frequency response is

$$H(\omega) = \frac{Xe^{i\omega t}}{Fe^{i\omega t}} = \frac{X}{F} \tag{3.16}$$

Note that $H(\omega)$ is the ratio of (complex) amplitudes and is complex and usually depends on frequency, but is *not* a function of time. It defines the system in terms of its response to harmonic input.

For a harmonically vibrating system, note that

$$V = i\omega X; \quad A = i\omega V = -\omega^2 X \tag{3.17}$$

where:
- X is the amplitude of the displacement
- V is the amplitude of the velocity
- A is the amplitude of the acceleration

In vibration, certain FRFs are given names as indicated in Table 3.1. For example, the mobility μ is the ratio of the (complex) amplitudes of velocity and force, when both are time harmonic and in the steady state.

Table 3.1 Definitions of FRFs

Receptance	$\alpha = \dfrac{X}{F}$	Dynamic stiffness	$K = \dfrac{F}{X}$
Mobility	$\mu = \dfrac{V}{F}$	Impedance	$Z = \dfrac{F}{V}$
Accelerance	$\dfrac{A}{F}$	Apparent mass	$M = \dfrac{F}{A}$

Note also from Equation 3.17 that, since displacement, velocity and acceleration are all related by factors of $i\omega$, then $\mu = i\omega\alpha$, where α is the receptance, while the dynamic stiffness and impedance are related by $K = i\omega Z$.

A *point FRF* is one where the input and output are at the same point and act in the same direction. For a *transfer FRF* the input and output are at different points or in different directions. Finally, FRFs may equally involve rotations or moments.

3.1.7 Stiffness and flexibility

Stiffness and flexibility (the reciprocal of stiffness) are properties of any structural component that deforms elastically under the action of forces. The *effective stiffness* of the component can be found by considering the force–deformation relation or the potential energy–deformation relation as described here.

A *spring* is represented by the symbol shown in Figure 3.5. When a force F is applied to the ends of a spring it extends by an amount x where

$$F = kx \tag{3.18}$$

and k is the *spring constant* (units N m^{-1}). It is a constant if the spring is linear. The work done by this force (Equation 3.4) when stretching the spring is

$$U = \int_0^x F \, dx = \int_0^x kx \, dx = \tfrac{1}{2} kx^2 \tag{3.19}$$

and is stored in the spring as potential energy

$$V = \tfrac{1}{2} kx^2 \tag{3.20}$$

In practice, springs are often nonlinear. The force–deformation relationship for a hardening spring is shown in Figure 3.6. For a small-amplitude vibration about some equilibrium position $x = x_0$, we can *linearise* the behaviour by a Taylor series expansion around x_0, that is,

$$F(x) = F(x_0) + \left.\frac{dF}{dx}\right|_{x_0} (x - x_0) \tag{3.21}$$

Figure 3.5 Force applied to a spring.

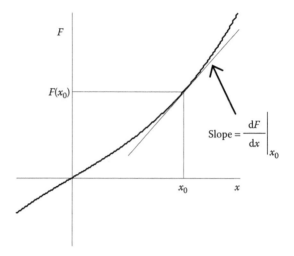

Figure 3.6 Force–deformation relation for hardening spring and linearisation around $x = x_0$.

so that the effective stiffness at x_0 is

$$k_{\text{eff}} = \left.\frac{dF}{dx}\right|_{x_0} \qquad (3.22)$$

Since vibration is often of a small amplitude, this linear model usually gives a good model of the response.

In systems of practical interest there is often more than one stiffness element and their effects combine to give the effective total stiffness. Two springs in *series* (Figure 3.7a) carry the same force F. If their extensions are x_1 and x_2, then

$$F = k_1 x_1; \qquad F = k_2 x_2 \qquad (3.23)$$

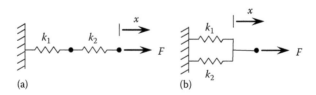

Figure 3.7 Two springs (a) in series and (b) in parallel.

Since the total extension $x = x_1 + x_2$, it follows that

$$F = k_{eff}x; \quad \frac{1}{k_{eff}} = \frac{1}{k_1} + \frac{1}{k_2}; \quad k_{eff} = \frac{k_1 k_2}{k_1 + k_2} \qquad (3.24)$$

On the other hand, two springs in *parallel* (Figure 3.7b) have the same extension x but carry forces F_1 and F_2, so that the total force $F = F_1 + F_2$. Consequently, the effective stiffness is such that

$$F = k_{eff}x; \quad k_{eff} = k_1 + k_2 \qquad (3.25)$$

The effective stiffnesses of more general deformable components can be found from strength of materials analysis. The analysis is beyond the scope of this chapter, and we will only quote results. For example, under the action of a force F at its tip, the cantilever beam in Figure 3.8a deforms in bending by an amount x such that

$$F = \left(\frac{3EI}{L^3}\right)x \qquad (3.26)$$

where:
  E is the elastic modulus
  I is the second moment of area of the cross-section
  L is the length of the cantilever

For a rectangular cross-section, $I = bh^3/12$, where b and h are the width and thickness of the beam, respectively. The effective stiffness of the cantilever at its tip is thus $k_{eff} = 3EI/L^3$. As a second example, under the action of a torque T the shaft in Figure 3.8b rotates through an angle θ where

$$T = \frac{GJ}{L}\theta \qquad (3.27)$$

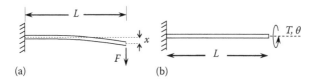

Figure 3.8 Deformation of (a) cantilever beam and (b) shaft in torsion.

where:
- G is the shear modulus of the material
- J is the polar moment of area of the cross-section
- L is the shaft length

Hence, $k_{eff} = GJ/L$. Table 3.2 lists various other common structural components and their effective stiffnesses.

3.2 FREE VIBRATION OF AN SDOF SYSTEM

In this section we will consider the free vibration of an SDOF system, first without damping and then with viscous damping. Although the discussion will concern a simple SDOF model, it is of central importance to vibration analysis, since the behaviour of all SDOF systems is similar and, as we will see in Section 3.5, the vibration of an MDOF system can be regarded as being composed of the response from a number of SDOF systems. Thus the SDOF system provides the building block for vibration analysis of complicated systems.

To anticipate the results, we will find that there is an *equilibrium position*, where the system can remain at rest, a *natural frequency*, the frequency at which an undamped system executes SHM about the equilibrium position if it is disturbed from equilibrium, and a *damping ratio*, which quantifies the effects of damping on the free vibration.

3.2.1 Undamped free vibration

Consider the SDOF system shown in Figure 3.9. The spring is assumed to be massless and hangs vertically, its free end when unloaded lying at $y = 0$ (Figure 3.9a). Suppose the mass is attached to the spring. The forces acting on the mass are shown in the free-body diagram, Figure 3.9d. If the mass is at rest and lies at the static equilibrium position (Figure 3.9b), the spring is stretched by an amount Δ, the static deflection. The end of the spring now lies at $y = \Delta$, or equally at $x = 0$, where $x = y - \Delta$.

Suppose now the mass is displaced from the equilibrium position (Figure 3.9c). The equation of motion is

$$m\ddot{y} = mg - ky \tag{3.28}$$

or alternatively

$$m\ddot{y} + ky = mg \tag{3.29}$$

Fundamentals of vibration 89

Table 3.2 Effective stiffnesses of structural components

System	Effective stiffness
Springs in series	$\dfrac{k_1 k_2}{k_1 + k_2}$
Springs in parallel	$k_1 + k_2$
Rod in axial extension	$\dfrac{EA}{L}$
Shaft in torsion	$\dfrac{GJ}{L}$
Cantilever beam (load at end)	$\dfrac{3EI}{L^3}$
Simply supported beam (load at centre)	$\dfrac{48EI}{L^3}$
Built-in at both ends (load at centre)	$\dfrac{192EI}{L^3}$
Built-in/simply supported (load at centre)	$\dfrac{768EI}{7L^3}$
Sway frame	$\dfrac{12EI}{L^3}$

Note: A is cross-sectional area.

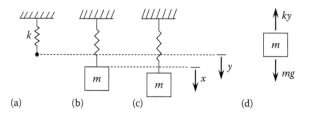

Figure 3.9 (a) Spring in unstretched position, SDOF system; (b) at rest in equilibrium position and (c) vibrating; (d) free-body diagram.

The equilibrium position, where $y = \Delta$ and $\ddot{y} = 0$, is at

$$\Delta = \frac{mg}{k} \tag{3.30}$$

Dividing Equation 3.29 by m leads to

$$\ddot{y} + \omega_n^2 y = g; \quad \omega_n = \sqrt{\frac{k}{m}} \tag{3.31}$$

or in terms of motions about the equilibrium position

$$\ddot{x} + \omega_n^2 x = 0; \quad \omega_n = \sqrt{\frac{k}{m}} \tag{3.32}$$

Equations 3.31 and 3.32 represent SHM about the equilibrium position at frequency $\omega_n = \sqrt{k/m}$ radians per second. A larger stiffness implies a higher natural frequency, while a larger mass implies a lower natural frequency. Note, also, that the static deflection is related to the natural frequency by $\Delta = g/\omega_n^2$.

The actual motion of the SDOF depends on the *initial conditions* that are applied when the motion is started. For example, suppose that at time $t = 0$, the mass is displaced by an amount x_0 and given a velocity v_0. Then, if we write the solution for SHM as (Equation 3.8)

$$x(t) = A \sin \omega_n t + B \cos \omega_n t \tag{3.33}$$

the initial conditions imply that

$$x(t) = \frac{v_0}{\omega_n} \sin \omega_n t + x_0 \cos \omega_n t \tag{3.34}$$

Figure 3.10 Cantilever with tip mass.

The importance of this is that, in principle, *all* undamped SDOF systems vibrate in this way: an equilibrium position (defined by the static extension of the spring, Equation 3.30) about which time-harmonic motion at frequency ω_n occurs, as given in Equation 3.32. The only differences are that for different systems – a building swaying in the wind, a car bouncing on its suspension, a child on a swing – the numerical values for ω_n or f_n will be different and the coordinate x might represent displacement or rotation and so forth. We might also need to put in much more effort to find the equation of motion for a complex system, but the fundamentals of vibration will remain the same: an undamped SDOF system vibrates time-harmonically at a natural frequency $\omega_n = \sqrt{k/m}$.

As an example, consider the cantilever beam supporting a mass m at its tip as shown in Figure 3.10. This might represent a bracket supporting a machine, or someone standing at the end of a diving board. The effective stiffness of the beam was found to be $k_{\text{eff}} = 3EI/L^3$ in Section 3.1.7. If we ignore the mass of the beam, then the static deflection will be $\Delta = mgL^3/3EI$, while the natural frequency is $\omega_n = \sqrt{3EI/mL^3}$.

As a second example, consider a pendulum of length L carrying a bob of mass m on the end of a massless string, as shown in Figure 3.11. If the pendulum is inextensible and rotates freely about point O, then taking moments about point O gives

$$mL^2\ddot{\theta} = -mgL\sin\theta \tag{3.35}$$

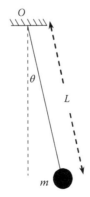

Figure 3.11 Simple pendulum.

or equally

$$\ddot{\theta} + \frac{g}{L}\sin\theta = 0 \tag{3.36}$$

This equation differs from Equation 3.32 because it is nonlinear. However, if we are looking for *small-amplitude* vibration, then we can *linearise* the equation of motion by a Taylor series expansion about the equilibrium position $\theta = 0$ (when the pendulum bob hangs vertically downwards at rest). If we note that $\sin\theta = \theta - \theta^3/3! + \ldots$ and ignore all the higher-order terms (i.e. assume $\theta^2 \ll 1$), then Equation 3.36 becomes

$$\ddot{\theta} + \frac{g}{L}\theta = 0 \tag{3.37}$$

Hence, the pendulum vibrates with a natural frequency $\omega_n = \sqrt{g/L}$. (One word of warning: sometimes the nonlinear terms, although small, can cause qualitatively quite different behaviour, especially over larger periods of time. One example would be if the pendulum could stretch. This is, however, outside the scope of this chapter.)

3.2.2 Energy: effective mass and stiffness

While we can use Newton's laws, as in Section 3.2.1, to find the equations of motion, energy provides a systematic alternative approach. It is especially useful for complex systems with many mass and/or stiffness elements but with many constraints, that is, many relations between the motions of the different parts of the system, and hence few DOFs (here, just a single DOF).

In an undamped free vibration, there are no external forces and hence no work is done on the system. There is no damping, and thus no dissipation of energy. Hence, the total energy $T + V$, the sum of the kinetic and potential energies, must be constant. Therefore,

$$\frac{d}{dt}(T+V) = 0 \tag{3.38}$$

For a linear system the kinetic and potential energies of an SDOF system can (usually) be written as

$$T = \frac{1}{2}m_{\text{eff}}\dot{x}^2$$

$$V = \frac{1}{2}k_{\text{eff}}x^2 + Ax + B \tag{3.39}$$

where A and B are constants and the potential energy has gravitational and elastic (Equation 3.20) contributions. Hence,

$$\frac{d}{dt}(T+V) = (m_{\text{eff}}\ddot{x} + k_{\text{eff}}x + A)\dot{x} = 0 \tag{3.40}$$

This implies that either $\dot{x} = 0$ (a trivial solution) or $m_{\text{eff}}\ddot{x} + k_{\text{eff}}x + A = 0$. This gives the equilibrium position at $x = -A/k_{\text{eff}}$ and the natural frequency $\omega_n = \sqrt{k_{\text{eff}}/m_{\text{eff}}}$.

Note that the kinetic and potential energies both change with time, but their sum is constant: there is thus a continual oscillatory exchange of energy between potential and kinetic. If, for example, $x(t) = C \sin \omega_n t$, then (with $A = B = 0$)

$$V = \frac{1}{2}k_{\text{eff}}x^2 = \frac{1}{2}k_{\text{eff}}C^2 \sin^2 \omega_n t = \frac{1}{4}k_{\text{eff}}C^2(1 - \cos 2\omega_n t)$$

$$T = \frac{1}{2}m_{\text{eff}}\dot{x}^2 = \frac{1}{2}m_{\text{eff}}\omega_n^2 C^2 \cos^2 \omega_n t = \frac{1}{4}m_{\text{eff}}\omega_n^2 C^2(1 + \cos 2\omega_n t) \tag{3.41}$$

The kinetic energy is a maximum when the potential energy is zero (at $x = 0$), while the potential energy is maximum when $\dot{x} = 0$.

As an example, consider the SDOF system in Figure 3.9. Taking $y = 0$ as the datum for gravitational potential energy, this gives

$$T = \frac{1}{2}m\dot{y}^2; \quad V = \frac{1}{2}ky^2 - mgy \tag{3.42}$$

Summing, differentiating and equating to zero gives the equation of motion, Equation 3.29. For the pendulum of Figure 3.11, in terms of θ,

$$T = \frac{1}{2}m(L\dot{\theta})^2; \quad V = mgL(1 - \cos\theta) \tag{3.43}$$

Noting that, for small θ, $\cos\theta = 1 - \theta^2/2 + ...$; Equation 3.37 follows as the equation of motion.

Consider, finally, the more complicated example shown in Figure 3.12. The bar is rigid, of length L, mass m, and its second moment of mass about the pivot point A is I. At its midpoint a spring of stiffness k is attached, while at the lower end there is a mass m_2. Three coordinates are shown in the figure: the rotation θ and the displacements x and y, but they are not independent: there is only one DOF. Assume the vibration is small so that the coordinates are related by *constraint equations*

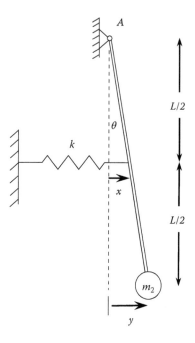

Figure 3.12 Energy method: free vibration of a bar. Note: three coordinates but only one DOF.

$$y = L\sin\theta \approx L\theta; \quad x = \frac{L\sin\theta}{2} \approx \frac{L\theta}{2} \qquad (3.44)$$

The kinetic energy is then

$$T = \frac{1}{2}I\dot\theta^2 + \frac{1}{2}m_2\dot y^2 \qquad (3.45)$$

with terms arising from each of the mass elements in the system. The potential energy is

$$V = \frac{1}{2}kx^2 + m_1 g(1-\cos\theta)\frac{L}{2} + m_2 g(1-\cos\theta)L \qquad (3.46)$$

and has terms from the strain energy in the spring and gravity. Note, also, for small vibrations, $\cos\theta \approx 1 - \theta^2/2$, so that

$$V = \frac{1}{2}kx^2 + (m_1 + 2m_2)\frac{gL\theta^2}{4} \qquad (3.47)$$

Because of the constraints of Equation 3.44, we can eliminate two of the coordinates and write the equation of motion in terms of the remaining coordinate, our chosen DOF. The natural frequency is the same, irrespective of which we choose, but the effective mass and stiffness depend on our reference point. Suppose we choose y as our DOF. Then $x = y/2$ and $\theta = y/L$. The energies become

$$T = \frac{1}{2}\left(m_2 + \frac{I}{L^2}\right)\dot{y}^2; \quad V = \frac{1}{2}\left(\frac{k}{4} + \frac{(m_1 + 2m_2)g}{4L}\right)y^2 \tag{3.48}$$

and hence the effective mass and stiffness *at point B* are

$$m_{\text{eff}} = \left(m_2 + \frac{I}{L^2}\right); \quad k_{\text{eff}} = \left(\frac{k}{4} + \frac{(m_1 + 2m_2)g}{4L}\right) \tag{3.49}$$

with the natural frequency of the system being $\omega_n = \sqrt{k_{\text{eff}}/m_{\text{eff}}}$.

3.2.3 Viscously damped SDOF system

Figure 3.13a shows a schematic of a *viscous damper*, or dashpot. Under the action of a force F_d, a viscous damper opens at a rate $\dot{x}$ such that

$$F_d = c\dot{x} \tag{3.50}$$

where c is the *damping coefficient* and has units of N s m^{-1}. The damper represents a piston moving in a cylinder filled with a viscous fluid and the force is associated with laminar flow of the fluid. The damper dissipates *vibrational energy* into heat; as a result, the free vibration decays with time.

The SDOF system and free-body diagram are shown in Figure 3.13b and c respectively. The equation of motion is

$$m\ddot{x} = -c\dot{x} - kx \tag{3.51}$$

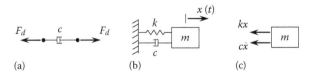

Figure 3.13 Viscously damped SDOF system: (a) viscous damper; (b) system; (c) free-body diagram.

or alternatively

$$m\ddot{x} + c\dot{x} + kx = 0 \tag{3.52}$$

Dividing by m gives

$$\ddot{x} + 2\zeta\omega_n\dot{x} + \omega_n^2 x = 0 \tag{3.53}$$

where the *natural frequency* and *damping ratio* are

$$\omega_n = \sqrt{\frac{k}{m}}; \quad \zeta = \frac{c}{2\sqrt{km}} = \frac{c}{2m\omega_n} \tag{3.54}$$

Thus, the free vibration depends on two parameters, ζ and ω_n. The solution to the equation of motion depends on the value of the damping ratio ζ. If $\zeta < 1$, the system is said to be *underdamped*. In fact, the large majority of engineering structures are underdamped, with the damping being small, typically $\zeta \sim 0.01–0.1$ or so. As a result, the free vibration is a decaying oscillation given by

$$x(t) = Ae^{-\zeta\omega_n t}\cos(\omega_d t + \theta) \tag{3.55}$$

which represents oscillations at a frequency ω_d which decay at the rate $\exp(-\zeta\omega_n t)$. A and θ are constants that depend on the initial conditions and

$$\omega_d = \omega_n\sqrt{1-\zeta^2} \tag{3.56}$$

is the *frequency of damped vibration*. If ζ is small (typically if $\zeta < 0.1$ so that $\zeta^2 \ll 1$: this is the case for the large majority of practical situations) then $\omega_d \approx \omega_n$. The free vibration is shown in Figure 3.14 for various values of ζ.

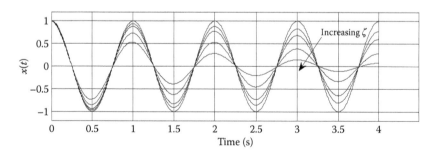

Figure 3.14 Free vibration of SDOF system ($x_0 = 1$; $v_0 = 0$): $\zeta = 0, 0.01, 0.02, 0.05, 0.1$.

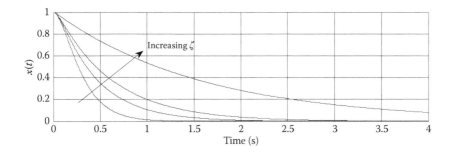

Figure 3.15 Free vibration of critically damped and overdamped SDOF systems ($x_0 = 1$; $v_0 = 0$): $\zeta = 1$, 1.5, 2.5.

The oscillations are at approximately the same frequency ω_n, while the decay is more rapid as ζ increases.

If $\zeta > 1$, the system is *overdamped*, while if $\zeta = 1$, the system is *critically damped*; in both cases the free vibration decays *without oscillation*. The free vibration is given by

$$x(t) = Ae^{\left(-\zeta+\sqrt{\zeta^2-1}\right)\omega_n t} + Be^{\left(-\zeta-\sqrt{\zeta^2-1}\right)\omega_n t}; \quad \zeta > 1$$

$$x(t) = (A + Bt)e^{-\omega_n t}; \quad \zeta = 1 \tag{3.57}$$

illustrated in Figure 3.15.

3.2.4 Estimation of SDOF-system parameters from measured free vibration

We often want to estimate ω_n and ζ, and one way to do this is from the decay of a free vibration. Assuming the system is underdamped so that the response is oscillatory, a free vibration is initiated (hit the structure, or displace it and let go) and the resulting vibration is measured (Figure 3.16).

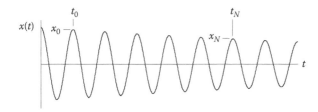

Figure 3.16 Estimation of parameters from free vibration of undamped SDOF system.

Two peaks are selected, N cycles apart, which occur at times $t = t_0$ and $t = t_N$ and have amplitudes x_0 and x_N. The period of the decaying oscillation (Equation 3.55) is then

$$\tau = \frac{t_N - t_0}{N} = \frac{2\pi}{\omega_d} \approx \frac{2\pi}{\omega_n} \tag{3.58}$$

where the approximation is valid if $\zeta^2 \ll 1$. The amplitudes of the peaks are related by

$$\frac{x_N}{x_0} = e^{-\zeta\omega_n(t_N - t_0)} = e^{-\zeta\omega_n N\tau} \tag{3.59}$$

from which

$$\zeta = \frac{1}{\omega_n N\tau} \ln \frac{x_0}{x_N} \approx \frac{1}{2\pi N} \ln \frac{x_0}{x_N} \tag{3.60}$$

Sometimes the decay is expressed in terms of the *log decrement*

$$\delta = \ln \frac{x_0}{x_1} = \frac{1}{N} \ln \frac{x_0}{x_N} \tag{3.61}$$

so that $\zeta = \delta/2\pi$.

3.3 TIME-HARMONIC FORCED VIBRATION OF AN SDOF SYSTEM

Suppose now that the SDOF system is excited by a force $f(t)$ as shown in Figure 3.17. In particular, in this section the force is time harmonic, so that $f(t) = F \exp(i\omega t)$. Such forces are common and may be produced by rotating or reciprocating machines running at a constant speed (although generally the forces from such machines are not strictly SHM). We will also use the response to harmonic excitation as a building block for the response to more general excitation in Section 3.4.

The equation of motion is now

$$m\ddot{x} + c\dot{x} + kx = f(t) = Fe^{i\omega t} \tag{3.62}$$

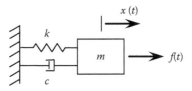

Figure 3.17 Forced SDOF system.

The full solution is the sum of two parts:

1. The *complementary function*: the solution to $m\ddot{x} + c\dot{x} + kx = 0$. This is just a free vibration as discussed in Section 3.2.4, the solution of which, if $\zeta < 1$, is a decaying, damped oscillation given by Equation 3.55. This motion decays to zero as the time increases: it is a *transient motion*.
2. The *particular integral*. This is a solution for $x(t)$ that satisfies the right-hand side, $F \exp(i\omega t)$. We might suspect (and will see in Equations 3.63 and 3.64) that this is also time harmonic and of the form $x(t) = X \exp(i\omega t)$, a harmonic motion at the forcing frequency ω. This does not decay with time but keeps going forever: it is the *steady-state motion*, that is, the motion that remains as $t \to \infty$. This steady-state motion is the motion of interest to us in this section.

Thus, if, for example, the force $f(t)$ is time harmonic and commences at $t=0$, the response is not purely harmonic but contains a transient component. To find the steady-state motion, we assume that $x(t) = X \exp(i\omega t)$ and substitute this into Equation 3.41. This gives

$$-\omega^2 m X e^{i\omega t} + i\omega c X e^{i\omega t} + k X e^{i\omega t} = F e^{i\omega t} \quad (3.63)$$

Note that the term $\exp(i\omega t)$ can be cancelled out, removing the time-dependence and leaving only a relation between the amplitude of the force, the system properties m, c and k and the response amplitude. This is given by

$$X = \frac{1}{k - \omega^2 m + i\omega c} F \quad (3.64)$$

and consequently the receptance (the displacement per unit force) is

$$\alpha = \frac{X}{F} = \frac{1}{k - \omega^2 m + i\omega c} = \frac{1}{k} \frac{1}{\left(1 - \omega^2/\omega_n^2\right) + i\left(2\zeta\omega/\omega_n\right)} \quad (3.65)$$

Note that this is *complex*: the response thus has a *phase* with respect to the applied force. It is also *frequency dependent* and in particular depends on the *frequency ratio* ω/ω_n. The behaviour thus depends not so much on the excitation frequency, but rather on its value compared with the natural frequency. It also depends on the damping ratio ζ which, as we have noted, is usually small.

The response X divided by the response to a static force F (i.e. F/k) is shown in Figures 3.18 and 3.19 for various levels of damping. The magnitude and phase are

$$\left|\frac{X}{F/k}\right| = \sqrt{\frac{1}{\left(1-\omega^2/\omega_n^2\right)^2 + \left(2\zeta\omega/\omega_n\right)^2}}; \quad \phi = -\tan^{-1}\left(\frac{2\zeta\omega/\omega_n}{1-\omega^2/\omega_n^2}\right) \quad (3.66)$$

Note that the phase is negative: X *lags* F. The most obvious feature is the large *resonance* peak when $\omega \approx \omega_n$. This peak is very sharp if the damping is light: as a result, the linear scales of Figure 3.18 cannot capture the behaviour well – the *dynamic range* is too large. Consequently, we usually

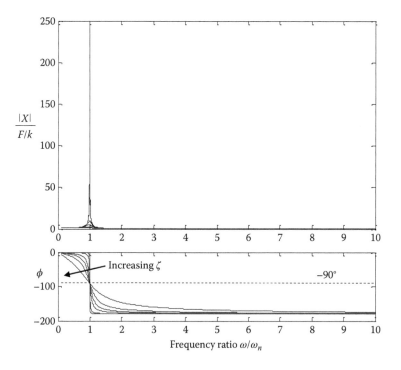

Figure 3.18 Non-dimensional response of an SDOF system to a time-harmonic force: $\zeta = 0.01, 0.02, 0.05, 0.1, 0.2$; linear axes. Phase in degrees.

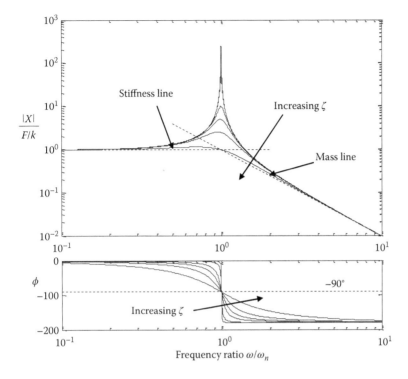

Figure 3.19 Non-dimensional response of an SDOF system to a time-harmonic force: $\zeta = 0.01, 0.02, 0.05, 0.1, 0.2$; mass and stiffness asymptotes; logarithmic axes. Phase in degrees.

plot FRFs on log axes as in Figure 3.19. This also highlights the asymptotes of the response at low and high frequencies. The general features of the response are:

- At *low frequencies*, defined such that $\omega \ll \omega_n$, $X \approx F/k$, and the system responds quasi-statically. The response is dominated by the stiffness k and is said to be *stiffness controlled*. The low frequency asymptote, the *stiffness line*, is marked in Figure 3.19.
- At *high frequencies*, when $\omega \gg \omega_n$, $X \approx -F/\omega^2 m$, and the system is *mass controlled*. The *mass line* is also indicated in Figure 3.19. It decreases as $(1/\omega^2)$, or at -40 dB per decade. Note that, at high frequencies, the phase lag is close to 180°.
- Around *resonance*, when $\omega \approx \omega_n$, the response is *damping controlled* and depends sensitively on the level of damping and the frequency. If $\omega = \omega_n$, then $X = -iF/\omega c = -iF/2\zeta k$, the amplitude of the resonance peak is inversely proportional to the damping ratio ζ and the

response lags the force by 90°. (The peak value actually occurs at $\omega = \omega_n\sqrt{1-2\zeta^2}$ or at $\omega \approx \omega_n$ if the damping is small.) Note also that there is a very rapid change in the phase around resonance.

As a final comment, the imaginary part of the receptance is Im$\{\alpha\} \leq 0$, while Re$\{\mu\} \leq $ Re$\{i\omega\alpha\} \geq 0$. This is a consequence of the fact that the SDOF system is *passive*: it does not create energy but can only dissipate it.

In summary, then, the response of a lightly damped SDOF system shows a sharp resonance peak. If the excitation frequency is close to the natural frequency, then a small force can produce a large response. The magnitude and phase of the response change rapidly with frequency around resonance. The response in a narrow frequency range around resonance is shown in Figure 3.20. The peak response X_{max} occurs when $\omega \approx \omega_n$. The two *half-power points*, at $\omega \approx \omega_1$ and $\omega \approx \omega_2$, occur when $|X| = X_{max}/\sqrt{2}$ (−3 dB with respect to the peak or equivalently $|X|^2 = X_{max}^2/2$, hence the name 'half-power point'). The *bandwidth* of the resonance is $\Delta\omega = \omega_2 - \omega_1$, and if the damping is light is given by

$$\Delta\omega = 2\zeta\omega_n \tag{3.67}$$

Because the interesting features of the resonance are compressed into a narrow frequency range, we sometimes display the FRF as a *polar plot*, sometimes called a *Nyquist plot*. This is a plot of the imaginary part of the FRF against the real part. The effect of displaying the FRF as a polar plot is

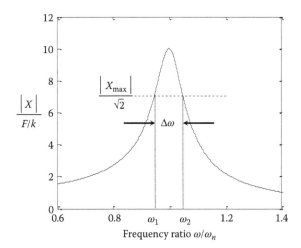

Figure 3.20 Response of an SDOF system to a time-harmonic force: $\zeta = 0.05$; half-power points and bandwidth.

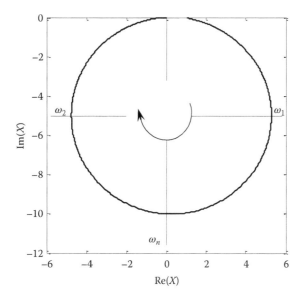

Figure 3.21 Polar plot of the mobility Xk/F of an SDOF system to a time-harmonic force: $\zeta = 0.05$. Arrow indicates increasing frequency ω.

to enlarge the part of the plot associated with the narrow frequency range around resonance. Figure 3.21 shows the polar plot of the displacement response Xk/F. The locus of this curve as frequency increases is approximately a circle. It starts at the origin when $\omega = 0$ (and when the phase is 0°), and sweeps clockwise around a circular locus. Its magnitude is a maximum around $\omega = \omega_n$ (see the peak in Figure 3.20), the phase being −90°, and tends again to the origin as $\omega \to \infty$, where the phase is −180°. The 'half-power points' are those frequencies where the square of the magnitude of the response (a measure of the 'power' of the response) is one-half the peak value squared: in the polar plot, the two half-power points lie on opposite ends of a diameter, the phases there being −45° and −135°.

3.3.1 Estimation of SDOF-system parameters from measured FRFs

There are various ways to estimate parameters from a measured FRF. The simplest is the *peak value* method. The response around resonance is measured (Figure 3.20). The frequency at which the peak occurs gives an estimate of ω_n. The half-power bandwidth is then estimated. From Equation 3.67, this gives an estimate for $\zeta = \Delta\omega/2\omega_n$. In Figure 3.20, $\Delta\omega \approx 0.1\omega_n$ and hence, $\zeta \approx 0.05$.

3.3.2 Vibration isolation

Vibrating sources, such as machines, can transmit vibration to their surroundings and this is often undesirable, for example causing discomfort, annoyance, fatigue and damage. We are often interested in isolating the vibrating source from its surroundings. In a simple SDOF-system model (Figure 3.22), the mass represents the machine which, when running, causes harmonic forces $f(t) = F \exp(i\omega t)$ to be produced. These forces produce vibration and consequently transmit forces through the supports of the machine, the spring and the damper, and hence to the surrounding structure. Vibration isolation involves the design of the supports so that these transmitted forces are acceptably small.

The equation of motion of the mass is given by Equation 3.62. The total force transmitted to the structure, which is assumed to be rigid (Figure 3.22b), is

$$f_{TR}(t) = kx + c\dot{x} \tag{3.68}$$

where the two terms represent the forces transmitted by the spring and damper respectively. If we assume time-harmonic motion then in the steady state $x(t) = X \exp(i\omega t)$ and $f_{TR}(t) = F_{TR} \exp(i\omega t)$. Substituting these into Equations 3.62 and 3.68 gives

$$(k - \omega^2 m + i\omega c)X = F;$$

$$F_{TR} = (k + i\omega c)X \tag{3.69}$$

The *transmissibility* TR is defined as the magnitude of the ratio F_{TR}/F, and vibration isolation involves making TR acceptably small. From Equation 3.69:

$$TR = \left|\frac{F_{TR}}{F}\right| = \left|\frac{k + i\omega c}{k - \omega^2 m + i\omega c}\right| = \left|\frac{1 + i2\zeta\omega/\omega_n}{1 - \omega^2/\omega_n^2 + i2\zeta\omega/\omega_n}\right| \tag{3.70}$$

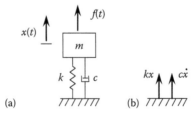

Figure 3.22 (a) Machine on supports on a rigid base as an SDOF system and (b) transmitted forces.

Fundamentals of vibration

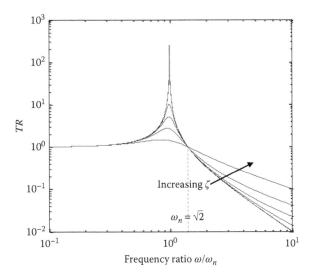

Figure 3.23 Transmissibility of SDOF system: $\zeta = 0.01, 0.02, 0.05, 0.1, 0.2$.

The transmissibility is shown in Figure 3.23 for various damping ratios. For lower frequencies ($\omega \ll \omega_n$), then $TR \approx 1$ and no isolation occurs, while around resonance ($\omega \approx \omega_n$), there is a resonance peak and TR can be large, the transmitted force being larger than the exciting force F generated in the machine. At higher frequencies, TR decreases as frequency increases. It can be shown that $TR = 1$ when $\omega = \sqrt{2}\omega_n$, whatever the value of ζ. If the damping is zero, then $TR \to \infty$ at resonance and decreases as $1/\omega^2$ at high frequencies, that is, at -40 dB per decade. Increasing the damping decreases the peak at resonance, but increases the transmitted forces at high frequencies.

What these observations lead us to conclude is that, to obtain some vibration isolation, we require $TR < 1$. Consequently, we must choose the stiffness k of the supports so that the natural frequency $\omega_n = \sqrt{k/m}$ is small enough compared with the operating frequency ω, which is often fixed by the operation of the machine. Thus, we require ω/ω_n to be large (and of course ω/ω_n must be greater than $\sqrt{2}$ to achieve *any* isolation) and hence, we need flexible mounts, that is, mounts with low stiffness, to give a low natural frequency. It might be intuitive to try to bolt the machine down tightly to the structure, but that will usually make things worse: we need soft mounts, typically metal springs, rubber mounts or pads made of cork, felt or rubber to introduce flexibility. If we neglect damping, then at high enough frequencies (i.e. if $\omega \gg \omega_n$, so that $TR < 1$) then

$$TR \approx \frac{\omega_n^2}{\omega^2} \tag{3.71}$$

This provides a guide as to how stiff to make the supports. Increasing the damping then increases TR, thus increasing the transmitted forces. Damping thus has an undesirable effect at high frequencies. Of course, if the damping in the supports is light, the response at resonance will be large, so some damping may be required to limit the resonant response, which might be unavoidable if, for example, the speed of the machines passes through resonance as the machine speeds up.

The analysis above involved various simplifying assumptions; for example, that the structure on which the machine is mounted is infinitely stiff and that the supports are massless. In practice, these and other effects are often important and affect the transmitted forces, but these are outside the scope of this chapter.

3.3.3 Other applications

Two other applications are worthy of brief mention. First, vibration isolation is also desirable when equipment (a delicate instrument, etc.) is placed in a vibrating environment (Figure 3.24a). The equation of motion is

$$m\ddot{x} + c\dot{x} + kx = c\dot{y} + ky \tag{3.72}$$

Motion of the surroundings, $y(t) = Y \exp(i\omega t)$, then causes vibration, $x(t) = X \exp(i\omega t)$, of the instrument, which we may want to reduce. The *displacement transmissibility*

$$TR = \left|\frac{X}{Y}\right| \tag{3.73}$$

is given by exactly the same expression as the force transmissibility in Equation 3.47. The same conclusions hold: good isolation requires flexible supports.

As a second example, forces generated as a machine runs are often caused by some out-of-balance rotation. Suppose the machine in Figure 3.24b has a total mass m, with an out-of-balance mass m_0 rotating at a frequency ω

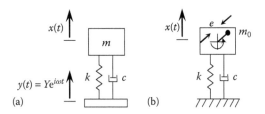

Figure 3.24 SDOF system with (a) base excitation and (b) rotating imbalance.

with an eccentricity e. If only vertical motion occurs, then the equation of motion becomes

$$m\ddot{x} + c\dot{x} + kx = m_0 e\omega^2 e^{i\omega t} \tag{3.74}$$

so that the force that excites the vibration, $m_o e\omega^2$, now increases with frequency.

3.3.4 Damping and loss factor

There are many mechanisms of damping, each of which dissipates vibrational energy. Many damping mechanisms are nonlinear, but if the damping is light, then approximating the behaviour as linear results in small errors. So far, we have considered only *viscous damping*, in which the damping force is proportional to velocity. This is associated with the laminar flow of viscous fluids in dampers, dashpots and shock absorbers. It is linear and hence a common choice, even though not common in practice: the assumption of linearity simplifies the analysis substantially.

There are many causes of *hysteresis* when structures are repeatedly stressed, including viscoelastic, thermal and micromechanical effects and internal friction. Such damping mechanisms are grouped together in what is termed *structural, material* or *hysteretic damping*. Measurements indicate that the energy loss per cycle is approximately proportional to the square of the displacement. This implies that the damping force is proportional to the displacement, but, of course, it must act with a phase that opposes the velocity of the structure, so that it always does negative work on the structure and hence dissipates energy. For time-harmonic motion, this can be allowed for by introducing a complex stiffness, $k(1 + i\eta)$, into the SDOF system, where η is the *loss factor*. This is often approximately independent of frequency, at least over moderate frequency ranges. The presence of the term $i\eta$ ensures that the imaginary part of the spring force, $i\eta kx$, is in phase with the velocity $i\omega X$ and, hence, the complex stiffness dissipates energy. With structural damping, the equation of motion for time-harmonic vibration becomes

$$m\ddot{x} + k(1 + i\eta)x = Fe^{i\omega t} \tag{3.75}$$

so that the response is $x(t) = X \exp(i\omega t)$ where

$$\frac{X}{F} = \frac{1}{k - \omega^2 m + i\eta k} = \frac{1}{k} \frac{1}{1 - \omega^2/\omega_n^2 + i\eta} \tag{3.76}$$

The receptance again has a resonance peak when $\omega \approx \omega_n$ and this is sharp if $\eta \ll 1$. There is also low- and high-frequency stiffness and mass-controlled regions. The bandwidth of the resonance peak is now $\Delta\omega = \omega\eta$, so that we can regard η as being equivalent to 2ζ. The polar plot of the receptance is a circle. In practical applications, η is typically 0.01 or so, usually lying in the range $0.001 < \eta < 0.1$.

Other forms of damping include *velocity-squared damping* $\left(f_d \propto \dot{x}^2\right)$, which is associated with turbulent fluid flow, wind resistance, drag and so on. *Dry friction*, or *Coulomb damping*, occurs due to the sliding of unlubricated surfaces, slippage in joints and so forth. The damping force is often approximately constant when slip occurs and is determined by the coefficient of friction between the surfaces. Finally, vibrating structures produce sound in the surrounding air, and sound energy radiates away from the structure, thus acting as damping.

These forms of damping are often nonlinear, but usually light, so that nonlinear effects are small. In practice, especially for harmonic motion, we can regard them as being equivalent to an amount of viscous damping (or structural damping) with the effective damping ratio (or loss factor) being chosen, so that the amount of energy dissipated by the actual damping mechanism is equal to that dissipated by the equivalent viscous damper (or structural damper). Damping constants and loss factors can rarely be predicted; they are usually measured. For a more detailed discussion of damping, see Nashif et al. (1985).

3.4 RESPONSE OF AN SDOF SYSTEM TO GENERAL EXCITATION

In practice, excitation is frequently not time harmonic. It may be transient, that is, nonzero, only for a finite time, or more generally, it may be made up of the superposition of a number of time-harmonic components at different frequencies. In this section, we will consider the response to more general excitation using two approaches. One approach is in the *time domain*, and represents the response as the sum of the time responses to impulses, while the other is in the *frequency domain*, and involves *frequency response*. More details of the theory behind the approaches can be found in Chapter 4. Both make use of the fact that the SDOF system is *linear* so that *superposition* applies (see also Section 2.2.9). In summary, this means that if a force $f_1(t)$ would produce a response $x_1(t)$, while a second force $f_2(t)$ would produce a response $x_2(t)$, then if a force $f(t) = f_1(t) + f_2(t)$, that is, the sum of the two individual forces, was applied, the response would be $x(t) = x_1(t) + x_2(t)$, that is, the sum of the responses each force component would individually produce. Equally, if a harmonic force $F_1 \exp(i\omega_1 t)$ at frequency ω_1 were to produce a response $X_1 \exp(i\omega_1 t)$, while a force at a second frequency ω_2 produced a response $X_2 \exp(i\omega_2 t)$, then the two harmonic forces applied simultaneously would produce a response $x(t) = X_1 \exp(i\omega_1 t) + X_2 \exp(i\omega_2 t)$, that is, the sum of the two individual harmonics.

3.4.1 Transient excitation: impulse response

The *impulse* of a force is defined by Equation 3.5 in Section 3.1.3, and is related to the change in momentum of a mass by Equation 3.6. An *impulsive force* is one with a very short duration, but a finite impulse. We might think of the *unit impulse* applied at time t_0 as the pulse shown in Figure 3.25: it is nonzero for a finite time ε and the magnitude equals $1/\varepsilon$, so that the impulse = 1. In the limit $\varepsilon \to 0$, this becomes the *delta function* at $t = t_0$, which has the properties

$$\delta(t-t_0) = 0, t \neq t_0; \quad \int_{-\infty}^{\infty} \delta(t-t_0)dt = 1; \quad \int_{-\infty}^{\infty} x(t)\delta(t-t_0)dt = x(t_0) \quad (3.77)$$

Note that the delta function 'sifts out' the value of $x(t)$ at $t = t_0$.

The *impulse response* $h(t)$ of a system is defined as the response of the system to a unit impulse applied at $t=0$, assuming the system is at rest and at the equilibrium position before the impulse is applied. The response of an undamped SDOF system to a unit impulse can be found by noting (see Equation 3.6) that a unit impulse produces a unit change in momentum, so that immediately after the impulse is applied, the velocity v_0 is such that $mv_0 = 1$, the velocity for $t < 0$ being zero. The response is given by Equation 3.21 with $x_0 = 0$, $v_0 = 1/m$, so that the impulse response for an undamped SDOF system is

$$h(t) = \frac{1}{m\omega_n} \sin \omega_n t \quad (3.78)$$

Similarly, from Equation 3.55 for an underdamped SDOF system,

$$h(t) = \frac{1}{m\omega_d} e^{-\zeta\omega_n t} \sin \omega_d t \quad (3.79)$$

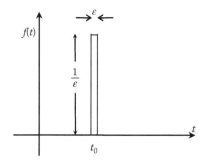

Figure 3.25 Unit impulse at time t_0.

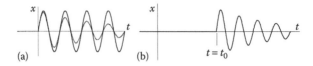

Figure 3.26 Impulse response of (a) undamped and underdamped ($\zeta = 0.05$) SDOF systems and (b) response to impulse I at time t_0.

These impulse responses are shown in Figure 3.26.

Now suppose an impulse I is applied at time t_0, as in Figure 3.25. The response $x(t)$ will then be

$$x(t) = Ih(t - t_0) \tag{3.80}$$

that is, the impulse response $h(t)$, shifted in time by an amount t_0, and I times as large. This is shown in Figure 3.26b.

3.4.2 Response to general excitation: convolution

Now, suppose a transient force $f(t)$ is applied to an SDOF system. We can regard the force as being made up of a series of impulses. Figure 3.27a shows the force $f(t)$, with the impulse at time $t = \tau$ and of width $\delta\tau$ (shaded). This impulse is

$$\delta I = f(\tau)\delta\tau \tag{3.81}$$

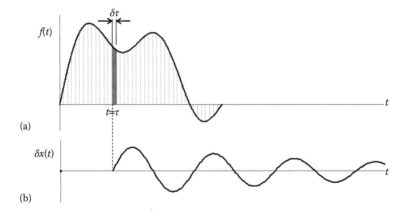

Figure 3.27 (a) Transient force as a series of impulses, with the impulse at $t = \tau$, of width $\delta\tau$ (shaded); (b) response $\delta x(t)$ to impulse $f(\tau)\delta\tau$ applied at time $t = \tau$.

The response to this impulse is therefore $\delta x(t) = \delta I\, h(t - \tau) = f(\tau)h(t - \tau)\delta\tau$. The total response is therefore

$$x(t) = \int_{-\infty}^{t} f(\tau)h(t - \tau)\,d\tau \tag{3.82}$$

(where we note that the response to an impulse applied at $\tau > t$ is zero). If $f(\tau) = 0$ for $\tau < 0$, then this becomes

$$x(t) = \int_{0}^{t} f(\tau)h(t - \tau)\,d\tau \tag{3.83}$$

This is the *convolution integral*. For idealised pulse shapes, such as a half-sine pulse associated with a bump or hitting a structure with a hammer, and for simple systems, such as an SDOF system, we might be able to integrate this analytically; otherwise, we would use numerical integration.

Structures can be subjected to base excitation, for example due to earthquake loading on a building, or if a package containing a delicate item is dropped on the floor. For an SDOF system, the situation is as shown in Figure 3.24a. The *relative displacement* of the mass with respect to the base is $z = x - y$. Substituting this into the equation of motion (Equation 3.72) gives

$$m\ddot{z} + c\dot{z} + kz = -m\ddot{y}; \quad \Rightarrow \ddot{z} + 2\zeta\omega_n\dot{z} + \omega_n^2 z = -\ddot{y} \tag{3.84}$$

Thus it appears as if a 'reversed inertia force' equal to $-m\ddot{y}$ acts on the SDOF system. The response, by convolution, is

$$z(t) = \int_{0}^{t} (-m\ddot{y})h(t - \tau)\,d\tau \tag{3.85}$$

3.4.3 Periodic excitation

A periodic excitation is one which repeats continuously after equal intervals of time T, called the period, so that $f(t + T) = f(t)$. Such forces arise from the repetitive operation of rotating or reciprocating machines. An example is shown in Figure 3.28. A periodic force can be written as a sum of time harmonics at frequencies which are integer multiples of the *fundamental frequency* $\omega_0 = 2\pi/T$. The amplitudes of the harmonics can be found by decomposing $f(t)$

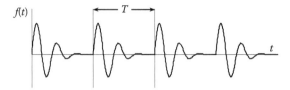

Figure 3.28 Periodic force: period T.

into a *Fourier series*.* It is out of place here to discuss Fourier series analysis in depth. However, the principles of *harmonic analysis* or *spectral analysis* are as follows (see also Figure 3.29). For more details, see Chapter 4.

- The force $f(t)$ is *periodic* in time.
- It can be expressed in terms of frequency components as $f(t) = \sum_{n=-\infty}^{\infty} F_n e^{in\omega_0 t}$.
- The response to $F_n e^{in\omega_0 t}$ is $X_n e^{in\omega_0 t} = H(n\omega_0) F_n e^{in\omega_0 t}$, where $H(n\omega_0)$ is the *frequency response* for frequency $(n\omega_0)$, for example, the receptance $\alpha(n\omega_0)$ in Equation 3.65.
- The response in the time domain is the sum of these harmonics, that is, $x(t) = \sum_{n=-\infty}^{\infty} X_n e^{in\omega_0 t} = \sum_{n=-\infty}^{\infty} H(n\omega_0) F_n e^{in\omega_0 t}$. This is also periodic.

Note: digital spectral analysis of nonperiodic signals is implemented by selecting sections of a signal and analysing them as if they were periodic. The artefacts so generated can be minimised by suitable signal processing (see Chapter 4).

3.4.4 Response to general excitation: Fourier transform and frequency response

The response to a general excitation can be found in a similar manner using the *Fourier transform*. The force can be written as a superposition of time harmonics at frequency ω as

$$f(t) = \int_{-\infty}^{\infty} F(\omega) e^{i\omega t} d\omega \tag{3.88}$$

---

* If $f(t)$ is periodic with a period T, then it can be written as the complex Fourier series

$$f(t) = \sum_{n=-\infty}^{\infty} F_n e^{in\omega_0 t}; \quad \omega_0 = 2\pi/T \tag{3.86}$$

where

$$F_n = \frac{1}{T} \int_0^T f(t) e^{-in\omega_0 t} dt \tag{3.87}$$

An alternative is to express $f(t)$ as a sum of harmonics $\sin n\omega_0 t$ and $\cos n\omega_0 t$, but since we found the response to a time-harmonic force $F\exp(i\omega t)$ in Section 3.3, the complex series is preferable.

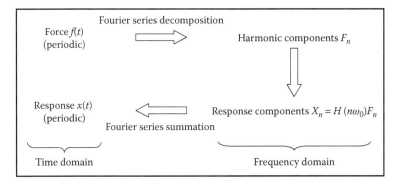

Figure 3.29 Harmonic analysis: response to a periodic force.

where the *spectrum* $F(\omega)$ is the *Fourier transform* of $f(t)$, given by

$$F(\omega) = \frac{1}{2\pi} \int_{-\infty}^{\infty} f(t) e^{-i\omega t} dt \qquad (3.89)$$

Note that, in this case, the spectrum is continuous: $F(\omega)$ is a continuous function of frequency. Formally $f(t)$ and $F(\omega)$ are a *Fourier transform pair*. (The factor of 2π can be included in many different ways.)

Again, it is not the place here to explain Fourier transform methods in detail. Instead, we should point out how the response $x(t)$ to the force $f(t)$ can be found by transforming from the *time domain* to the *frequency domain*, writing the response to each frequency component of the force $F(\omega)\exp(i\omega t)$ in terms of the frequency response as $X(\omega) = H(\omega) F(\omega)$, then taking the inverse Fourier transform to find $x(t)$. The process is illustrated in Figure 3.30. Also shown is the direct solution in the time domain found by convolution of the force with the impulse response $h(t)$ as discussed in Section 3.4.2. In fact, $h(t)$ and $H(\omega)$ are a Fourier transform pair (except for a factor of 2π).

3.4.5 Response to random excitation

Vibration excitation is often random: wind loading, turbulent boundary layers, a rough road or an earthquake. Random signals are considered in some detail in Chapter 4. Here we will just consider a few consequences for the random excitation of an SDOF system. We will also only consider stationary, ergodic,[*] random forces.

---

[*] See definition in Chapter 4.

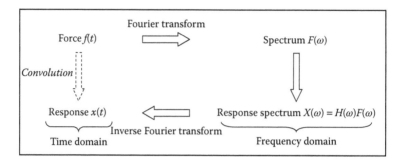

Figure 3.30 Spectral analysis and Fourier transform.

Suppose the force is random and described by its power spectral density* (psd) $S_f(\omega)$ (see Section 4.5.6). By definition, the mean-square force is

$$\overline{f^2(t)} = \int_{-\infty}^{\infty} S_f(\omega)\,d\omega \qquad (3.90)$$

For an SDOF system excited by such a force, the psd of the displacement is

$$S_x(\omega) = |H(\omega)|^2 S_f(\omega) \qquad (3.91)$$

where

$$H(\omega) = \frac{1}{k - \omega^2 m + i\omega c} \qquad (3.92)$$

is the frequency response. The mean-square displacement $\overline{x^2(t)}$ is therefore

$$\overline{x^2(t)} = \int_{-\infty}^{\infty} S_x(\omega)\,d\omega = \int_{-\infty}^{\infty} |H(\omega)|^2 S_f(\omega)\,d\omega \qquad (3.93)$$

Now, usually, the damping is light, so that the FRF has a sharp resonance peak around the natural frequency. Consequently, $|H(\omega)|^2$ has an extremely sharp resonance peak.

---

* The psd $S(\omega)$ is *double-sided*, since it is defined for both positive and negative frequencies (although $S(-\omega)$ is equal to $S(\omega)$). Often, in practice, we might prefer to deal with a *single-sided* psd $G(f)$, a function of frequency f in hertz and defined only for $f \geq 0$. They are related by $G(f) = 4\pi S(\omega)$.

Consider now the case of *white noise* excitation, for which S_f is uniform. The psd $S_x(\omega)$ is therefore relatively extremely large over a narrow frequency band around ω_n. The response $x(t)$ is therefore random and *narrowband* – its spectrum has significant frequency components only in a narrow frequency band around resonance – in contrast to the broadband input. In the time domain, $x(t)$ appears as a sinusoid of frequency ω_n with a slowly varying amplitude and phase.

The mean-square response from Equation 3.61 is then

$$\overline{x^2(t)} = \int_{-\infty}^{\infty} \frac{S_f(\omega)}{\left(k-\omega^2 m\right)^2 + (\omega c)^2} d\omega = \frac{S_f}{k^2} \int_{-\infty}^{\infty} \frac{1}{\left(1-\omega^2/\omega_n^2\right) + \left(2\zeta\omega/\omega_n\right)^2} d\omega \qquad (3.94)$$

The last integral is

$$\int_{-\infty}^{\infty} \frac{1}{\left(1-\omega^2/\omega_n^2\right) + \left(2\zeta\omega/\omega_n\right)^2} d\omega = \frac{\pi\omega_n}{2\zeta} \qquad (3.95)$$

so that

$$\overline{x^2(t)} = \frac{\pi\omega_n}{2k^2\zeta} S_f = \frac{\pi}{kc} S_f \qquad (3.96)$$

Thus, the mean-square response is inversely proportional to both the stiffness and the damping coefficient.

A similar result follows if $S_f(\omega)$ varies with frequency but its bandwidth includes the natural frequency ω_n. Since $|H(\omega)|^2$ is so large around resonance, the response $x(t)$ is still narrowband and its psd $S_x(\omega)$ is dominated by the resonant contributions. To a very good approximation, the response to $S_f(\omega)$ is the same as the response to white noise whose psd is $S_f(\omega_n)$, that is, the value of $S_f(\omega)$ at the natural frequency. The mean-square response is again given by Equation 3.96, with S_f replaced by $S_f(\omega_n)$.

Finally, since the psd of the velocity $S_v(\omega) = \omega^2 S_x(\omega)$, and since $S_v(\omega)$ is also narrowband and dominated by the resonance peaks, then

$$\overline{v^2(t)} \approx \omega_n^2 \overline{x^2(t)} \qquad (3.97)$$

Note that $v(t)$ is also approximately sinusoidal with frequency ω_n and with slowly varying magnitude and phase.

3.5 MULTIPLE DEGREES OF FREEDOM SYSTEMS

Many vibration problems can be modelled using only a single DOF. Many more, however, require a model with more than one DOF. Two examples of 2DOF systems are shown in Figure 3.1b and c: the first allows for 'wheel hop' behaviour as well as vertical car-body vibration, while the second allows for car-body pitch as well as bounce. The wheel rim in Figure 3.1d is modelled by finite elements and has many thousands of DOFs. Indeed, a full FE model of a car might have many millions of DOFs.

This section concerns the vibration of MDOF systems with n DOFs, where n could be any integer. We start by looking at a system with two DOFs, and go on to analyse MDOF systems using a matrix approach. The advantage of this is generality: the matrix description can be used irrespective of the value of n. Moreover, the fundamentals of the vibration of MDOF systems are the same irrespective of the value of n. In summary, there are n *modes of vibration*, n *natural frequencies* and n resonances. We will see that each of the modes of vibration behaves just like an SDOF system: indeed, this is partly why understanding the vibration of an SDOF system is so important.

In this section, we will assume that the equations of motion are known or are easy to find (this is usually not the case in practice). We will start with the 2DOF system shown in Figure 3.31. The masses are assumed to vibrate horizontally with displacements $x_1(t)$ and $x_2(t)$. We will first consider undamped free vibration, and then damped and undamped forced vibration.

3.5.1 Undamped free vibration of 2DOF systems

Figure 3.31 also shows the free-body diagrams of the two masses. Note that the forces exerted on the masses by the central spring are proportional to the extension of the spring $(x_2 - x_1)$ and, hence, they *couple* the motion: the motion of mass 2 produces a force on mass 1, and vice versa. The equations of motion are

$$m_1 \ddot{x}_1 + (k_1 + k_2) x_1 - k_2 x_2 = 0;$$
$$m_2 \ddot{x}_2 + (k_2 + k_3) x_2 - k_2 x_1 = 0 \qquad (3.98)$$

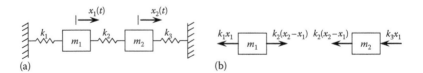

Figure 3.31 (a) 2DOF system and (b) free-body diagrams.

Note that the stiffness terms involving x_1 in the equation involving $\ddot{x}_1$ are both positive: this is because all the springs generate restoring forces that act to return the mass to its equilibrium position.

We can write Equation 3.98 in *matrix notation* as

$$\mathbf{M}\ddot{\mathbf{x}} + \mathbf{K}\mathbf{x} = 0; \quad \mathbf{x} = \begin{Bmatrix} x_1 \\ x_2 \end{Bmatrix}; \quad \mathbf{M} = \begin{bmatrix} m_1 & 0 \\ 0 & m_2 \end{bmatrix}; \quad \mathbf{K} = \begin{bmatrix} k_1 + k_2 & -k_2 \\ -k_2 & k_2 + k_3 \end{bmatrix}$$
(3.99)

where **x** is a vector of displacements at the DOFs, and **M** and **K** are the *mass and stiffness matrices*. They are real, symmetric and positive definite* (or sometimes semi-definite). The off-diagonal terms in **K** couple the two DOFs: they represent static or stiffness coupling. Sometimes, there are off-diagonal terms in the mass matrix, representing dynamic or mass coupling. The coupling means that motion in x_1 implies forces on x_2 and vice versa.

In Section 3.2, we saw that the free vibration of all undamped SDOF systems takes the same form: time-harmonic motion. We might then ask ourselves this question: is free (unforced) time-harmonic motion at frequency ω possible for the 2DOF system? If so, then

$$x_1(t) = X_1 e^{i\omega t}; \quad x_2(t) = X_2 e^{i\omega t}$$

where X_1 and X_2 are the complex amplitudes. Substituting these into the equations of motion gives

$$\left(k_1 + k_2 - \omega^2 m_1\right) X_1 - k_2 X_2 = 0$$
$$-k_2 X_1 + \left(k_2 + k_3 - \omega^2 m_2\right) X_2 = 0$$
(3.100)

The first equation can be solved for X_2 in terms of X_1. Substituting this into the second equation leads to

$$[(k_1 + k_2 - m_1 \omega^2)(k_2 + k_3 - m_2 \omega^2) - k_2^2] X_1 = 0$$
(3.101)

The solutions to this equation are either $X_1 = 0$ (this is a *trivial solution*, for which the system is at rest) or else the term in square brackets is zero, that is,

$$m_1 m_2 \omega^4 - [m_1(k_1 + k_2) + m_2(k_2 + k_3)]\omega^2 + [(k_1 + k_2)(k_2 + k_3) - k_2^2] = 0 \quad (3.102)$$

---

* A real-valued matrix **A** is positive definite if $\mathbf{z}^T \mathbf{A} \mathbf{z} > 0$ for every nonzero real column vector **z**, with $\mathbf{z}^T$ the transpose of **z**. A matrix is positive semi-definite if $\mathbf{z}^T \mathbf{A} \mathbf{z} \geq 0$. The implication of this for the stiffness matrix **K** is that $\mathbf{x} = 0$ is a position of minimum potential energy. Similarly, for the mass matrix **M**, $\dot{\mathbf{x}} = 0$ results in minimum kinetic energy.

This is the *characteristic equation* of the system. It is a quadratic equation in ω^2. It thus has two solutions for ω^2, that is, ω_1^2 and ω_2^2, where $\omega_1 \leq \omega_2$. These are the two *natural frequencies* of the system, with ω_1^2, the lower, being the square of the fundamental.

So the answer to our question – is free time-harmonic motion at frequency ω possible? – is yes, it is possible, but *only* if ω equals one of the two natural frequencies ω_1 and ω_2. The natural frequencies depend on the masses and stiffnesses of the system. Note that the algebra to find these frequencies, while easy, is quite lengthy: one can begin to imagine the difficulties with three, four or more DOFs. In general, however, the free vibration will be the superposition of sinusoidal motions in each of the two natural frequencies (see Section 3.5.3 for a discussion of this).

For simplicity, let us assume henceforth that the system properties are $m_1 = m$, $m_2 = 2m$, $k_1 = k_2 = k$ and $k_3 = 2k$. The natural frequencies are then

$$\omega_1 = \sqrt{\frac{k}{m}}; \quad \omega_2 = \sqrt{\frac{5k}{2m}} \tag{3.103}$$

The first of Equations 3.101 relates the amplitudes by

$$X_2 = \frac{k_1 + k_2 - \omega^2 m_1}{k_2} X_1 = \frac{2k - m\omega^2}{k} X_1 \tag{3.104}$$

Thus, whenever the system vibrates time-harmonically at one of the natural frequencies, the amplitudes of motion of the two masses must be related by this equation. Specifically,

Harmonic motion at ω_1: $X_2 = X_1$
Harmonic motion at ω_2: $X_2 = -0.5 X_1$

Note that, for the second natural frequency, the displacement of m_2 is of the opposite sign to that of m_1 – they vibrate in antiphase. (In general, if the system is disturbed from equilibrium and released, it will vibrate with components in both of its natural frequencies simultaneously and the ensuing motion is not SHM and not even periodic – see Section 3.5.3.) We can write these in a vector form as

$$\text{Mode 1:} \quad \begin{Bmatrix} X_1 \\ X_2 \end{Bmatrix} \propto \phi^{(1)} = \begin{Bmatrix} 1 \\ 1 \end{Bmatrix}$$

$$\text{Mode 2:} \quad \begin{Bmatrix} X_1 \\ X_2 \end{Bmatrix} \propto \phi^{(2)} = \begin{Bmatrix} 1 \\ -0.5 \end{Bmatrix} \tag{3.105}$$

where $\phi^{(1)}$ and $\phi^{(2)}$ are the *mode shapes* of the *modes of vibration*. They define the *shape* of the motion at each of the natural frequencies. The vector of the actual amplitudes X_1 and X_2 must be proportioned to the mode shape if the system vibrates in that mode. Note that if ϕ is a mode shape, then any constant times ϕ is an identical description of the *shape* of the motion, and hence could be used to define the mode shape. This allows us to *normalise* the mode shapes, that is, to scale ϕ so that the mode shapes satisfy some condition; in Equation 3.105 they are normalised so that $X_1 = 1$. An alternative will be described in Section 3.5.4 (Equation 3.117). The same conclusions follow for *any* 2DOF system: there are two natural frequencies, two modes of vibration and two mode shapes.

Some systems can have natural frequencies equal to zero. They are called *degenerate* or *unrestrained* systems, and the corresponding mode is a *rigid-body motion* in translation or rotation. This is not harmonic motion as such, but involves constant velocity or angular velocity motion without deformation of the system. For example, a car rolling freely along a road has a rigid body mode associated with pure translational motion in the direction of motion.

3.5.2 Matrix notation: *n* arbitrary

We can follow through a similar analysis for a general MDOF system with n DOFs. The equations of motion in matrix form are

$$\mathbf{M\ddot{x}} + \mathbf{Kx} = 0; \quad \mathbf{x}(t) = \begin{Bmatrix} x_1(t) \\ x_2(t) \\ \vdots \\ x_n(t) \end{Bmatrix} \quad (3.106)$$

Can free time-harmonic motion at frequency ω exist? If so, then assume

$$\mathbf{x}(t) = \mathbf{X} e^{i\omega t}; \quad \mathbf{X} = \begin{Bmatrix} X_1 \\ X_2 \\ \vdots \\ X_n \end{Bmatrix} \quad (3.107)$$

where $\mathbf{X}$ is a vector of amplitudes of the DOFs. Substituting into Equation 3.70 gives

$$\left[\mathbf{K} - \omega^2 \mathbf{M} \right] \mathbf{X} = 0 \quad (3.108)$$

which means that either $\mathbf{X}=0$ (trivial) or the determinant

$$\left|\mathbf{K} - \omega^2 \mathbf{M}\right| = 0 \tag{3.109}$$

This is an *eigenvalue problem*. The matrices $\mathbf{K}$ and $\mathbf{M}$ are both symmetric and of size $n \times n$. There are n eigenvalues ω_j^2 for $j = 1, 2, \ldots, n$; ω_j are the *natural frequencies*, ordered so that $\omega_j \leq \omega_{j+1}$. (There may be rigid body modes for which $\omega_j = 0$.)

Corresponding to each natural frequency is a mode shape $\phi^{(j)}$, an *eigenvector* of Equation 3.108, such that

$$\left[\mathbf{K} - \omega_j^2 \mathbf{M}\right] \phi^{(j)} = 0; \quad \phi^{(j)} = \begin{Bmatrix} \phi_1^{(j)} \\ \phi_2^{(j)} \\ \vdots \\ \phi_n^{(j)} \end{Bmatrix} \tag{3.110}$$

where $\phi_k^{(j)}$ is the amplitude of the jth mode shape at the kth DOF.

Thus the answer to our question is: Yes, free time-harmonic motion of the nDOF system *can* occur, but only at each of the n natural frequencies, when the motion will be proportional to the corresponding mode shape ϕ.

3.5.3 General free vibration: modal superposition

In Section 3.2.1, we saw that the free vibration of an undamped SDOF system was SHM, as in Equation 3.8. Now, free vibration in the jth mode is also time harmonic, so we can write the motion in the jth mode as

$$y_j(t) = A_j \sin \omega_j t + B_j \cos \omega_j t \tag{3.111}$$

where A_j and B_j are constants. During the free vibration of the jth mode, the physical coordinates $\mathbf{x}(t)$ have amplitudes which are in the ratios defined by the jth mode shape $\phi^{(j)}$. Therefore, in free vibration in mode j, the physical DOFs are

$$\mathbf{x}(t) = \phi^{(j)} y_j(t) = \phi^{(j)} \left(A_j \sin \omega_j t + B_j \cos \omega_j t \right) \tag{3.112}$$

The *general* free vibration will be a superposition of all the modes of vibration, so that

$$x(t) = \sum_{j=1}^{n} \phi^{(j)} y_j(t) \tag{3.113}$$

which can be written as

$$x(t) = Py(t); \quad y(t) = \begin{Bmatrix} y_1(t) \\ y_2(t) \\ \vdots \\ y_n(t) \end{Bmatrix}; \quad P = \begin{bmatrix} \phi^{(1)} & \phi^{(2)} & \cdots & \phi^{(n)} \end{bmatrix} \tag{3.114}$$

The matrix P is the *modal matrix*. The jth column is the jth mode shape $\phi^{(j)}$. The vector $y(t)$ is a vector of *modal coordinates*. Equation 3.114 defines a *transformation* between physical coordinates x and modal coordinates y, and describes x as a superposition of modal responses. For free vibration of the form of Equation 3.111, the constants A_j and B_j, $j = 1, 2, \ldots n$ can be found from the initial conditions.

3.5.4 Orthogonality

The mode shapes are *orthogonal* to the mass and stiffness matrices, so that*

$$\phi^{(i)T} K \phi^{(j)} = \begin{cases} 0, & i \neq j \\ K_j, & i = j \end{cases}$$

$$\phi^{(i)T} M \phi^{(j)} = \begin{cases} 0, & i \neq j \\ M_j, & i = j \end{cases} \tag{3.115}$$

where the superscript ^{T} denotes the transpose. The constants K_j and M_j are the *modal stiffness* and *modal mass* of mode j. They are such that $K_j = \omega_j^2 M_j$. Equations 3.116 mean that the mode shapes are orthogonal, independent and uncoupled. In terms of the modal matrix P, the orthogonality conditions are

$$P^T M P = \text{diag}(M_j); \quad P^T K P = \text{diag}(K_j) \tag{3.116}$$

---

* By definition, for distinct natural frequencies ($\omega_i \neq \omega_j$)
$(K - \omega_i^2 M) \phi^{(i)} = 0; \quad (K - \omega_j^2 M) \phi^{(j)} = 0$
Premultiply the first by $\phi^{(j)T}$ and the second by $\phi^{(i)T}$, then take the transpose of the second equation, noting that M and K are symmetric (and hence $(\phi^{(j)T} K \phi^{(i)})^T = \phi^{(i)T} K \phi^{(j)}$) and then subtract from the first equation. This leads to
$(\omega_j^2 - \omega_i^2) \phi^{(i)T} M \phi^{(j)} = 0; \quad \Rightarrow \phi^{(i)T} M \phi^{(j)} = 0$

where diag(M_j) is a diagonal matrix whose elements are the modal masses.

It is often convenient to normalise the mode shapes so that the modal masses $M_j = 1$. This results in *mass normalised modes*. The orthogonality conditions then become

$$\phi^{(i)T} \mathbf{K} \phi^{(j)} = \begin{cases} 0, & i \neq j \\ \omega_j^2, & i = j \end{cases}$$

$$\phi^{(i)T} \mathbf{M} \phi^{(j)} = \begin{cases} 0, & i \neq j \\ 1, & i = j \end{cases} \qquad (3.117)$$

3.5.5 Modal decomposition: uncoupling the equations of motion

The MDOF system has modes of vibration which satisfy the orthogonality conditions. We can use this information to write the equations of motion in terms of the modal coordinates **y**. In terms of the physical coordinates **x**, the equations of motion are

$$\mathbf{M}\ddot{\mathbf{x}} + \mathbf{K}\mathbf{x} = 0 \qquad (3.118)$$

where there are off-diagonal elements in **M** and **K** which couple the DOFs **x**. Now write $\mathbf{x} = \mathbf{P}\mathbf{y}$ and premultiply by $\mathbf{P}^T$ to give

$$\mathbf{P}^T \mathbf{M} \mathbf{P} \ddot{\mathbf{y}} + \mathbf{P}^T \mathbf{K} \mathbf{P} \mathbf{y} = 0 \qquad (3.119)$$

We can now use the orthogonality conditions of Equation 3.78 to get

$$\mathrm{diag}(M_j)\ddot{\mathbf{y}} + \mathrm{diag}(K_j)\mathbf{y} = 0 \qquad (3.120)$$

Note that there are *no nonzero off-diagonal terms*, so that the equations of motion in terms of the modal coordinates are *uncoupled*. The *j*th row is

$$M_j \ddot{y}_j + K_j y_j = 0; \quad \omega_j^2 = \frac{K_j}{M_j}; \quad j = 1, 2, \ldots, n \qquad (3.121)$$

This is the equation of motion of an SDOF system with mass M_j and stiffness K_j. Using the *transformation* $\mathbf{x} = \mathbf{P}\mathbf{y}$ allows us to *uncouple* the equations of motion; this is called *modal decomposition*. We decompose the motion of the *n*DOF system in coordinates **x** into the motion of *n* SDOF systems described by the coordinates **y** (Figure 3.32). The modal decomposition

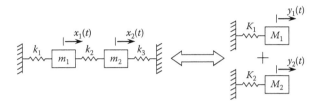

Figure 3.32 Modal decomposition: a 2DOF system is decomposed into two SDOF systems by the modal transformation $\mathbf{x} = \mathbf{Py}$.

which changes Equation 3.118 into n SDOF systems in Equation 3.121 is important for many reasons. First, we can write the response of an SDOF system using the results of Section 3.2 in a straightforward way. We can then find the physical response by adding up these modal responses because $\mathbf{x} = \mathbf{Py}$. Second, n could be a large number; the system may have many thousands or millions of DOFs. However, usually only the lowest few modes are important – we do not need to find *all* the modes, only the *important* ones. We need, then, to keep only some of the modes when forming the modal matrix $\mathbf{P}$. This matrix need not be square: we could keep only m modes, so that $\mathbf{P}$ is an $n \times m$ matrix. There would then be m equations of motion for each mode (Equation 3.121) with $m < n$. A model with a large number of DOFs can be described by just a few modes of vibration. Finally, software for solving the eigenvalue problem is widely available (e.g. in MATLAB®).

3.6 FORCED RESPONSE OF MDOF SYSTEMS

Suppose that a force $f_j(t)$ acts on the jth DOF $x_j(t)$. The equations of motion are

$$\mathbf{M\ddot{x}} + \mathbf{Kx} = \mathbf{f}(t); \quad \mathbf{f}(t) = \begin{Bmatrix} f_1(t) \\ f_2(t) \\ \vdots \\ f_n(t) \end{Bmatrix} \quad (3.122)$$

Following the same modal decomposition procedure of Section 3.5.5 (substitute $\mathbf{x} = \mathbf{Py}$; premultiply by $\mathbf{P}^T$; use the orthogonality conditions) leads to

$$\text{diag}(M_j)\mathbf{\ddot{y}} + \text{diag}(K_j)\mathbf{y} = \mathbf{q}(t); \quad \mathbf{q}(t) = \mathbf{P}^T\mathbf{f}(t) \quad (3.123)$$

where $\mathbf{q}(t)$ is the vector of *modal forces*. The jth element, $q_j(t)$, indicates how much loading the excitation $\mathbf{f}(t)$ applies to the jth mode. It is given by

$$q_j(t) = \phi^{(j)T}\mathbf{f}(t) = \sum_{i=1}^{n} \phi_i^{(j)} f_i(t) \qquad (3.124)$$

A force applied at a DOF i where the mode shape $\phi_i^{(j)}$ is large gives a large contribution to the modal force. Taking the jth row of Equation 3.123 gives

$$M_j \ddot{y}_j + K_j y_j = q_j(t); \quad j = 1, 2, \ldots, n \qquad (3.125)$$

The modes are uncoupled: the nDOF system is decomposed into n SDOF systems (Figure 3.32). The jth mode again behaves as an SDOF system. Its forced response follows from the results we developed in Sections 3.3 and 3.4.

3.6.1 Damped MDOF systems

Now, let us assume that there is viscous damping. The equations of motion are now

$$\mathbf{M}\ddot{\mathbf{x}} + \mathbf{C}\dot{\mathbf{x}} + \mathbf{K}\mathbf{x} = \mathbf{f} \qquad (3.126)$$

where $\mathbf{C}$ is a matrix of viscous damping coefficients. We can again try to uncouple these equations, but a problem arises. First, let us solve for the *undamped* modes of vibration. When we then follow the modal decomposition procedure, the damping terms lead to the matrix product $\mathbf{P}^T\mathbf{C}\mathbf{P}$. In general, this will *not* be a diagonal matrix, so that the undamped modes become coupled through the damping terms. However, the product will be diagonal if the damping matrix is proportional to $\mathbf{M}$ or $\mathbf{K}$ or, more generally, is of the form $\mathbf{C} = \alpha\mathbf{M} + \beta\mathbf{K}$, where α and β are constants. This form of damping is called *Rayleigh damping*.

If we assume we have *proportional damping*, then $\mathbf{P}^T\mathbf{C}\mathbf{P} = \text{diag}(C_j)$, a diagonal matrix, and hence the equations of motion uncouple. The motion of mode j is described by

$$M_j \ddot{y}_j + C_j \dot{y}_j + K_j y_j = q_j(t); \quad j = 1, 2, \ldots, n \qquad (3.127)$$

With *structural damping*, the damping term becomes $i\eta_j K_j$. Proportional damping means that the damping is distributed throughout the system in a similar manner to the distribution of mass or stiffness. The modes are uncoupled; each mode decays independently and is not coupled through the damping to the other modes. The procedure for solving for the forced vibration of the nDOF system is then summarised in Figure 3.33: find the undamped modes; transform to modal coordinates to get n SDOF

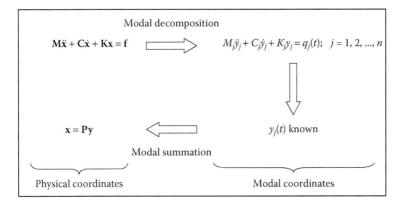

Figure 3.33 Forced response of MDOF systems: transformation to modal coordinates, solution for the modes of vibration $y_j(t)$, then summation of modes to find the response in physical coordinates.

systems; solve for the motion of each mode, using the SDOF results from Sections 3.3 and 3.4; sum the response in each mode to find the response in physical coordinates, $\mathbf{x}(t)$. We solve n SDOF problems rather than one nDOF problem.

In practice, damping is rarely proportional, but is usually light, and the errors introduced by assuming that the damping is proportional are usually negligible. We then deal with the (undamped) modes and allow each mode to have its own damping ratio ζ_j or loss factor η_j. Henceforth, we will assume proportional damping. In practice, if the damping is not proportional, then the mode shapes become complex; there is a phase difference between the motion of different points and the eigenvalues become complex, the imaginary parts representing the decay due to damping.

3.6.2 Response to harmonic forces: FRF

If the forces are time harmonic, then $\mathbf{f}(t) = \mathbf{F}\exp(i\omega t)$ in Equation 3.126, with $\mathbf{F}$ being a vector of force amplitudes. In the steady state, the response is time harmonic, so that $\mathbf{x}(t) = \mathbf{X}\exp(i\omega t)$. We could try to solve Equation 3.126 directly by noting that

$$\left[\mathbf{K} - \omega^2\mathbf{M} + i\omega\mathbf{C}\right]\mathbf{X} = \mathbf{F} \tag{3.128}$$

so that

$$\mathbf{X} = \alpha\mathbf{F}; \quad \alpha = \left[\mathbf{K} - \omega^2\mathbf{M} + i\omega\mathbf{C}\right]^{-1} \tag{3.129}$$

where α is a matrix of receptances. The element in row r, column e, α_{re}, is the response at DOF r when the excitation is applied at DOF e. This approach to finding the response, by direct matrix inversion, is potentially computationally expensive, since we must invert an $n \times n$ matrix at every frequency of interest and n could be very large. Furthermore, it gives little understanding regarding the nature of the response, the natural frequencies and so on.

As an alternative, we can solve for the response using modal decomposition. Note that the modal forces $\mathbf{q}(t) = \mathbf{Q}\exp(i\omega t)$ and modal responses $\mathbf{y}(t) = \mathbf{Y}\exp(i\omega t)$ are also time harmonic. The response for the jth mode (Equation 3.127) is then

$$Y_j = \frac{1}{K_j - \omega^2 M_j + i\omega C_j} Q_j \tag{3.130}$$

The physical response is then

$$\mathbf{X} = \mathbf{P}\mathbf{Y} = \left[\mathbf{P}\,\mathrm{diag}\!\left(\frac{1}{K_j - \omega^2 M_j + i\omega C_j}\right)\mathbf{P}^{\mathrm{T}}\right]\mathbf{F} \tag{3.131}$$

with the expression in brackets being the receptance matrix. In particular, the element $\alpha_{re} = X_r/F_e$ is

$$\alpha_{re} = \sum_{j=1}^{n}\frac{\phi_r^{(j)}\phi_e^{(j)}}{K_j - \omega^2 M_j + i\omega C_j} = \sum_{j=1}^{n}\frac{1}{K_j}\frac{\phi_r^{(j)}\phi_e^{(j)}}{1 - \omega^2/\omega_j^2 + i2\zeta_j\omega/\omega_j} \tag{3.132}$$

This receptance is the sum of terms arising from each mode j. Each term is the product of the mode shape at the response point and the mode shape at the excitation point, while the term in the denominator indicates how close mode j is to resonance: mode j resonates if $\omega \approx \omega_j$ where the response is governed by the loss factor ζ_j of mode j; the response in mode j is dominated by stiffness if $\omega \ll \omega_j$ or mass if $\omega \gg \omega_j$, respectively.

The receptance of *any* nDOF system can be written as Equation 3.132. The cost of calculating the response as a modal sum involves solving the free-vibration problem, but there are many benefits: the responses at different frequencies can be found at very little extra cost; the natural frequencies are known; we can interpret and understand the response – it is the sum of the modal terms, each of which has its own resonance peak. Finally, very often only a *few* modes are important. Suppose we are interested in the response up to some maximum frequency. We might then only keep the first m modes, where the natural frequency of the highest mode retained, ω_m, is (say) twice the highest frequency of interest. The modal sum in Equation 3.132 can then be truncated at m: the sum runs from 1 to m rather than 1 to n. The value of this is that the vibration of a very large model (there might be millions of DOFs) can be described by only a few modes (maybe 10 or so).

3.6.3 Reciprocity

It can be seen from Equation 3.132 that for any pair of points 1 and 2, then

$$a_{12} = a_{21} \tag{3.133}$$

This is an example of the *principle of reciprocity*, or *Betti's reciprocity theorem*. In this case, the principle states that the response at point 2 per unit force applied at point 1 (i.e. X_2/F_1) is equal to the response at point 1 per unit force applied at point 2 (i.e. X_1/F_2). Indeed, the principle holds for each mode j, since $\phi^{(j)}_{r=1}\phi^{(j)}_{e=2} = \phi^{(j)}_{e=1}\phi^{(j)}_{r=2}$.

The principle of reciprocity holds widely in the fields of elastodynamics and acoustics, and indeed physics more generally: it is valid for *any* linear system and is not restricted to the vibration of MDOF systems. The excitation F_1 could be a force, a moment or an acoustic pressure, while X_1 is the corresponding response quantity: this must be such that the integral $\int F_1 dX_1$ gives the work done by F_1. Thus, for a point force, the response X_1 must be the displacement in the direction and along the line of action of the force, while for a moment, X_1 must be the rotation at that point and around the same axis about which the moment is applied. Finally, for acoustic pressure, the reciprocal quantity is the volume velocity of the source.

The principle has many applications, including the estimation of parameters such as the coupling loss factors in statistical energy analysis and determining relations between the coefficients that describe the reflection of waves. One application of particular relevance in this book concerns the measurement of FRFs as described in Chapter 9.

3.6.4 Characteristics of FRFs

The fact that the receptances of all MDOF systems can be expressed by the modal sum of Equation 3.132 means that they share certain characteristics. These will be described in this section, assuming that the damping is light and that the natural frequencies are well separated. Typical receptances for a 2DOF system are shown in Figure 3.34, and as a polar plot in Figure 3.35. Refer also to Figures 3.19 and 3.21 for equivalent figures for an SDOF system.

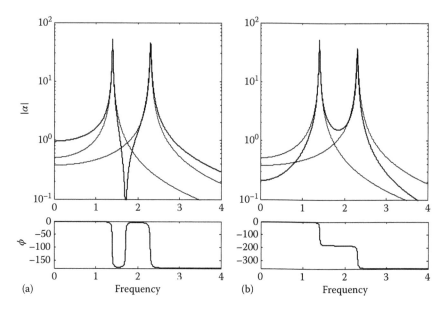

Figure 3.34 Typical receptances of a 2DOF system and contributions from each mode: (a) input receptance α_{11} and (b) transfer receptance α_{12}.

First, because the damping is light, one mode dominates the response at each resonance. The whole structure resonates and each FRF has a resonance peak at the same frequency (with two exceptions as described here). Figure 3.34 also shows the response of each mode, the total response being their sum. The phase of the FRF changes very rapidly around resonance, changing by 180° over a narrow frequency band.

The input FRF has an *antiresonance* between successive resonance peaks. This exists for the following reason. Suppose there is no damping. At each of the two resonances, a single mode dominates the response. It has a positive real response below resonance and a negative real response above resonance. As the frequency increases from one resonance frequency to the other, the response changes from being negative (dominated by the response of the first resonance) to positive. Thus, at some frequency between the two resonances, the total response becomes zero: this is the antiresonance frequency.

Note that for the input FRF the input and response point are identical so that the *modal constant* $\phi_r^{(j)} \phi_e^{(j)} = \phi_r^{(j)2}$ for every mode is positive. As the frequency increases, the contribution of each mode changes from positive to negative as the frequency passes through the natural frequency (note that the phase changes from 0° to −180°). Thus, between the first and the second resonances, mode 1 gives a negative contribution to the total response which decreases with increasing frequency, while mode 2 gives an increasing and positive contribution. At some frequency the magnitudes of these contributions are equal but their phases are (nearly) opposite so that they almost cancel each other out, causing the antiresonance. The presence of small amounts of damping simply blunts the antiresonance somewhat.

For a transfer FRF there may or may not be an antiresonance between two resonances. There will be one if the modal constants of adjacent modes have the same sign, while there will not be an antiresonance if the modal constants have opposite signs.

A *node* of a mode shape is a point where the mode shape equals zero. If the excitation is applied at a node of a particular mode (i.e. $\phi_e^{(j)} = 0$) then that mode will not be excited and a resonance will not be present in any FRF at the natural frequency of that mode. Similarly, if we measure the response at a node point for a particular mode (i.e. $\phi_r^{(j)} = 0$), a resonance peak for that mode will not be seen. Small levels of damping will blunt the resonances and antiresonances somewhat.

The polar plots of the FRFs are shown in Figure 3.35. We can see two modal circles in each FRF, one circle for each mode. The circle appears to the right of the imaginary axis if the modal constant of the mode is positive and to the left if it is negative. The circular form of the response offers a means of identifying individual modal responses of MDOF systems provided that their natural frequencies are not too close together.

3.6.5 Vibration absorber

A *vibration absorber*, or *neutraliser*, is a spring-mass system, tuned to a particular frequency, which exploits the existence of an antiresonance.

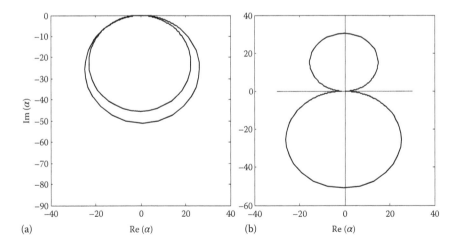

Figure 3.35 Polar plots of typical receptances of a 2DOF system: (a) input receptance α_{11} and (b) transfer receptance α_{12}.

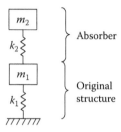

Figure 3.36 Vibration absorber attached to host structure.

Figure 3.36 shows an original system which may have a troublesome resonance at some frequency ω_a, or perhaps may run at a speed which generates forces at this frequency. Synchronous machines are an example where the excitation frequency may be known. The absorber is attached, changing the system from an SDOF system to a 2DOF system. Of course, there will now be two resonances, but an antiresonance will be created at a frequency $\omega_a = \sqrt{k_2/m_2}$. The FRFs of the system with and without the absorber are shown in Figure 3.37. The absorber needs to be carefully tuned to be effective, and damping can be included to give broadband reduction. Details on the analysis and design can be found in, for example, Den Hartog (2003).

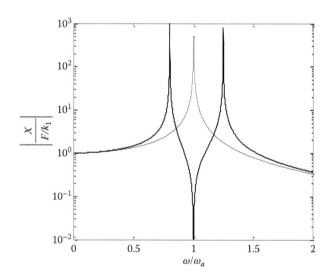

Figure 3.37 FRF of a system with a vibration absorber: $m_2/m_1 = 0.2$, $\omega_a = \sqrt{k_2/m_2} = \sqrt{k_1/m_1}$; (a) (- - -) original system; (b) (—) system with absorber.

3.7 CONTINUOUS SYSTEMS

So far, we have considered models of the vibration of structures described by idealised discrete rigid masses and massless springs and with a finite number of DOFs. We saw that their vibration can be described by natural frequencies, mode shapes and so on. Physical structures, however, have mass and stiffness distributed continuously through the structure. When they vibrate, the deformation of the structure is continuous in space. The purpose of this section is to investigate the vibration of such continuous systems. We might think of a continuous system as being made up of an infinite number of infinitesimal mass particles and hence having an infinite number of DOFs. We will not go into any great depth of analysis, but we will see that continuous systems do indeed have natural frequencies and modes and, as in the case of MDOF systems, these modes behave like SDOF systems. We will consider just two examples – axial vibration of a rod and bending vibration of a beam – but the principles are applicable to more general systems. Further analysis and details can be found in more advanced textbooks (e.g. Rao 2010; Meirovitch 1997; Den Hartog 2003).

3.7.1 Axial vibration of a rod

Consider a rod made of an elastic material with density ρ and elastic modulus E. It has a cross-sectional area A and lies along the x-axis. We will consider the example shown in Figure 3.38, where the rod is of length L, one end is fixed and the other end is free. The derivation of the equation of motion can be found in many texts (e.g. Meirovitch 1997; Rao 2010) and only a brief summary will be given here.

Suppose we consider a small element of length δx. When the rod vibrates axially, this element will have a displacement $u(x)$ in the x-direction and

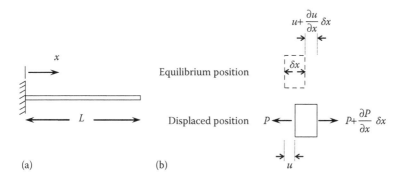

Figure 3.38 Fixed–free rod undergoing axial vibration and (b) element δx.

carry a stress $\sigma(x)$. The internal tension $P(x) = \sigma A$. From Hooke's law, $\sigma = E\varepsilon$, where $\varepsilon = \partial u/\partial x$ is the strain. Therefore,

$$P = EA\frac{\partial u}{\partial x} \tag{3.134}$$

From Newton's second law for the element,

$$(\rho A \delta x)\frac{\partial^2 u}{\partial t^2} = \left(P + \frac{\partial P}{\partial x}\delta x\right) - P = \frac{\partial P}{\partial x}\delta x \tag{3.135}$$

where the term in parentheses on the left-hand side of the equation is the mass of the element. If the rod is uniform, the area A is constant and the equation of motion becomes

$$\frac{\partial^2 u}{\partial t^2} = c^2 \frac{\partial^2 u}{\partial x^2}; \quad c^2 = \sqrt{\frac{E}{\rho}} \tag{3.136}$$

This is the *wave equation* (see also Section 2.2).

We might ask ourselves the same question: can the rod undergo time-harmonic free vibration? If so, we might assume that $u(x,t) = U(x)\exp(i\omega t)$. The equation of motion then becomes

$$\frac{d^2 U}{dx^2} + \frac{\omega^2}{c^2}U = 0 \tag{3.137}$$

This has the general solution

$$U(x) = B\sin\frac{\omega x}{c} + C\cos\frac{\omega x}{c} \tag{3.138}$$

where B and C are arbitrary constants. Of course, the motion must also satisfy the *boundary conditions* at the ends of the rod. The end at $x=0$ is fixed, so the displacement there must satisfy the *geometric* or *imposed* boundary condition $U(0)=0$. Substituting this in the expression for $U(x)$ gives $C=0$. On the other hand, the end at $x=L$ is free, so that the axial tension P must be zero – this is a *natural* or *dynamic* boundary condition. Noting Equation 3.134 for P, we get

$$EA\frac{dU}{dx}\bigg|_L = \frac{\omega B}{c}\cos\frac{\omega L}{c} = 0 \tag{3.139}$$

Thus either $B=0$ – a trivial solution – or the frequency must be such that

$$\cos\frac{\omega L}{c} = 0 \quad \Rightarrow \omega_n = \frac{(2n-1)\pi c}{2L}, \quad n = 1, 2, \ldots, \infty \quad (3.140)$$

This is the *characteristic equation* that gives the natural frequencies ω_n for any integral value of n.

So the answer to our question is this: yes, the system *can* undergo time-harmonic free vibration, but only at the natural frequencies ω_n given by Equation 3.140. Corresponding to each natural frequency there is the mode shape

$$\phi_n(x) = \sin\frac{\omega_n x}{c} = \sin\frac{(2n-1)\pi x}{2L} \quad (3.141)$$

as illustrated in Figure 3.39. This satisfies not only the equation of motion but also the boundary conditions. Thus, there is an infinite number of modes but the natural frequencies are distinct, and well separated, so that there is a finite number of modes below any given frequency. Consequently, if there is a given maximum frequency of interest, we can describe the vibration in terms of a finite number of modes of vibration.

The fundamentals of the vibration of the continuous system are, therefore, similar to those of an MDOF system: there are modes of vibration; the general free vibration is the superposition of these modes of vibration; the mode shapes satisfy orthogonality conditions and the forced response can be written as the sum of responses in each mode, with each mode behaving

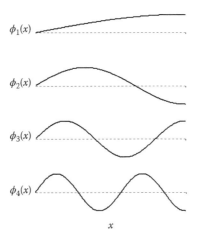

Figure 3.39 First four mode shapes of fixed–free rod. Note increasing number of nodes.

as an SDOF system. Once again, this emphasises the importance of understanding the behaviour of the SDOF system.

3.7.2 Bending vibration of a beam

Beams and plates are perhaps the most common structural component. Examples include panels in aircraft, cars and ships, columns and floors in buildings and stiffening frames in cars. Here we consider beams, and in particular thin beams for which Euler–Bernoulli bending theory applies. (It is assumed that plane cross-sections remain plane when deformed, that shear deformations can be neglected and that rotary inertia is negligible.) Thin plates can be modelled by a similar theory, but it is more complicated because they are two-dimensional. Beams and plates can carry transverse loads (acting perpendicularly to their centre lines) and also in-plane loads, similarly to the axial vibration of the rod discussed in the previous section. Apart from the ability to carry loads, bending vibrations of flat plates and curved shells are important because they couple well with any surrounding fluid, hence radiating sound into the fluid.

The theory describing bending vibration can be found elsewhere (e.g. Meirovitch 1997; Rao 2010). When at rest at the equilibrium position, the beam (Figure 3.40) lies along the x-axis. It has density ρ, cross-sectional area A and bending stiffness EI (see Section 3.1.7), assumed here to be constant along the length of the beam. When vibrating it has a displacement $w(x,t)$ and a slope $\theta(x,t) = \partial w/\partial x$. The equation of motion for free vibration is

$$EI\frac{\partial^4 w}{\partial x^4} + \rho A\frac{\partial^2 w}{\partial t^2} = 0 \tag{3.142}$$

while the internal forces (the *bending moment M* and *shear force V*, see Figure 3.40b) are

$$M = EI\frac{\partial^2 w}{\partial x^2}; \quad V = -EI\frac{\partial^3 w}{\partial x^3} \tag{3.143}$$

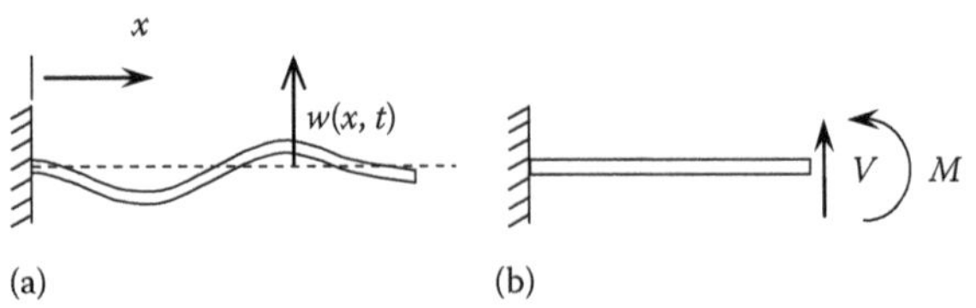

Figure 3.40 (a) Cantilever beam and (b) definitions of positive shear force and bending moment.

The equation of motion differs from the wave equation for a rod. It is of fourth order, and there are two deformation DOFs at any point (w, θ) and two corresponding internal force components (V, M).

Can we have time-harmonic free vibration? Well, let us start by assuming $w(x,t) = W(x)\exp(i\omega t)$. The equation of motion becomes

$$EI\frac{d^4W}{dx^4} - \rho A \omega^2 W = 0 \tag{3.144}$$

and hence the general solution is

$$W(x) = B\sin kx + C\cos kx + D\sinh kx + F\cosh kx; \quad k = \sqrt[4]{\frac{\rho A \omega^2}{EI}} \tag{3.145}$$

Of course, the general solution must also satisfy the boundary conditions. Common boundary conditions for a beam include free ($M = V = 0$), simply supported ($w = M = 0$), clamped ($w = \partial w/\partial x = 0$) and sliding ($V = \partial w/\partial x = 0$). For the case shown in Figure 3.40, the cantilever is clamped at $x = 0$ and free at $x = L$, so that the boundary conditions are

$$x = 0: \ w = 0, \ \frac{\partial w}{\partial x} = 0; \quad x = L: \ \frac{\partial^2 w}{\partial x^2} = 0, \ \frac{\partial^3 w}{\partial x^3} = 0 \tag{3.146}$$

Substituting these into the general solution (Equation 3.145) leads, after some arithmetic, to the answer to our question: yes, we *can* have time-harmonic free vibration, but only if

$$\cos kL \cosh kL = 0; \quad k = \sqrt[4]{\frac{\rho A \omega^2}{EI}} \tag{3.147}$$

This is the *characteristic equation*, to which we need to find numerical solutions for kL and hence for ω. These are listed in Tables 3.3 and 3.4 for beams and rods with some common boundary conditions. Corresponding to each natural frequency is a mode shape, for which the constants B, C, D and F in Equation 3.145 have special values, also indicated in Tables 3.3 and 3.4. (Note that simple analytical solutions only exist for beams with simply supported or sliding boundaries, while approximate expressions exist for large n whatever the boundary conditions. Note, also, that for some boundary conditions there are rigid body translational or rotational modes.)

Table 3.3 Natural frequencies and mode shapes for beams

Beams: boundary conditions:

▭ simply supported ($w = M = 0$)
▭ clamped ($w = \theta = 0$)
▭ free ($V = M = 0$)

Boundary conditions	Characteristic equation	Mode shapes	k_1L	K_2L	K_3L	k_nL (n large)
▭▭	$\sin kL = 0$	$\sin kx$	π	2π	3π	$n\pi$
▭	$\tan kL - \tan hkL = 0$	$\sin kx + A\sinh kx$ $A = \cos kL/\cosh kL$	3.927^a	7.069	10.210	$(n+\tfrac{1}{4})\pi$
▭	$\cos kL \cos hkL = 1$	$A(\cos kx + \cosh kx)$ $+ (\sin kx + \sinh kx)$ $A = (\cos kL - \cosh kL)/$ $(\sin kL + \sinh kL)$	4.730^a	7.853	10.996	$(n+\tfrac{1}{2})\pi$

Fundamentals of vibration   137

⊥ (clamped-free)	$\cos kL \cosh kL = -1$	$A(\cos kx - \cosh kx)$ $+ (\sin kx - \sinh kx)$ $A = (\cos kL + \cosh kL)/$ $(\sin kL - \sinh kL)$	1.875	4.694	7.855	$(n - \tfrac{1}{2})\pi$
(free-free)	$\cos kL \cosh kL = 1$	As clamped–free but with A as free–free	4.730	7.853	10.996	$(n + \tfrac{1}{2})\pi$
(clamped-clamped)	$\tan kL - \tanh kL = 0$	As clamped–free but $A = -(\sin kL + \sinh kL)/$ $(\cos kL + \cosh kL)$	3.927	7.069	10.210	$(n + \tfrac{1}{4})\pi$

[a] Existence of rigid body modes.

Table 3.4 Natural frequencies and mode shapes for rods

Boundary conditions	Characteristic equation	Mode shapes	Natural frequencies
fixed–fixed	$\cos\dfrac{\omega L}{c}=0$	$\sin\dfrac{(2n+1)\pi x}{2L}$	$\dfrac{(2n-1)\pi c}{L}$
free–free	$\sin\dfrac{\omega L}{c}=0$	$\cos\dfrac{n\pi x}{L}$	$\dfrac{n\pi c}{L}$ [a]
fixed–free	$\sin\dfrac{\omega L}{c}=0$	$\cos\dfrac{n\pi x}{L}$	$\dfrac{n\pi c}{L}$

Note: Boundary conditions ▨▨▨ fixed ($u=0$) and ▭▭▭ free ($P=0$).

[a] Existence of rigid body modes.

Clearly the algebra can be quite complicated, but the principles are clear: as for axial vibration of a rod or for an MDOF system, there are natural frequencies and modes, each of which has a unique mode shape. The free and forced vibrations can be written as a sum of modal terms; and, of course, each mode behaves like an SDOF system.

3.7.3 Waves in structures

Structural vibration can be represented in terms of waves as well as modes. In fact, the modes of distributed systems are simply the interference patterns (standing waves) formed by waves travelling around the structure. The wave interpretation is particularly useful at higher frequencies (e.g. higher than the first few natural frequencies) when we might want to describe the transmission of vibration from a source, such as a machine, through the structure to which the source is attached, to some receiver.

For the case of axial vibration in a rod, we note that we could also write the solution to the equation of motion, Equation 3.137, as

$$U(x) = Be^{-ikx} + Ce^{ikx}$$

so that the axial displacement is

$$u(x,t) = Be^{i(\omega t - kx)} + Ce^{i(\omega t + kx)} \tag{3.148}$$

The two terms represent waves which *propagate* in the positive and negative x-directions respectively. These are analogous to sound waves in air. In

Equation 3.148, k is the *wavenumber* (spatial frequency or phase change per unit distance), while the *wavelength* is $\lambda = 2\pi/k$. The waves travel with a speed $c = \omega/k = \sqrt{E/\rho}$. The situation is different for bending waves (Equation 3.144). In this case the displacement can be written as

$$w(x,t) = Be^{i(\omega t - kx)} + Ce^{i(\omega t + kx)} + De^{i\omega t}e^{-kx} + Fe^{i\omega t}e^{+kx} \qquad (3.149)$$

It comprises two propagating waves, represented by the first two terms, and two *nearfield* or *evanescent* waves which decay exponentially with distance, even in the absence of damping. For beams, the wavenumber k is defined by the last of Equations 3.145 and is proportional to $\sqrt{\omega}$. The wave motion is dispersive, which has important consequences for sound radiation and transmission (see Sections 5.5 and 5.6).

Waves propagate through a structure and are reflected and transmitted at boundaries and where the properties of the structure change. A more complete analysis is beyond the scope of this chapter.

3.8 CONCLUDING REMARKS

In this chapter, we have considered the vibration of mechanical structures, ranging from simple SDOF models to continuous systems. The essence of the behaviour – the fundamentals of vibration – is that all SDOF systems behave similarly, having a natural frequency and damping ratio and resonating if excited near their natural frequency. The natural frequency is determined by the stiffness and mass (Equation 3.32), while the free vibration, if underdamped, is a decaying oscillation (Equation 3.55). The response of the SDOF system can be found either in terms of the frequency response (Equation 3.64) or the impulse response (Equation 3.79). MDOF and continuous systems have modes of vibration. While the behaviour of the system might be more complicated, the modes themselves each behave as an SDOF system, so the SDOF results can be used to find the modal solution. Representing the behaviour as a sum of modes gives computational savings, but much more importantly, gives us insight into the behaviour of the systems.

Our analysis has considered relatively simple models for which the equations of motion are not difficult to find. For more complicated, real-life systems, the effort we need to spend to find the equations of motion and then to solve the eigenvalue problem – to find the natural frequencies and mode shapes – can be very much higher, and we will often need to use approximate numerical methods such as FE analysis. The underlying principles hold, however, and the fundamentals of vibration are the same.

3.9 QUESTIONS

1. Find the natural frequencies of the SDOF systems shown.

   (a)

   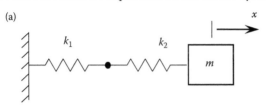

   (b)

   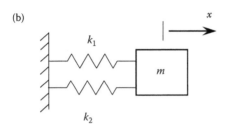

   (c) The second moment of mass of the disc shown is J. The string attached to the mass m is light and inextensible. Find also the effective masses and stiffnesses in terms of both the displacement y of the end of the spring and the rotation θ of the disc.

   $$\omega_n = \sqrt{kr_2^2/(J + mr_1^2)}$$

   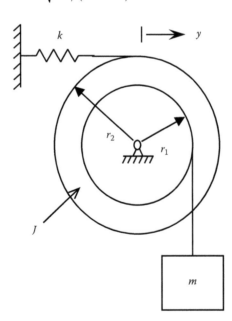

(d) The link AB is rigid and has negligible mass. Consider the two cases where (i) the link rotates in a horizontal plane, and (ii) the link rotates in a vertical plane (i.e. the motion is affected by gravity)

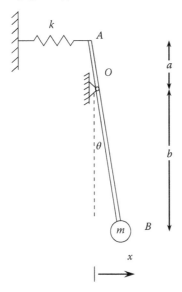

2. A machine of mass $m=500$ kg is mounted on a simply supported steel beam of length $l=2$ m having a rectangular cross-section (depth $d=0.1$ m, width $b=1.2$ m) and elastic modulus $E=2.06\times 10^{11}$ N m^{-2} as shown. To reduce the vertical deflection of the beam a spring of stiffness k is attached at mid-span. Determine the value of k needed to reduce the deflection of the beam to one-third of its original value. Note that the second moment of area of a rectangular cross-section is $I = bd^3/12$.
(Ans. 247 kN mm^{-1}.)

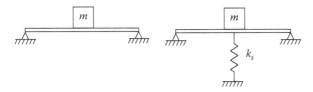

3. A compressor of mass 2000 kg is to be supported on four spring mounts of equal stiffness. The compressor runs at 1200 rpm (revolutions per minute). Find the natural frequency and the stiffness of each mount if the transmissibility is to be 10% or less. Neglect damping.

4. A piezoelectric accelerometer is a device used to measure the acceleration of a vibrating object. In principle, it consists of a disc of piezoelectric material (cross-sectional area A, thickness d and elastic modulus E) which behaves like a spring. Its top and bottom surfaces are covered by electrodes to which terminals are attached. On it is mounted a seismic mass m. The base is attached to the vibrating surface, which has an acceleration $\ddot{y}$. When the accelerometer vibrates, a force F is produced in the piezoelectric material, and a charge $Q = sF$ appears across the terminals, where s is a constant that depends on the piezoelectric constant for the material.

Find the stiffness of the piezoelectric disc and hence the equation of motion of the mass. Show that for frequencies which are much lower than the natural frequency of the accelerometer, the charge output from the terminals is indeed proportional to the acceleration of the base.

Up to what frequency (approximately – ignore damping) is the magnitude of the acceleration accurate to 1%? If the damping factor is $\zeta = 10\%$, up to what frequency is the phase of the acceleration accurate to 5°?

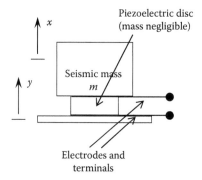

5. A delicate instrument is supported by some packaging. A simple model is an SDOF system with mass m (the instrument) supported by a stiffness k (the packaging). The package is dropped from a height h.

Show that on impact at $t = 0$ it has a velocity $v = \sqrt{2gh}$.

Measuring x from the position of m on impact, find an expression for the displacement $x(t)$ of the mass after impact, assuming that the duration of the impact is very short.

What is the maximum acceleration after impact? Express this answer in terms of the static deflection δ of the spring under the weight of the mass m.

What should this static deflection be if the instrument must be able to withstand a fall of 0.8 m, given that the acceleration must not exceed 100 m s^{-2}?

(Ans. $x(t) = \dfrac{\sqrt{2gh}}{\omega_n} \sin \omega_n t + \dfrac{g}{\omega_n^2}(1 - \cos \omega_n t); \quad \ddot{x}_{max} = -g\sqrt{\dfrac{2h}{\delta} + 1}$)

6. An aircraft engine can be modelled as an SDOF system with $m = 1500$ kg, $\omega_n = 30$ rad s^{-1} and $\zeta = 0.03$. During a particular manoeuvre it is excited by white noise with a constant psd (single-sided) of 5×10^4 N^{2}Hz^{-1}. Estimate the mean-square displacement and the r.m.s displacement, velocity and acceleration.
(Ans. 1.9 mm, 55.6 mm s^{-1}, 1.67 m s^{-2})

7. A two degree of freedom system is shown in this diagram with values of stiffness and mass in newtons per metre and kilograms indicated.

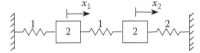

(a) Find the natural frequencies and mode shapes. Hence, find the modal matrix **B**.
(b) What are the values of the modal masses and modal stiffnesses?
(c) A time-harmonic force $f_1(t) = Fe^{i\omega t}$ acts on the left-hand mass. Find expressions for the modal forces.
(d) Using modal decomposition, find the receptances α_{11} and α_{21}, expressing them as the sum of two modal components.

8. A rod of length l undergoes axial vibration. The end at $x = 0$ is fixed, while the end at $x = l$ carries a mass M. Show that the characteristic equation is
$$a \tan a = \dfrac{\rho A l}{M}; \quad a = \dfrac{\omega l}{c}$$

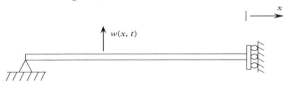

9. Find the natural frequencies and mode shapes of a beam in bending if it is simply supported at $x = 0$, while the opposite end slides freely in the vertical guide, but is not allowed to rotate.

REFERENCES

Den Hartog, J.P. (2003) *Mechanical Vibrations*, Dover Publications, New York.

Meirovitch, L. (1997) *Principles and Techniques of Vibrations*. Prentice Hall, Englewood Cliffs, NJ.

Meirovitch, L. (2000) *Fundamentals of Vibrations: International Student Edition*, McGraw-Hill, New York.

Meriam, J.L. and Kraige, L.G. (2003) *Engineering Mechanics: Dynamics* (5th edn), John Wiley & Sons, New York.

Nashif, A.D., Jones, D.I.G., and Henderson, J.P. (1985) *Vibration Damping*, John Wiley & Sons, New York.

Rao, S.S. (2010) *Mechanical Vibrations* (5th edn), Prentice Hall, Englewood Cliffs, NJ.

Chapter 4

Fundamentals of signal processing

Joe Hammond and Paul White

4.1 INTRODUCTION

Signal processing is the science of applying transformations to measurements to facilitate their use by an observer or a computer; and digital signal processing (DSP) is the enabling technology for applications across all disciplines and sectors. The study of sound and vibration is dependent on the use of special-purpose signal analysers or software packages which are easily installed on a variety of computing platforms. The accessibility and convenience of DSP analysis modules and procedures can sometimes create a deceptive air of simplicity in often complicated phenomena. It is important that practitioners, while availing themselves of the full range of DSP capabilities, should have a clear understanding of the fundamentals of the science of signal processing and so be fully aware of the assumptions, implications and limitations inherent in their analysis methods. The aim of this chapter is to provide an introduction to signal processing, emphasising the basis of the methodology.

Data analysis involves the three phases of data acquisition, processing and interpretation (of the results of the processing) and, of course, all three are linked in any application. The last phase is naturally very much related to the subject under investigation, but the first two may be discussed independently of specific applications. A vast body of theory and methodology has been built up as a consequence of the problems raised by the need for data analysis; this is often referred to as *signal analysis* or *time-series analysis*, depending on the context. This chapter concerns methods of analysis of time histories which have usually been obtained as measurements in the course of conducting an experiment, or when recording the behaviour of a physical system. We shall limit our considerations to 'basic' procedures; inevitably, the treatment will be superficial owing to the restricted scope afforded by a single chapter.

4.1.1 Data classification, generation and processing

Figure 4.1 illustrates a broad categorisation of time histories (Bendat and Piersol 2010). A basic distinction is the designation of a signal as 'random' or 'deterministic', where by 'random' we mean one that is not exactly predictable. Very often, processes are mixed, and the demarcations of Figure 4.1 are not easily applied. Additionally, it is often convenient to regard a time history as arising from some 'data generation model' which may be physical or conceptual (e.g. appropriately filtered white noise), as in Figure 4.2.

We shall neglect chaotic signals which may be generated by deterministic nonlinear processes but nevertheless have unpredictable behaviour. The processing strategy we employ may be significantly influenced by prior conceptions regarding the 'type' of signal and its mode of generation, leading to either *parametric* or *nonparametric* analysis.

These categorisations are a rather simplistic view of modern signal processing, which employs a variety of noise reduction, identification and inversion techniques based within optimisation theory, matrix theory and probabilistic methods. However, it is important to be aware in any analysis that procedures based on prior knowledge (or often merely assumptions) may bias the outcome. We should strive always to keep in mind the first principle of data reduction (Ables 1974): 'The result of any transformation imposed on the experimental data shall incorporate and be consistent with

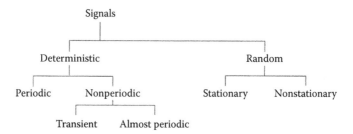

Figure 4.1 Types of signal.

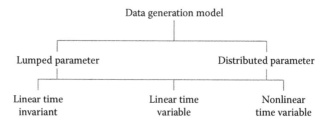

Figure 4.2 Data generation models.

all relevant data and be maximally noncommittal with regard to unavailable data'.

4.1.1.1 Objectives of data analysis

Generally speaking, the objective of data analysis is to highlight/extract information contained in a signal that direct observation may not reveal. Sometimes, only 'gross' characteristics of a signal may be required (e.g. mean value, mean-square level, total energy), in which case some simple processing will produce the required quantities, while ignoring other information contained in the signal which might only unnecessarily confuse matters. At other times, it may be necessary to probe more deeply in an attempt to extract features of interest.

4.1.1.2 Methods of data analysis

The signals that we shall consider are time histories (often electrical outputs from transducers), or equivalent (e.g. rough ground-height profiles where a space variable is the equivalent of time) and useful information extraction procedures may be performed while remaining in the 'time domain'. However, very often transform methods are employed, since they offer simplifications both in theory and physical interpretation, and 'frequency-domain methods' in particular are commonly used. The signals recorded from physical phenomena are generally 'continuous time' in nature (sometimes called analogue data), but developments in technology have now ensured that most signal processing is achieved digitally. This means the data are sampled, and the analysis of discrete (in time) data requires that the analyst should have a good grasp of the fundamentals of DSP, which is now a vast area of study. Sections 4.2 and 4.3 deal with elements of Fourier analysis of continuous time signals.

4.2 FOURIER ANALYSIS OF CONTINUOUS TIME SIGNALS

This section is concerned with the representation of signals as the sum of simpler components, namely sinusoids and cosinusoids. As a preliminary, we briefly discuss the (Dirac) delta function and the concept of convolution.

4.2.1 Preliminary: delta function

The delta function (sometimes called the unit-impulse function) is written $\delta(t)$ and is defined by

$\delta(t) = 0 \quad t \neq 0$

$$\int_{-\infty}^{\infty} \delta(t)\mathrm{d}t = 1 \quad\quad\quad (4.1)$$

Clearly, this is not a function in the ordinary sense. It belongs to the class of 'generalised' functions or distributions. We can intuitively think of it as a very narrow, tall spike at $t=0$; that is to say, consider a function (Figure 4.3)

$$\delta_\varepsilon(t) = \frac{1}{\varepsilon} \quad\quad -\frac{\varepsilon}{2} < t < \frac{\varepsilon}{2}$$

$$= 0 \quad\quad \text{elsewhere} \quad\quad\quad (4.2)$$

and so $\int_{-\infty}^{\infty} \delta_\varepsilon(t)\mathrm{d}t = 1$ for all ε. One can define the unit 'impulse' using $\delta(t) = \lim_{\varepsilon \to 0} \delta_\varepsilon(t)$.

Pictorially, the delta function is represented as an arrow whose height indicates the area under the delta function, referred to as the *strength* of the impulse (Figure 4.4).

Another interpretation may be given in terms of the unit (or Heaviside) step function, $u(t)$, defined as

$u(t) = 1 \quad t > 0$

$\quad\quad\, = 0 \quad t < 0 \quad\quad\quad (4.3)$

so that

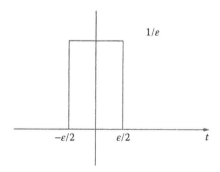

Figure 4.3 Impulse of unit magnitude.

Figure 4.4 Graphical representation of a delta function.

$$\delta(t) = \frac{d}{dt} u(t) \tag{4.4}$$

that is to say, the delta function is the derivative of the unit step function. The introduction of this concept permits us to differentiate functions having discontinuities. A further definition of the delta function, which plays an important role in frequency analysis, is

$$\delta(t) = \int_{-\infty}^{\infty} e^{\pm 2\pi i v t} dv = \frac{1}{2\pi} \int_{-\infty}^{\infty} e^{\pm i v t} dv \tag{4.5}$$

Consider a delta function located at $t=a$, which is written $\delta(t-a)$, and an ordinary function $f(t)$, then

$$\int_{-\infty}^{\infty} f(t)\delta(t-a)dt = f(a) \tag{4.6}$$

This referred to as the *sifting property* and is a key property of delta functions.

This exposition is rather heuristic and may be made more formal using *distribution theory*; see, for example, Zadeh and Desoer (1963).

4.2.2 Fourier series and Fourier integral

In Figure 4.1, we defined three types of deterministic signal, namely, periodic, transient and almost periodic. Periodic data are analysed using the Fourier series, transient data use the Fourier integral, and some classes of almost-periodic signals (and also damped signals) may be studied using the Prony series (Spitznogle and Quazi 1970; Davies 1983). The basic difference between the Prony series and the Fourier series is that in the former case the frequencies are not predetermined, and so both (complex) amplitudes and frequencies are computed, while in the Fourier method only complex amplitudes are calculated at a specified set of frequencies. The Prony method is computationally more difficult, but can be very attractive under certain circumstances (e.g. where data lengths are short and where the data match the model of a sum of damped oscillations).

If a signal repeats itself exactly every T_p seconds, then $x(t)$ may be represented as a sum of sinusoids and cosinusoids whose frequencies are integer multiples of the fundamental frequency $1/T_p$ Hz, together with a constant representing the 'offset' of the signal. Each harmonic component is characterised by its amplitude and phase. It is very convenient to express these oscillations in complex form which leads to expressing the signal $x(t)$ as

$$x(t) = \sum_{n=-\infty}^{\infty} C_n e^{i2\pi nt/T_p} \qquad (4.7)$$

The amplitude C_n is now complex, representing both the amplitude and the phase of each component, and the concept of a negative frequency has been introduced by the term n/T_p when n is negative. If $x(t)$ is a transient, then the discrete set of sinusoids and cosinusoids becomes a continuum and we write

$$x(t) = \int_{-\infty}^{\infty} X(f) e^{i2\pi ft} df = \mathbb{F}^{-1}\{X(f)\} \qquad (4.8)$$

$X(f)$ is given by

$$X(f) = \int_{-\infty}^{\infty} x(t) e^{-i2\pi ft} dt = \mathbb{F}\{x(t)\} \qquad (4.9)$$

The specification of C_n or $X(f)$ computed from $x(t)$ is equivalent to the original time history, since we can reconstruct the signal from these. The frequency f is in hertz and $\omega = 2\pi f$ is the frequency in radians per second. Note that Equation 4.8 differs from the form given in Section 3.4.4, because the integral is over f (hertz) in the former and over ω (radians per second) in the latter. We note that the properties of the delta function (particularly Equations 4.5 and 4.6) may be used to 'invert' Equation 4.8 to give Equation 4.9 and that the two equations also define the Fourier operator, $\mathbb{F}\{\ \}$, and its inverse, $\mathbb{F}^{-1}\{\ \}$.

4.2.2.1 Random signals and nonstationarity

It is a logical step to ask if the representations in Equations 4.7 and 4.8 may be generalised to allow the representation of random signals. 'Randomness' and the nature of the randomness (stationary or nonstationary) introduce some significant conceptual aspects (Priestley 1967). In summary, a stationary random signal may be expressed as

$$x(t) = \int_{-\infty}^{\infty} e^{i\omega t} dZ_x(\omega) \qquad (4.10)$$

which is a generalisation of Equation 4.8. (This is a Fourier–Stieltjes form – only reducing to the Riemann form (Equation 4.8) when $Z_x(\omega)$ is differentiable as for 'transients' in Section 4.2.2.) We will sidestep this mathematical correctness in Section 4.6 in our treatment of stationary random signals.

If the process is *nonstationary*, then while this form *may* still be appropriate, there are significant changes that occur to the corresponding spectra (due to the *nonorthogonality* of $dZ_x(\omega)$), and so concepts of two-dimensional (frequency–frequency or time–frequency) spectra have been introduced. The whole subject of time–frequency analysis has attracted considerable interest, but is beyond the scope of this introductory chapter, and the reader is referred to Priestley (1967), Cohen (1989) and Hammond and White (1996) for further details.

4.3 SOME RESULTS IN SIGNAL AND SYSTEM ANALYSIS

In this section, we establish some basic results in signal and system analysis that are needed in data analysis. Concepts that are described include convolution (in both time and frequency domains) and the Hilbert transform.

4.3.1 Effect of delay

If a signal $x(t)$ has Fourier transform $X(f)$, then a delayed signal $x(t - t_0)$ has Fourier transform $e^{-i2\pi f t_0} X(f)$. The effect of the delay in the frequency domain is the multiplier $e^{-i2\pi f t_0}$, which has unit magnitude and phase $\phi(f) = -2\pi f t_0$. The derivative of the phase with respect to f is $d\phi(f)/df = -2\pi t_0$, that is, proportional to delay t_0 for all frequencies. Written in terms of ω, one can say $d\phi/d\omega = -t_0$.

The quantity $-(d\phi/d\omega)$ is called the group delay of the signal. In the case of a pure delay, the group delay is the same for all frequencies: there is no dispersion. If $\phi(\omega)$ is a *nonlinear* function of the frequency ω, then the group delay depends on the frequency, and dispersion is present, causing phase distortion of the signal.

4.3.1.1 Echoes

If a simple echo is present in a signal, $y(t)$, this may be modelled as

$$y(t) = x(t) + ax(t - t_0) \qquad (4.11)$$

where $x(t)$ is the 'original' signal, a is the echo amplitude and t_0 is the 'epoch' of the echo. The Fourier transform of $y(t)$ is

$$Y(f) = \left(1 + ae^{-i2\pi d f t_0}\right) X(f) \qquad (4.12)$$

The effect of the echo is contained in the term $(1 + ae^{-i2\pi f t_0})$, which has an oscillatory form as a function of frequency in both the amplitude and phase. The squared amplitude is $(1 + a^2 + 2a \cos 2\pi f t_0)$, which imposes an oscillatory form on $|Y(f)|^2$. Accordingly, the appearance of such a feature in an energy (or power) spectrum is an indicator of an echo. Additional echoes and dispersion obviously complicate the picture. The analysis of such signals leads to the introduction of cepstral analysis (Bogert et al. 1963), which, in turn, was generalised to homomorphic deconvolution (Oppenheim and Schafer 1975).

4.3.2 Convolution

The *convolution* of two time histories is introduced via the concept of linear filtering. In Figure 4.5, $x(t)$ is the excitation of a linear system with an impulse-response function $h(t)$ producing the response $y(t)$.

With reference to Figure 4.5, since $h(t)$ is the response to a unit impulse $\delta(t)$, then for a time-invariant system, $h(t - t_1)$ is the response to $\delta(t - t_1)$. Splitting up $x(t)$ into elemental impulses $x(t_1)\Delta t_1$, the response at time t due to such an impulse is then $h(t - t_1)x(t_1)\Delta t_1$ if the system is linear (see Figure 4.6). Adding up all such responses and letting $\Delta t_1 \to 0$ yields

$$y(t) = \int_{-\infty}^{t} h(t - t_1) x(t_1) dt_1 \qquad (4.13)$$

Figure 4.5 Single input–single output system.

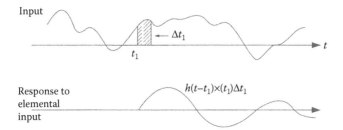

Figure 4.6 The response of a system to an 'elemental' impulse.

The upper limit is t because we assume the system is causal, that is to say, the impulse response $h(t)$ is such that $h(t)=0$ for all $t<0$, which physically means that the system does not respond until an input has been applied. Then, using the substitution $t-t_1=\tau$, this may be expressed as

$$y(t) = \int_0^\infty h(\tau)x(t-\tau)d\tau \tag{4.14}$$

where $h(\tau)$ now has the role of a weighting (or memory) function, as depicted in Figure 4.7.

The convolution of the functions h and x is often written $y=h*x$. This expression of the effect of a linear filter in the time domain is considerably simplified by taking the Fourier transform, which yields

$$Y(f) = H(f)X(f) \tag{4.15}$$

that is to say, convolutions transform to products. $H(f)$ is the frequency response of the linear filter and is the Fourier transform of the impulse-response function.

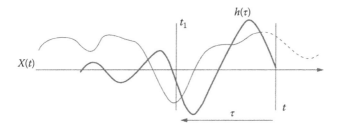

Figure 4.7 The impulse-response function considered as a weighting (memory) function.

4.3.3 Windowing

Suppose $x(t)$ is known only for $-T/2 < t < T/2$, or we choose to 'truncate' the signal to that length.

It is as though we are looking at the data through a 'window' $w(t)$, where, in this case

$$w(t) = 1 \quad |t| < T/2$$
$$= 0 \quad |t| > T/2 \quad (4.16)$$

giving the truncated signal $x_T(t) = w(t)\, x(t)$, as shown in Figure 4.8.

Fourier transforming $x_T(t)$, we obtain (for a general $w(t)$)

$$X_T(f) = \int_{-\infty}^{\infty} X(g) W(f-g) dg \quad (4.17)$$

that is to say, the Fourier transform of the *product* of two time functions is the convolution of their transforms (compare this with Equations 4.14 and 4.15).

The Fourier transform of the particular case of the 'rectangular' data window considered in this section is given by

$$W(f) = \frac{T \sin(\pi f T)}{\pi f T} \quad (4.18)$$

and is depicted in Figure 4.9. The convolution operation acts to widen the 'true' Fourier transform $X(f)$ and so distort it.

This distortion is sometimes called *smearing* (due to the main lobe) and *leakage*, since components of $X(f)$ at values other than $g=f$ 'leak' through the side lobes to contribute to the value of $X_T(f)$ at f. For example, suppose $x(t) = \cos 2\pi p t$, then

$$X(f) = \frac{1}{2}\left[\delta(f+p) + \delta(f-p)\right] \quad (4.19)$$

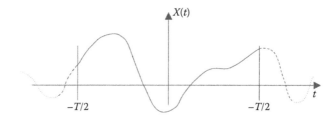

Figure 4.8 Data truncation with a rectangular window.

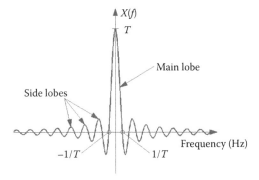

Figure 4.9 Fourier transform of the rectangular window.

and so

$$X_T(f) = \frac{1}{2}[W(f+p) + W(f-p)] \qquad (4.20)$$

showing that the shape of the spectral window (in the frequency domain) replaces the delta function. A comparison of theoretical and achieved spectra is depicted in Figure 4.10.

If two (or more) closely spaced harmonic components occur in a signal, then the distortion may mean that the components cannot (easily) be resolved (see Figure 4.11).

To separate two peaks at frequencies f_1, f_2, it is necessary to use a record length T of order $T \geq 2/(f_2-f_1)$ for a rectangular window.

We note that the rectangular window is a particularly 'poor' window as far as the side lobes are concerned (the highest side lobe is 13 dB below the main peak with an asymptotic roll-off of 6 dB per octave). The sharp

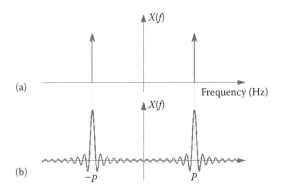

Figure 4.10 Spectra of a cosine wave: (a) theoretical; (b) windowed.

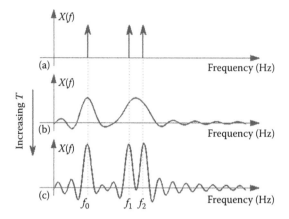

Figure 4.11 Effect of windowing on the modulus of the Fourier transform of three cosines.

corners of the data window result in this rippling effect. However, the main lobe is narrower than other windows.

Numerous alternatives have been suggested to reduce leakage in Fourier transform calculations. By tapering the windows to zero, the side lobes are reduced, but at the expense of widening the main lobe (increasing smearing); see Section 4.3.4. The frequency-domain behaviour of windows is often given in decibels (dB) where the characteristics have been normalised to unity gain (0 dB) at zero frequency, and some terms are defined in Figure 4.12, which relates to the rectangular window.

The properties of a few commonly used windows are summarised in Table 4.1 (see e.g. Childers and Durling (1975) for the appropriate formulae).

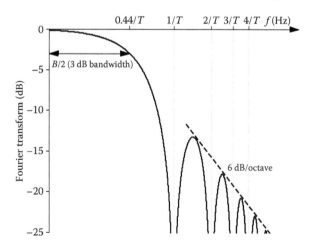

Figure 4.12 Frequency-domain characterisation of the rectangular window.

Table 4.1 Properties of a few commonly used windows

Window (length T)	Highest side lobe (dB)	3 dB bandwidth	Asymptotic roll-off (dB/octave)	First zero crossing
Rectangle	−13	0.88/T	6	1/T
Bartlett	−26	1.25/T	12	2/T
Hann(ing) or Tukey	−32	1.4/T	18	2/T
Hamming	−42	1.3/T	6	3/T

Further, the paper by Harris (1978) contains a comprehensive treatment of various windowing functions and their effects.

4.3.4 Uncertainty principle

A property of the Fourier transform of a signal is that the narrower a signal characterisation is in one domain, the wider is its characterisation in the other. An extreme example is a delta function $\delta(t)$ whose Fourier transform is a constant. Another is an oscillation $e^{i2\pi pt}$ whose Fourier transform is $\delta(f-p)$. This fundamental property of signals is summarised in the so-called *uncertainty principle*, as follows (Hsu 1970).

If $x(t)$ is a signal with finite energy, that is, $\|x\|^2 = \int_{-\infty}^{\infty} x^2(t)dt < \infty$, and has Fourier transform $X(f)$, then defining the following

$$\bar{t} = \frac{1}{\|x\|^2} \int_{-\infty}^{\infty} tx^2(t)dt \quad \Delta t^2 = \frac{1}{\|x\|^2} \int_{-\infty}^{\infty} (t-\bar{t})^2 x^2(t)dt \tag{4.21a}$$

$$\|X\|^2 = \|x\|^2 = \int_{-\infty}^{\infty} |X(f)|^2 df \quad \bar{f} = \frac{1}{\|X\|^2} \int_{-\infty}^{\infty} f|X(f)|^2 df \tag{4.21b}$$

and

$$(\Delta f)^2 = \frac{1}{\|X\|^2} \int_{-\infty}^{\infty} (f-\bar{f})^2 |X(f)|^2 df \tag{4.21c}$$

where $\bar{t}, \bar{f}$ are measures of location (i.e. the average time and frequency of the signal) and $\Delta t, \Delta f$ are measures of spread (i.e. the duration and bandwidth,* respectively), then it can be shown that

---

* Note that both Δf and B are measures of bandwidth, but are not the same quantity: Δf is called the root-mean-square bandwidth, to distinguish it from the 3 dB bandwidth, B.

$$\Delta f \Delta t \geq \frac{1}{4\pi} \qquad (4.22)$$

Thus, the bandwidth-time (BT) product of a signal has a lower bound of $1/4\pi$ where bandwidth and time here refer to the 'extent' of the signal in these two domains. Clearly, as Δf gets narrower, Δt must get wider (i.e. there is an inverse spreading property). It is this inequality that is at the heart of 'difficulties' in time–frequency analysis methods which are Fourier-based; that is, if one wants a 'local' Fourier transform, then increasing the 'localisation' in time leads to poorer resolution in frequency and vice versa, that is, one cannot simultaneously obtain arbitrarily fine 'resolution' in *both* frequency and time domains. It is noted that the Gaussian pulse e^{-at^2} (whose Fourier transform is another Gaussian pulse) has the minimum BT product of $1/4\pi$.

4.3.5 Hilbert transform

The Hilbert transform of a signal $x(t)$ is defined as the convolution of $x(t)$ with $1/\pi t$ and is denoted $\hat{x}(t)$, that is to say,

$$\hat{x}(t) = \frac{1}{\pi t} * x(t) \qquad (4.23)$$

This concept may be arrived at via analytic function theory. Despite the singular nature of the noncausal filter $1/\pi t$, many useful analytic results can be obtained. For example, the Hilbert transform of $\cos(2\pi f_0 t)$ is $\sin(2\pi f_0 t)$ (and that of $\sin(2\pi f_0 t)$ is $-\cos(2\pi f_0 t)$), and consequently, a Hilbert transform is sometimes referred to as a 90° phase shifter.

Transformation to the frequency domain is helpful, as the Fourier transform of the convolution operation can be expressed as

$$\hat{X}(f) = H(f)X(f) \qquad (4.24)$$

where

$$\begin{aligned} H(f) &= i & f < 0 \\ &= -i & f > 0 \\ &= 0 & f = 0 \end{aligned} \qquad (4.25)$$

This result may be used to obtain simple analytic results (rather than the time-domain convolution).

The Hilbert transform has considerable significance in signal-processing analysis. For our purposes here, we note that it is used in the definition of the 'instantaneous frequency' of a signal in the following way. Combining the original signal $x(t)$ with its Hilbert transform $\hat{x}(t)$ in the form

$$\tilde{x}(t) = x(t) + i\hat{x}(t) \tag{4.26}$$

creates what is referred to as the 'analytic' signal or 'preenvelope' signal $\tilde{x}(t)$, which is complex and hence expressible as

$$\tilde{x}(t) = A_x(t)e^{i\phi_x(t)} \tag{4.27}$$

where:

$A_x(t) = \sqrt{x^2 + \hat{x}^2}$ is the instantaneous amplitude
$\phi_x(t) = \tan^{-1}(\hat{x}(t)/x(t))$ is the instantaneous phase
$\dot{\phi}_x / 2\pi$ is the instantaneous frequency

For the trivial case $x(t) = a \cos(2\pi f_0 t)$, the analytic signal is $ae^{2\pi i f_0 t}$ and the instantaneous frequency is f_0, with instantaneous amplitude $|a|$, as one would expect. These definitions of instantaneous amplitude, phase and frequency have come to be accepted for some classes of signals, and are particularly relevant to amplitude-modulated and frequency-modulated signals, for example, for a chirp, when tracking a 'variable' frequency may be significant.

However, these definitions do fail to yield meaningful results when the signal being analysed comprises more than one component, that is, more than one modulated sinusoid is presented. Such signals are commonly referred to as multicomponent signals (this concept lacks a precise definition). The converse of a multicomponent signal, that is, a signal containing only one component, is referred to as a monocomponent signal.

4.3.6 Instantaneous frequency and group delay

In Section 4.3.1, the notion of group delay was introduced, and in Section 4.3.5 the idea of instantaneous frequency was noted. Both these notions are important as starting points for time–frequency analysis. One can conceive of a plot of group delay versus ω, that is, time versus frequency and, quite separately, a plot of $\dot{\phi}$ versus t, that is, (instantaneous) frequency versus time, yielding apparently two *different* ways of obtaining a time–frequency representation.

Both appear to offer the possibility of combining time and frequency in a plane to provide the basis of time–frequency analysis. The question arises as to whether the two time–frequency laws are the same or not; specifically, whether they are inverse functions of each other. In general they are not, and Boashash (1987) describes conditions that must be met by a signal $x(t)$ for these two time–frequency laws to be equivalent. These conditions are that the signal should be 'asymptotic'; that is, that the BT product (referred to in Section 4.3.4) should be 'large', and that the signal should be monocomponent. For such signals, the instantaneous frequency and group delay are equivalent, in which case there is no ambiguity in the definition of a time–frequency plot. The reader is referred to two review papers for time–frequency methods (Cohen 1989; Hammond and White 1996).

4.4 EFFECTS OF SAMPLING

This section introduces the concepts associated with converting a continuous time signal $x(t)$ into a sequence of numbers obtained by observing the signal every Δ seconds.

4.4.1 Analogue-to-digital conversion

An analogue-to-digital converter (ADC) is a device that operates on a continuous time history (input) and produces a sequence of numbers (output) that are sample values of the input. It is convenient to regard the process as consisting of two stages, namely 'sampling' and 'quantisation'. In Figure 4.13, $x(n\Delta)$ is the exact value attained by time history $x(t)$ at time $t = n\Delta$ (Δ is the sample interval for uniform-rate sampling). $\tilde{x}(n\Delta)$ is the representation of $x(n\Delta)$ in a computer, and differs from $x(n\Delta)$, since a finite number of bits are used to represent each number. Actual ADCs do not consist of two separate components (as in the 'conceptual' ADC depicted in Figure 4.13), and a variety of forms of ADC are available.

4.4.2 Fourier transform of a sequence

Since the calculations of Fourier coefficients and transforms are usually carried out digitally, it is necessary to reexamine the Fourier methods we described for data, but now in discrete form. We might treat a discrete version of Equation 4.8, for example, as an approximation to the 'correct' result. However, completely rigorous, exact theories can be developed for discrete data which are, of course, closely related to the continuous time counterparts, but there are fundamental differences. It is often more convenient to consider the problems of discrete data as self-contained, and relate the results to those of the continuous domain only as and when they are required, accounting at that stage for any errors or approximations that might have been incurred in using discrete methods in place of the

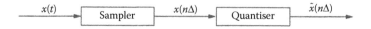

Figure 4.13 Conceptual model of analogue-to-digital conversion.

continuous operations. Having said this, we will, nevertheless, attempt to relate the Fourier analyses of continuous and discrete data.

4.4.2.1 Impulse train modulation

One way of introducing the Fourier transform of a sequence involves the use of the mathematical notion of 'ideal sampling' of a continuous wave. If an analogue signal $x(t)$ is to be sampled every Δ seconds, it is convenient to model the sampled signal as the product of the continuous signal with a 'train' of delta functions $\tilde{i}(t)$ where

$$\tilde{i}(t) = \sum_{n=-\infty}^{\infty} \delta(t - n\Delta) \tag{4.28}$$

so that the sampled signal, $x_s(t)$, is modelled as $x_s(t) = x(t)\tilde{i}(t)$. The Fourier transform of $x_s(t)$ is $X_s(f)$, and using the properties of the delta function (Equations 4.5 and 4.6), is written

$$X_s(f) = \sum_{n=-\infty}^{\infty} x(n\Delta) e^{-i2\pi f n \Delta} \tag{4.29}$$

Multiplying both sides by $e^{i2\pi f n \Delta}$ and integrating with respect to f from $-½\Delta$ to $½\Delta$ gives

$$x(n\Delta) = \Delta \int_{-1/2\Delta}^{1/2\Delta} X_s(f) e^{i2\pi f n \Delta} df \tag{4.30}$$

Equations 4.29 and 4.30 relate the sequence of numbers $x(n\Delta)$ to the quantity $X_s(f)$, often written $X(e^{i2\pi f/\Delta})$, which is termed the Fourier transform of the sequence.

A natural question to ask at this stage is 'How is $X_s(f)$ related to $X(f)$?' The answer to this can be obtained as follows. Since $\tilde{i}(t)$ is periodic, we represent it as a Fourier series, that is,

$$\tilde{i}(t) = \frac{1}{\Delta} \sum_{n=-\infty}^{\infty} e^{(2\pi i n t)/\Delta} \tag{4.31}$$

Now substituting this into $x_s(t) = x(t)\tilde{i}(t)$ and Fourier transforming gives an alternative right-hand side to Equation 4.29, namely

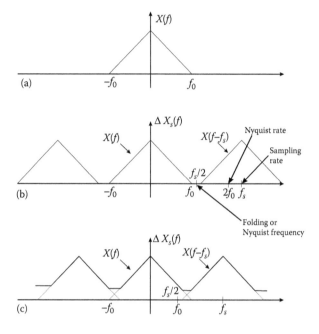

Figure 4.14 An example of the effect of sampling on the Fourier transform: (a) magnitude of the Fourier transform of an analogue signal with a triangular spectrum, with highest frequency f_0; (b) Fourier transform of the sampled signal, scaled by the sampling interval, Δ, when the sampling frequency is more than $2f_0$; (c) scaled Fourier transform of the sampled signal when sampling frequency is less than $2f_0$.

$$X_s(f) = \frac{1}{\Delta} \sum_{n=-\infty}^{\infty} X\left(f - \left(\frac{n}{\Delta}\right)\right) \qquad (4.32)$$

This important equation (depicted graphically in Figure 4.14) relates the Fourier transform of a continuous signal and the Fourier transform of the sequence formed by sampling the signal at equispaced intervals, and is commonly referred to as the Poisson sum formula. It will be further discussed in the explanation of 'aliasing' in Section 4.4.3.

4.4.2.2 z-transform

An alternative route to Equation 4.29 is to use the notion of the z-transform of a sequence (used in the solution of difference equations).

The z-transform of a sequence of numbers $x(n)$, say (where the notion of time is suppressed by setting the sampling interval, Δ, to unity) is

$$X(z) = \sum_{n=-\infty}^{\infty} x(n) z^{-n} \qquad (4.33)$$

z is a complex number and $X(z)$ is a function of a complex variable. If z is allowed to take values on the unit circle in the z-plane, that is, $z = e^{i\omega} = e^{i 2\pi f}$, then Equation 4.33 is

$$X(e^{i2\pi f}) = \sum_{n=-\infty}^{\infty} x(n) e^{-i 2\pi f n} \qquad (4.34)$$

which is seen to correspond identically with Equation 4.29, with $\Delta = 1$.

4.4.3 Aliasing

Equation 4.32 describes how the frequency components of the sampled waveform relate to that of the continuous waveform, and Figure 4.14 explains this pictorially.

In Figure 4.14a, it is assumed that $X(f) = 0$ for $|f| > f_0$ (say), and Figure 4.14b is a plot of $\Delta X(e^{i2\pi/\Delta})$ assuming that the sampling rate $f_s = 1/\Delta$ is such that $f_0 < 1/2\Delta$. Some commonly used terms are defined within the diagram.

If $f_0 > f_s/2$, there is an overlapping of the shifted versions of $X(f)$, resulting in a distortion of the frequency description for $|f| < 1/2\Delta$ as in Figure 4.14c. This 'distortion' is due to the fact that high frequencies in the data are indistinguishable from lower frequencies, owing to the sampling rate not being fast enough. Specifically, consider $x(t) = \cos 2\pi p t$ sampled at $f_s = 1/\Delta$, that is, at $t = n\Delta$ (with, say, $p < \tfrac{1}{2}\Delta$). This may be written

$$x(n\Delta) = \frac{1}{2}\left(e^{i 2\pi n p \Delta} + e^{-i 2\pi n p \Delta}\right) \qquad (4.35)$$

Now, letting frequency p above be replaced by $k/\Delta \pm p$, for any integer k, results in an identical sequence. So, if a frequency component were detected at p Hz, any one of these higher frequencies might have been responsible for this, rather than a 'true' component at p Hz. This phenomenon is called *aliasing*. The values $k/\Delta \pm p$ can be seen graphically for some p between 0 and $\tfrac{1}{2}\Delta$ by 'pleating' the frequency axis as in Figure 4.15.

To avoid aliasing, the sample rate must be chosen to be greater than twice the highest frequency contained in the signal (see the discussion in Section 4.4.6 regarding the use of 'antialias' filters).

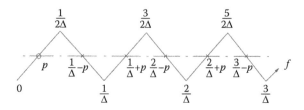

Figure 4.15 Possible aliases of frequency p Hz.

4.4.4 Discrete Fourier transform

Equation 4.29 is still not in a form suitable for implementation on a digital computer. Since, in practice, any sequence will be of finite duration, it is appropriate to consider a finite summation which, for an N-point sequence, could be written

$$X_s(f) = \sum_{n=0}^{N-1} x(n\Delta) e^{-i2\pi f n \Delta} \tag{4.36}$$

Obviously, if this involved the truncation (windowing) of data, then there would be some distortion introduced as discussed in Section 4.3.3.

One can think of simplifying the task by only evaluating f at specific points, say at $k/N\Delta$ Hz, $k = 0, ..., N-1$. Then,

$$X\left(e^{i2\pi k/N}\right) = \sum_{n=0}^{N-1} x(n\Delta) e^{-i(2\pi/N)nk} \tag{4.37}$$

Or, adopting the more usual notation (and setting Δ to unity for convenience),

$$X(k) = \sum_{n=0}^{N-1} x(n) e^{-i(2\pi/N)nk} \tag{4.38}$$

Multiplying both sides by $e^{i(2\pi/N)mk}$ and summing over k gives

$$x(n) = \frac{1}{N} \sum_{n=0}^{N-1} X(k) e^{i(2\pi/N)nk} \tag{4.39}$$

Equations 4.38 and 4.39 constitute the discrete Fourier transform (DFT) and are the forms that are suitable for machine computation.

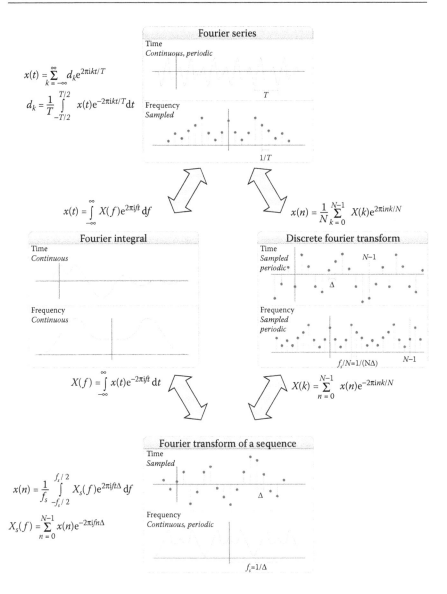

Figure 4.16 Summary of Fourier transforms.

It is important to realise that even though the original sequence $x(n\Delta)$ may have been zero for n outside the range $0 \to N-1$, the act of 'sampling in frequency' has imposed a *periodic structure* to the sequence, that is, it follows from Equation 4.39 that $x(n+N) = x(n)$.

We are now in a position to summarise the various Fourier transforms we have defined and their differences, and this is done in Figures 4.16 and 4.17

Continuous Fourier transform
The original continuous time signal and its Fourier transform.

Fourier transform of the truncated signal
Truncation of the continuous time signal introduces oscillations into the Fourier transform.

Fourier transform of the sampled signal
Sampling of the time domain signal makes the Fourier transform periodic. Sample rate 10 Hz.

Discrete Fourier transform
The Fourier transform in the frequency domain, at an interval of $1/T$ Hz, introduces periodicity into the time domain.

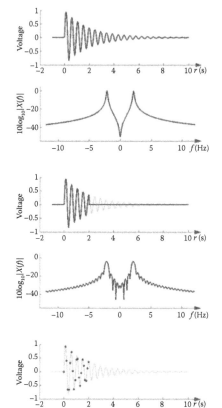
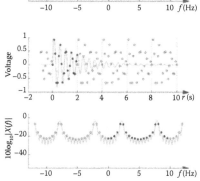

Figure 4.17 Relationship of a DFT to the Fourier integral.

(see also Randall 1977). It is instructive to follow (diagrammatically) the interpretation of a DFT of an aperiodic function that is truncated, as in Figure 4.17 (see also Brigham 1974).

4.4.5 Fast Fourier transform

The evaluation of the DFT can be accomplished efficiently by a set of algorithms known as the fast Fourier transform (FFT). This section briefly outlines the basic 'decimation-in-time' method for a radix 2 FFT and follows Oppenheim and Schafer (1975) and Rabiner and Gold (1975).

4.4.5.1 Radix 2 FFT

Writing $e^{-i(2\pi/N)}$ as W_N, then the DFT, Equation 4.27, can be expressed as

$$X(k) = \sum_{n=0}^{N-1} x(n) W_N^{nk} \tag{4.40}$$

The FFT algorithms essentially exploit the periodicity and symmetry properties of the DFT, and so reduce the number of multiply and add operations needed to calculate $X(k)$ from about N^2 to approximately $N \log_2 N$ which, for example, equates to a reduction by a factor of about 100 when $N = 1024$.

There are a considerable range of FFT algorithms that have been proposed. We shall outline the principles of one; a radix 2 decimation-in-time algorithm. For a radix 2 scheme, N has to be a power of two ($N = 2^v$ with v an integer). Decimation-in-time algorithms divide the sequence $x(n)$ into successively smaller subsequences. In particular, two $N/2$ point sequences $x_1(n), x_2(n)$ are defined as the even and odd members of $x(n)$, so that

$$X(k) = \sum_{n=0}^{N/2-1} x(2n) W_N^{2nk} + \sum_{n=0}^{N/2-1} x(2n+1) W_N^{(2n-1)k} \tag{4.41}$$

Then, noting that $W_N^{2nk} = W_{N/2}$, this can be written

$$X(k) = \sum_{n=0}^{N/2-1} x_1(n) W_{N/2}^{2nk} + W_N^k \sum_{n=0}^{N/2-1} x_2(n) W_{N/2}^{nk} \tag{4.42}$$

or

$$X(k) = X_1(k) + W_N^{nk} X_2(k) \tag{4.43a}$$

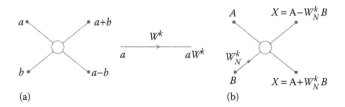

(a)   (b)

Figure 4.18 Signal-flow terminology.

Further, it can be shown that

$$X(k + N/2) = X_1(k) - W_N^{nk} X_2(k) \qquad (4.43b)$$

where $X_1(k)$, $X_2(k)$ are $N/2$ point DFTs of $x_1(n)$, $x_2(n)$. Thus, using Equation 4.43, one can compute $X(k)$ for all k, once their $X_1(k)$, $X_2(k)$ have been determined. The computation of $X_1(k)$ and $X_2(k)$ requires roughly $(N/2)^2 + (N/2)^2 = N^2/2$ operations, that is, roughly half the computation needed to compute $X(k)$ directly.

Some signal-flow graph terminology is defined in Figures 4.18 and 4.19, showing the first stage of decomposition for the case $N = 8$ based on two 4-point transforms. This principle can then be applied to compute the 4-point transforms by dividing each of $x_1(n)$, $x_2(n)$ into two sequences of even and odd members, resulting in four 2-point transforms.

Following Rabiner and Gold (1975), the overall scheme is shown for an 8-point transform in Figure 4.20, where the process has been recursively applied until the inputs correspond to the raw time series.

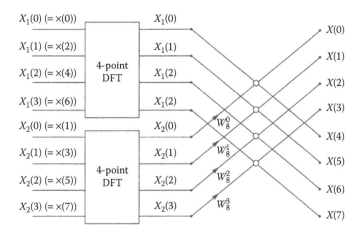

Figure 4.19 An 8-point DFT in terms of two 4-point DFTs.

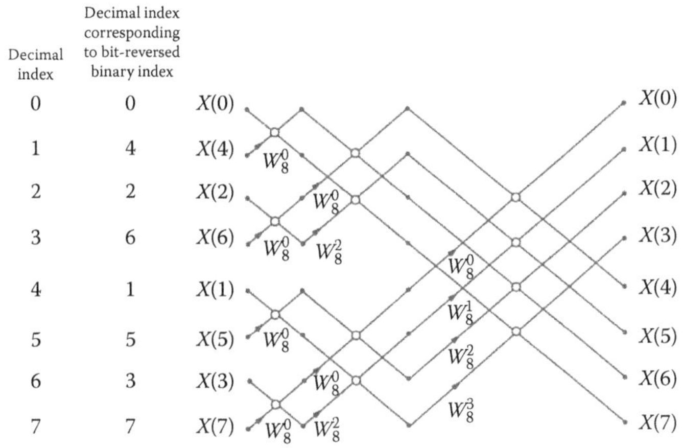

Fundamentals of signal processing 169

Figure 4.20 Decimation-in-time FFT algorithm.

There are $N/2$ butterflies at each stage, and if $N=2^v$ there are $v=\log_2 N$ stages, and with approximately two multiply/add operations per butterfly we have of the order of $N \log_2 N$ operations in all. (Note that the output array $X(k)$ is in the correct order, but the input array is shuffled. In fact, it is in 'bit-reversed' order.)

4.4.6 Practical issues associated with analogue-to-digital conversion

As stated in Section 4.4.1, the use of an ADC represents each input value as one of a finite set of values, effectively rounding the input voltage to the nearest discrete value. The errors introduced by this process are called quantisation errors (or quantisation noise). Quantisation errors are analysed in more detail in Oppenheim and Schafer (1975), Rabiner and Gold (1975), Childers and Durling (1975) and Otnes and Enochson (1978).

The output $\tilde{x}(n\Delta)$ can be written as $\tilde{x}(n\Delta) = x(n\Delta) + e(n\Delta)$, where $e(n\Delta)$ describes the errors introduced through finite word-length effects; these errors are treated as random 'noise'. The probability distributions describing the error depend on the particular way in which the quantisation is implemented. Often, the error is ascribed a uniform distribution (with zero mean) over one quantisation step and furthermore is assumed to be stationary and 'white'. For a b-bit word length, and where the ADC corresponds to a range of A volts, say, then the variance of the error is $\sigma_e^2 = (A/2^b)^2/12$. Further, if the data x is assumed to be a random signal, and σ_x^2 denotes the variance of $x(n\Delta)$, then a measure of the signal-to-noise ratio is $S/N = 10\log_{10}(\sigma_x^2/\sigma_e^2)$. Choosing $\sigma_x = A/4$, to limit the effects of clipping, then $S/N = 6b - 1.24$ dB gives a value of about 70 dB for a 12-bit ADC. This is reduced further by practical considerations (see Otnes and Enochson 1978) relating to the quality of the acquisition system.

It is vital that the rate at which $x(t)$ is sampled should be high enough to avoid aliasing (see Section 4.4.3). This usually means that $x(t)$ should be low-pass filtered using analogue filters preceding the ADC. The cut-off frequency and cut-off rate of these 'antialias' filters should be chosen with particular applications in mind, but very roughly speaking, if the 3 dB point is a quarter of the sampling frequency and the cut-off rate better than 48 dB per octave, then this at least ensures a 40–50 dB reduction at the folding frequency (half the sampling frequency; see Section 4.4.2), and so probably an acceptable level of aliasing (though we must emphasise that this may not be adequate for some applications). The selection of the sample rate is very important. While aliasing is to be avoided, unnecessarily high sample rates are wasteful. The specific application (the bandwidth of interest) and the characteristics of the antialias filter should dictate the selection of an 'optimal' rate.

4.5 RANDOM PROCESSES

4.5.1 Introduction

In previous sections, we discussed the analysis of deterministic signals using Fourier methods. This section is devoted to the treatment of nondeterministic signals. We may visualise a sample of such a signal as shown in Figure 4.21.

An example might be the record of wave height at a point at sea. The essential feature of such a signal is that its value cannot be predicted *exactly* in advance. We shall briefly summarise the way in which such random processes are described.

4.5.2 Probability theory

The mathematical theory underlying models describing *uncertain* situations is probability theory. It is often explained by considering examples taken from games of chance involving coins, dice and so forth. We need to define a few terms.

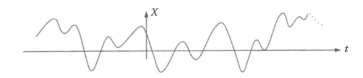

Figure 4.21 Sample time history.

1. **Experiment of chance:** An experiment whose outcome is not predictable is called an experiment of chance.
2. **Sample space:** The collection (set) of all possible outcomes of an experiment is called its sample space (denoted Ω), with individual elementary outcomes as ω (not to be confused in this context with frequency).
3. **Event:** Any collection (subset) of points in the sample space is an event. Individual events in the sample space are called *elementary events*. We see that *events* are collections of elementary events.

To each event E in Ω, we now wish to assign a number which measures our belief that E will occur. This is the *probability* of event E and it is written $Prob(E)$ or $P(E)$ and satisfies $0 \leq P[E] \leq 1$.

4.5.3 Random variables and probability distributions

In many cases, we may not wish to define the outcome of an experiment in terms of elements of the sample space but in terms of a convenient set of numbers. This leads to the definition of a random variable. A *random variable* is a function defined on a sample space. We write $X\{\omega\} = x_j$ to denote that the random variable X takes the value x_j for elements ω that comprise the event. If the x_j are countable, the random variable is *discrete*, and if not it is *continuous*. Some processes may be mixed.

4.5.3.1 Probability distributions for discrete random variables

Definition: If random variable X takes on only a discrete set of values x_1, x_2, ..., then its probability distribution is characterised by specifying the probabilities with which X assumes these values, that is to say

$$P[X = x] \text{ for } x = x_1, x_2, ...,$$

$P[X=x]$ is called the *probability function* of random variable X.

4.5.3.2 Probability distribution function

Definition: The distribution function $F(x)$ of a random variable X is defined as

$$F(x) = P[X \leq x] = \text{Prob } [X \text{ taking on a value up to and including } x]$$

For a discrete process, there are jumps (discontinuities) in the function.

4.5.3.3 Continuous distributions

For a continuous random variable, it is not possible to enumerate all possible outcomes; that is to say, the number of elements in the sample space is infinite and uncountable and so $P[X=x]=0$, that is, the probability that X takes an particular value is zero. A more useful approach is to think of the probability of X lying within *intervals* on the x axis, that is to say,

$$P[X > a], \quad P[a \leq X \leq b]$$

Starting from the distribution function $F(x) = P[X \leq x]$, we are led to the *probability-density function* as follows.

4.5.3.4 Probability-density functions

Consider the probability of X lying in an interval x to $x + \delta x$: then

$$P[x < X \leq x + \delta x] = F(x + \delta x) - F(x) \tag{4.44}$$

Then $(P[x < X \leq x + \delta x])/(\delta x)$ is a measure of the probability/unit length near x, and if $\lim_{\delta x \to 0} (P[x < X \leq x + \delta x])/(\delta x)$ exists, it is called the probability-density function (pdf) and is denoted by $p(x)$ and satisfies

$$p(x) = \frac{dF(x)}{dx} \tag{4.45}$$

so that

$$F(x) = \int_{-\infty}^{x} p(u) du \tag{4.46}$$

It follows that

1. $p(x) \geq 0$
2. $\int_{-\infty}^{\infty} p(x) dx = 1$
3. $P[a < x < b] = \int_{a}^{b} p(x) dx$

Note: We can define the probability-density function for discrete random variables if we include delta functions.

4.5.3.5 Joint distributions

Now, let us extend our considerations to two or more random variables (say X, Y, i.e. a bivariate process). Then, the probability of $X<x$ *jointly occurring* with $Y \leq y$ is $P[X \leq x, Y \leq y] = F(x, y)$.

Just as for the univariate case, we can define a joint density function

$$p(x,y) = \lim_{\substack{\delta x \to 0 \\ \delta y \to 0}} \frac{P[x \leq X \leq x+\delta x, y \leq Y \leq y+\delta y]}{\delta x \delta y}$$

$$= \frac{\partial^2 F(x,y)}{\partial x \partial y} \quad (4.47)$$

4.5.4 Expectations of functions of a random variable

Let X be a random variable. The *expected value* of X, $\mathbb{E}[X]$, is defined as

$$\mathbb{E}[X] = \int_{-\infty}^{\infty} xp(x)dx \quad (4.48)$$

We use the notation $\mu_x = \int_{-\infty}^{\infty} xp(x)dx$ for this *mean value* or *first moment* of X.

More generally, if Y is a function of X, that is, if $Y = g(X)$, then the average value of Y is

$$\mathbb{E}[Y] = \mathbb{E}[g(X)] = \int_{-\infty}^{\infty} g(x)p(x)dx \quad (4.49)$$

This may be generalised to functions of several random variables. In a bivariate case (X, Y) with random variables (X, Y), and if W is a function of X and Y, that is, $W = g(X, Y)$, then the average value of W is

$$\mathbb{E}[W] = \int_{-\infty}^{\infty}\int_{-\infty}^{\infty} g(x,y)p(x,y)dxdy \quad (4.50)$$

4.5.4.1 Moments of a random variable

The moments of a random variable are often taken as relatively simple indicators of the behaviour of a process. The first few moments are noted as follows:

First moment (mean) $\mathbb{E}[X] = \mu_x = \int_{-\infty}^{\infty} xp(x)dx$ (4.51)

This is a measure of location.

Second moment (mean square) $\mathbb{E}[X^2] = \int_{-\infty}^{\infty} x^2 p(x)dx$ (4.52)

This is the measure of spread relative to the origin.

We usually consider 'central moments', that is, moments *about the mean*. The second moment about the mean is the *variance*:

$$\mathbb{V}[X] = \sigma_x^2 = \mathbb{E}\left[(X - \mu_x)^2\right] = \int_{-\infty}^{\infty} (x - \mu_x)^2 p(x)dx \quad (4.53)$$

This is a measure of spread relative to the mean. The standard deviation is $\sigma_x = \sqrt{\mathbb{V}[X]}$.

4.5.4.2 Normal distribution (Gaussian distribution)

A random variable is normally distributed (or has a Gaussian distribution) if it has a probability-density function given by

$$p(x) = \frac{1}{\sigma_x \sqrt{2\pi}} e^{-(x-\mu_x)^2/2\sigma_x^2} \quad (4.54)$$

where

$$E[X] = \mu_x, \quad \sigma_x^2 = \int_{-\infty}^{\infty} (x - \mu_x)^2 p(x)dx \quad (4.55)$$

(see Figure 4.22).

This distribution is completely characterised by the two parameters μ_x and σ_x. The importance of this distribution is stressed by the *central limit theorem*, which (roughly speaking) says that the sum of independent random variables $S_n = \sum_{k=1}^{n} X_k$ tends to a Gaussian random variable as n gets large, regardless of the individual distributions of X_k.

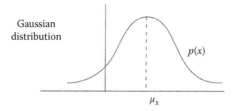

Figure 4.22 Gaussian probability-density function.

4.5.4.3 Higher moments

Higher moments of random variables give useful information about the probability-density function of the process. For example, a measure of the departure from symmetry or 'skewness' of a distribution is the third moment about the mean $\mathbb{E}[(X-\mu_x)^3]$. This is sometimes measured in terms of the coefficient of skewness

$$\gamma_1 = \frac{\mathbb{E}\left[(X-\mu_x)^3\right]}{\sigma_x^3} \tag{4.56}$$

$\gamma_1 = 0$ for a Gaussian distribution.

The degree of *peakiness* of a pdf near its mean is sometimes measured by the coefficient of *kurtosis*

$$\gamma_2 = \frac{\mathbb{E}\left[(X-\mu_x)^4\right]}{\sigma_x^4} - 3 \quad \text{(fourth central moment)}$$

$\gamma_2 = 0 \quad$ for a Gaussian distribution $\tag{4.57}$

If $\gamma_2 > 0$, the distribution is more peaky than Gaussian (super-Gaussian or leptokurtic).

If $\gamma_2 < 0$, the distribution is more flattened than Gaussian (sub-Gaussian or platykurtic).

Sometimes, the kurtosis is defined as $(\mathbb{E}[(X-\mu_x)^4])/(\sigma_x^4)$ and its value is considered relative to 3.

If a process is Gaussian, then $\gamma_1 = \gamma_2 = 0$; consequently, monitoring the third and fourth moments allows detection of non-Gaussianity. Since many *nonlinear* systems generate signals which are non-Gaussian, then γ_1 and γ_2 can also be used to detect (and perhaps characterise) nonlinearity. The fourth moment (kurtosis) is used as an indicator of damage in machinery condition monitoring. Early damage in rotating-machinery rolling elements often results in accelerometer signals whose kurtosis increases significantly above zero owing to the impact arising from faults.

4.5.4.4 Bivariate processes

The concept of a moment may be generalised to jointly distributed random variables; for example, the second joint moment is

$$\mathbb{E}[(X - \mu_x)(Y - \mu_y)] = \int\!\!\!\int_{-\infty}^{\infty} (x - \mu_x)(y - \mu_y) p(x,y) \, dx \, dy \qquad (4.58)$$

and is called the *covariance* function relating X and Y. Some definitions relating to bivariate processes (X, Y) are:

1. *Uncorrelated* if $\mathbb{E}[XY] = \mathbb{E}[X]\mathbb{E}[Y]$
2. *Orthogonal* if $\mathbb{E}[XY] = 0$
3. *Independent* if $p(x, y) = p(x) p(y)$

A joint distribution for Gaussian (bivariate) process (X, Y) is expressed as follows. If the means of these processes are defined as $\mathbb{E}[X] = \mu_x$; $\mathbb{E}[Y] = \mu_y$, with variances $\mathbb{V}[X] = \sigma_x^2$; $\mathbb{V}[Y] = \sigma_y^2$; and covariance $\mathbb{E}[(X - \mu_x)(Y - \mu_y)] = \sigma_{xy}$, then the joint pdf for this bivariate Gaussian process is given by:

$$p(x, y) = \frac{1}{2\pi} \frac{1}{|S|^{1/2}} \exp\left\{ -\frac{1}{2} (\underline{v} - \underline{\mu}_v)^T S^{-1} (\underline{v} - \underline{\mu}_v) \right\} \qquad (4.59)$$

where the covariance matrix S is defined as

$$\begin{bmatrix} \sigma_x^2 & \sigma_{xy} \\ \sigma_{xy} & \sigma_y^2 \end{bmatrix} \text{ and } \underline{v} = \begin{bmatrix} x \\ y \end{bmatrix}, \underline{\mu}_v = \begin{bmatrix} \mu_x \\ \mu_y \end{bmatrix}.$$

4.5.5 Stochastic processes

So far, time has played no part in our description of random processes. Now, consider a random variable (defined on a sample space) which is time dependent. Such a random variable is denoted as $X(t)$, and implies that for each point in time t in the range $-\infty < t < \infty$ the random variable $X(t)$ exists and has a sample space $-\infty < X(t) < \infty$, with an associated probability-density function denoted as $p(x(t))$. A continuous time series can be described as an ordered set of random variables $X(t)$, $-\infty < t < \infty$ and is referred to as a *stochastic process*. To completely describe a stochastic process, it is necessary to specify the joint probability-density functions associated with an arbitrary set of times $t_1, t_2, \ldots, t_n$.

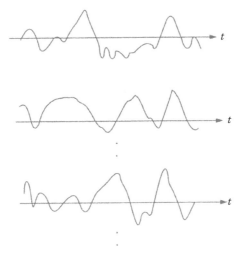

Figure 4.23 Ensemble of time histories.

4.5.5.1 Ensemble

An observed time series $x(t)$ is one realisation of the stochastic process $X(t)$. It can be regarded as one of an infinite number of functions which might have been observed. The set of outcomes may be visualised as being generated by a large number of independent repetitions of the same experiment under conditions which are essentially identical (see Figure 4.23). Such a set of a large number of outcomes is referred to as an ensemble. Theoretical quantities relating to $X(t)$ should be considered with respect to this ensemble; for example, the expectation of $X(t)$, $\mathbb{E}[X(t)]$, corresponds to the mean value of $X(t)$ when it is averaged across the ensemble, a process referred to as *ensemble averaging*.

4.5.5.2 Probability distributions associated with a stochastic process

We conceive of a probability-density function for a stochastic process as follows. Let the distribution function be $F_1(x(t))$, where $F_1(x(t)) = P[X(t) \leq x(t)]$. So

$$P[x(t) < X(t) \leq x(t) + \delta x] = F_1(x(t) + \delta x) - F_1(x(t)) \qquad (4.60)$$

Therefore,

$$\lim_{\delta x \to 0} \frac{P[x(t) < X(t) \leq x(t) + \delta x]}{\delta x} = \frac{dF_1(x(t))}{dx}$$

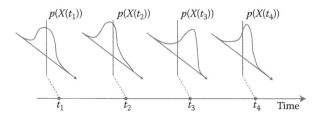

Figure 4.24 Evolving probability-density of a nonstationary process.

$$p_1(x(t)) = \frac{dF_1(x(t))}{dx} \quad (4.61)$$

that is to say, $p_1(x(t))$ is a probability-density function which evolves with time (Figure 4.24).

Similarly, we may define joint probability-density functions, relating $x(t)$ at times t_1 and t_2. If $F_2(x(t_1),x(t_2)) = \text{Prob}[X(t_1) \leq x(t_1)$ jointly with $X(t_2) \leq x(t_1)]$, then the joint probability-density function is

$$p_2(x(t_1),x(t_2)) = \frac{\partial^2 F_2(x(t_1),x(t_2))}{\partial x(t_1)\partial x(t_2)} \quad (4.62)$$

In general, a univariate stochastic process is described by the nth order joint probability-density function $p_n(x(t_1), x(t_2), \ldots, x(t_n))$. For our purposes, we shall only consider p_1, p_2.

4.5.5.3 Moments of a stochastic process

We define moments for stochastic processes just as we did for random variables, noting now that time is involved. Each moment is, in general, now time dependent; for example,

$$\text{Mean } \mu_x(t) = \mathbb{E}[X(t)] = \int_{-\infty}^{\infty} x p(x,t) dx \quad \text{(1st moment)}$$

$$\text{Mean square } \mathbb{E}[X^2(t)] = \int_{-\infty}^{\infty} x^2 p(x,t) dx \quad \text{(2nd moment)}$$

$$\text{Variance } \sigma_x^2(t) = \mathbb{V}[X(t)] = \int_{-\infty}^{\infty} (x - \mu_x)^2 p(x,t) dx \quad (4.63)$$

4.5.5.4 Stationarity

In general, the properties of a stochastic process are time dependent. To simplify matters, we often assume a limited form of 'steady state' has been reached, in the sense that the statistical properties are unchanged under a shift in time. Specifically, two conditions are assumed: (a) $p_1(x(t))$ does not vary with time, that is, $p_1(x(t)) = p(x)$ – this implies that μ_x and σ_x^2 are constant; and (b) $p_2(x(t_1), x(t_2))$ is a function of the interval between measurement, that is, it depends on $(t_2 - t_1)$ only and not t_1 and t_2 explicitly.

If conditions (a) and (b) hold, the process is said to be *weakly stationary* or simply *stationary*. If $p_n(x(t_1), x(t_2), \ldots, x(t_n)) = p_n(x(t_1 + T), x(t_2 + T), \ldots, x(t_n + T))$ for all n and all T, that is, p_n is invariant under a time shift, then we say the process is *completely* stationary. In most applications, only the verification of conditions (a) and (b) is feasible, and weak stationarity is commonly adopted as a proxy for complete stationarity. Note that, over sufficiently large timescales, all practical processes evolve in time, that is to say, they are nonstationary; the assumption of stationarity is an idealisation. However, we shall see that it has important consequences. For example, in many situations we may only have a single data record available for analysis (rather than a set of records) and so, rather than averaging across a finite ensemble, we will average along time, that is, we will use *time averages* in place of *ensemble averages*. Clearly, *stationarity* is a necessary condition if this is to be useful.

4.5.5.5 Covariance (correlation) functions

For a random process $X(t)$, a widely used quantity is the *autocovariance function*, which is defined as

$$\mathbb{E}\big[(X(t_1) - \mu_x(t_1))(X(t_2) - \mu_x(t_2))\big] = R_{xx}(t_1, t_2). \quad (4.64)$$

This is a measure of the 'degree of association' of the signal at time t_1 with *itself* at time $t_2 (t_2 \geq t_1)$. Commonly, in the engineering literature, the mean value is not explicitly subtracted and one computes $\mathbb{E}[X(t_1)X(t_2)]$, which is referred to as the *autocorrelation function*. (We note that the term 'autocorrelation function' is sometimes used to refer to the *normalised* autocovariance function, which is also called the autocorrelation coefficient – so, beware of terminology!)

Since we shall deal mainly with stationary processes, we shall simplify Equation 4.64 by noting that the statistical properties are unchanged under a time shift, so

$$\mathbb{E}\big[(X(t_1) - \mu_x)(X(t_2) - \mu_x)\big] = R_{xx}(t_2 - t_1). \quad (4.65)$$

that is, a function of the difference (t_2-t_1) only. We usually write $t_1=t$ and $t_2=t+\tau$, so

$$\mathbb{E}[(X(t)-\mu_x)(X(t+\tau)-\mu_x)] = R_{xx}(\tau) \qquad (4.66)$$

τ is called the lag.

When $\tau=0$, $R_{xx}(0)=\mathbb{V}[X(t)]$; also, as τ gets large, it seems reasonable that for random processes the average $\mathbb{E}[X(t)X(t+\tau)]$ should approach zero, since the values of $X(t)$ and $X(t+\tau)$ for large time separations are 'less associated'; that is, $R_{xx}(\tau)$ has a shape typified by that shown in Figure 4.25.

$R_{xx}(\tau)$ is an *even* function of τ, since $\mathbb{E}[x(t)x(t+\tau)]=\mathbb{E}[x(t-\tau)x(t)]$.

4.5.5.6 Cross-covariance (cross-correlation) function

When two random processes are involved, $X(t)$, $Y(t)$, for example, input–output processes, then we generalise the equations in Section 4.5.5.5 and define the cross-covariance function as

$$R_{xy}(t_1,t_2) = \mathbb{E}\bigl[(X(t_1)-\mu_x(t_1))(Y(t_2)-\mu_y(t_2))\bigr] \qquad (4.67)$$

This is a measure of the degree of association of signal X at t_1 with signal Y at t_2. If we assume stationarity of both signals and the relationship between them, then $R_{xy}(\tau)$ is a function of $\tau=t_2-t_1$.

If one assumes the means of the two processes are zero, so that one can consider the autocorrelation function, then for a stationary process

$$R_{xy}(\tau) = \mathbb{E}[X(t)Y(t+\tau)] \qquad (4.68)$$

In general, $R_{xy}(\tau)$ is *not* an even function. In fact, $R_{xy}(\tau)=R_{yx}(-\tau)$. We list some properties of covariance functions.

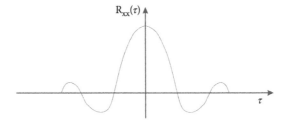

Figure 4.25 Autocorrelation function.

4.5.5.7 Summary of the properties of covariance functions

4.5.5.7.1 Autocovariance function

Define $\rho_{xx}(\tau) = (R_{xx}(\tau))/(R_{xx}(0)) = (R_{xx}(\tau))/(\sigma_x^2)$; $\rho_{xx}(\tau)$ is the normalised (non-dimensional) form of the autocovariance function. Then

1. $R_{xx}(\tau) = R_{xx}(-\tau)$, that is, the autocovariance function is even.
2. $\rho_{xx}(0) = 1$; $R_{xx}(0) = \sigma_x^2$.
3. $|R_{xx}(\tau)| \leq R_{xx}(0)$ so $-1 \leq \rho_{xx}(\tau) \leq 1$.
4. $R_{xx}(\tau)$ is a continuous function of τ for a continuous stochastic process.

4.5.5.7.2 Cross-covariance function

Defining $\rho_{xy}(\tau) = \left(R_{xy}(\tau)\right) / \left(\sqrt{R_{xx}(0) R_{yy}(0)}\right) = \left(R_{xy}(\tau)\right) / \left(\sigma_x \sigma_y\right)$ (the cross-correlation coefficient), then

1. $R_{xy}(-\tau) = R_{yx}(\tau)$, so that there is no general form of symmetry.
2. $\left|R_{xy}(\tau)\right|^2 \leq R_{xx}(0)\, R_{yy}(0)$.
3. $-1 \leq \rho_{xy}(\tau) \leq 1$.

Note: The cross-correlation coefficient is particularly useful when the two quantities involved are different in scale and dimension.

4.5.5.8 Ergodic processes and time averages

The definitions of the moments in Section 4.5.5.3 have all involved the underlying theoretical probability distributions of the stochastic processes and should be interpreted as involving ensemble averages. Often, we find that we have only a single realisation (record) of a limited length from which to make estimates of these averages we have defined. This motivates the question as to whether averages *along the record*, that is, *time averages*, might be used in place of ensemble averages. We have noted this already in Section 4.5.5.4 and stated that *stationarity* is a prerequisite if we are to follow this up. In fact, it turns out that many processes have the desirable property that *some ensemble averages* are equivalent to *time averages* over one record. Such an average is said to be *ergodic*. Sometimes processes are termed *weakly ergodic* if the mean-value and covariance functions may be calculated by performing time averages in place of ensemble averages.

We add that while the concept of ergodicity is given prominence in many texts, it can only be approached correctly through estimation theory (Section 4.7) and it must be related directly to the particular average in question; for example, mean value, autocovariance function and so forth. We *cannot* simply refer to a *process* as ergodic.

Nevertheless, we shall anticipate that many processes are ergodic with respect to the mean and covariance functions, implying that we can write the following definitions for these averages. The mean value is

$$\mu_x = \lim_{T \to \infty} \frac{1}{T} \int_0^T x(t) dt \quad \text{or} \quad \lim_{T \to \infty} \frac{1}{T} \int_{-T/2}^{T/2} x(t) dt \qquad (4.69)$$

that is, we get the same value for the time average over any single member of the ensemble as $\mathbb{E}[X(t)]$.

Similar arguments hold for the cross-covariance function

$$R_{xy}(\tau) = \lim_{T \to \infty} \int_0^T \left(x(t) - \mu_x\right)\left(y(t + \tau) - \mu_y\right) dt \qquad (4.70)$$

4.5.5.9 Examples

4.5.5.9.1 Autocovariance function of a sinusoid

Let $x(t) = A \sin(\omega_0 t + \theta)$, where θ is a random variable with uniform probability-density function

$$p(\theta) = \frac{1}{2\pi} \quad 0 \le \theta \le 2\pi$$

$$= 0 \quad \text{elsewhere}$$

From the results of earlier sections,

$$\mu_x = \mathbb{E}[x(t)] = \int_{-\infty}^{\infty} A \sin(\omega_0 t + \theta) p(\theta) d\theta = 0$$

$$R_{xx}(\tau) = \mathbb{E}[x(t)x(t + \tau)] = \int_{-\infty}^{\infty} A^2 \sin(\omega_0 t + 0) \sin(\omega_0 t + \tau + \theta) p(\theta) d\theta = \frac{A^2}{2} \cos \omega_0 \tau$$

Note that the autocovariance function is 'phase blind'; that is, it is independent of θ. Note also that the same result can be obtained using a time average over a single record of the process (and allowing the record length T to get large).

4.5.5.9.2 Asynchronous random telegraph signal

Consider a time function that switches between two voltage levels $+a$, $-a$ (see Figure 4.26).

The crossing times t_i are random, and we shall assume the times t_i conform to a Poisson process with parameter λ, that is, the probability of k crossings in time τ is

$$P_k = \frac{e^{-\lambda|\tau|}(\lambda|\tau|)^k}{k!}$$

where λ is the number of crossings divided by unit time. The mean value of $x(t)$ is zero, that is, $\mu_x = 0$; we proceed to compute the autocorrelation function. The product $x(t)x(t+\tau)$ can only take one of two values, a^2 or $-a^2$. The product is a^2 if an even number of crossings occur in time τ and the product is $-a^2$ if an odd number of crossings occur. To compute $R_{xx}(\tau)$, one evaluates:

$$\mathbb{E}\big[x(t)x(t+\tau)\big] = a^2 P[k\,\text{even}] - a^2 P[k\,\text{odd}]$$

leading to

$$R_{xx}(\tau) = \sum_{k=0}^{\infty}\big[a^2 P_{2k} - a^2 P_{2k+1}\big]$$

$$= a^2 e^{-\lambda|\tau|}\left[\sum_{k=0}^{\infty}\frac{(\lambda|\tau|)^{2k}}{2k!} - \frac{(\lambda|\tau|)^{2k+1}}{(2k+1)!}\right]$$

$$R_{xx}(\tau) = a^2 e^{-2\lambda|\tau|}$$

that is, the autocovariance function is an exponentially decaying function where λ controls the decay rate.

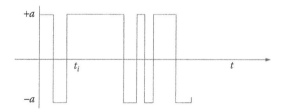

Figure 4.26 Asynchronous random telegraph signal.

4.5.5.9.3 White noise

This is a theoretical concept, but one which is, nevertheless, very useful. We can lead up to it by considering a limiting form of the equations in Section 4.5.5.9.2. As $\lambda \to \infty$, $R_{xx}(\tau)$ gets very narrow; that is, the process is very erratic. In order that $R_{xx}(\tau)$ does not 'disappear' completely, we intuitively let the parameter a get large to compensate. We see that $R_{xx}(\tau) \to k\delta(\tau)$. This leads to the idea that a 'completely erratic' random process will have a delta function-like autocovariance function; that is, $R_{xx}(\tau) = k\delta(\tau)$. Processes with this property are called 'white noise'. In practice, we shall only be able to approximate such processes and often refer to 'band-limited white noise', which anticipates the spectral description of such processes.

Note: Processes with a variety of probability distributions can have a delta function autocorrelation, so simply saying a process is 'white' does not convey its probabilistic structure; it only describes its temporal structure.

4.5.5.9.4 Cross-covariance (correlation) function

Consider two functions:

$$x(t) = A\sin(\omega_0 t + \theta_1) + B$$

$$y(t) = C\sin(\omega_0 t + \theta_2) + D\sin(n\omega_0 t + \theta_3)$$

Using the time-averaged form, the cross-correlation is

$$R_{xy}(\tau) = \frac{1}{2}AC\cos(\omega_0\tau + (\theta_1 - \theta_3))$$

Note: The cross-correlation function seeks out the part of $y(t)$ that 'matches' or fits $x(t)$ *and it retains the relative phase* $(\theta_1 - \theta_3)$, that is, the cross-correlation detects the delay relating the similar parts of x and y.

4.5.5.9.5 Signal embedded in noise

Let us assume the noise and signal are uncorrelated; for example, $y(t) = s(t) + n(t)$. If $s(t)$ is a sine wave and $n(t)$ is broadband noise (uncorrelated with s), then

$$R_{yy}(\tau) = E\big[(s(t) + n(t))(s(t+\tau) + n(t+\tau))\big]$$

$$= R_{ss}(\tau) + R_{nn}(\tau)$$

Fundamentals of signal processing 185

Since $R_{nn}(\tau)$ decays quickly, then $R_{ss}(\tau)$ dominates the result for larger values of τ, and so this provides a way of identifying the sine component as a noisy signal.

4.5.6 Spectra

So far, our considerations of stochastic processes have been restricted to the time domain. We now introduce the frequency-domain counterparts of the concepts. We have seen that Fourier methods may be applied to 'periodic' and 'transient' phenomena; we shall now see that Fourier methods may be applied to random (stationary) processes. We shall draw on the work in Sections 4.2 and 4.3.

Consider a sample of a stationary random process $x(t)$, and let us look at a truncated version $x_T(t)$ (Figure 4.27).

$$x_T(t) = x(t), \quad |t| < \frac{T}{2}$$
$$= 0 \quad \text{elsewhere} \qquad (4.71)$$

We shall study the frequency decomposition of the power of this signal. Since $x_T(t)$ is 'pulse-like' (i.e. transient), it has a Fourier representation

$$x_T(t) = \int_{-\infty}^{\infty} X_T(f) e^{i2\pi ft} df \qquad (4.72)$$

The total energy of the signal is $\int_{-\infty}^{\infty} x_T^2(t) dt$, which tends to infinity as T gets large, and so we consider the average power, which is the variance,

$$\frac{1}{T}\int_{-\infty}^{\infty} x_T^2(t) dt$$

By Parseval's theorem

$$\frac{1}{T}\int_{-\infty}^{\infty} x_T^2(t) dt = \frac{1}{T}\int_{-T/2}^{T/2} x^2(t) dt = \frac{1}{T}\int_{-\infty}^{\infty} |X_T(f)|^2 df \qquad (4.73)$$

Figure 4.27 Truncated sample function.

Now, as $T \to \infty$,

$$\lim_{T \to \infty} \frac{1}{T} \int_{-T/2}^{T/2} x^2(t) dt = \int_{-\infty}^{\infty} \lim_{T \to \infty} \frac{|X_T(f)|^2}{T} df \qquad (4.74)$$

The quantity $|X_T(f)|^2/T$ is called the sample (or 'raw'; see Section 4.7.4) spectral density, and is given the notation $\hat{S}_{xx}(f)$. Since the left-hand side of Equation 4.74 is the average power of the sample function it is tempting to define $\lim_{T \to \infty} (|X_T(f)|^2)/T$ as the power-spectral density. Unfortunately, this is not useful, since this quantity does not converge in a statistical sense. When we come to estimation theory (Section 4.7), we shall see that if this quantity is evaluated from a data length $T' > T$ the answers are just as erratic as for the shorter data length; that is to say, the use of 'more data' does not 'improve' our estimate. However, we shall see that a very useful theoretical function may be introduced by averaging this quantity to 'remove' the erratic behaviour; that is, consider the average

$$\sigma_x^2 = \mathbb{E}\left[\lim_{T \to \infty} \frac{1}{T} \int_{-T/2}^{T/2} x^2(t) dt\right] = \mathbb{E}\left[\int_{-\infty}^{\infty} \lim_{T \to \infty} \frac{|X_T(f)|^2}{T} df\right] = \int_{-\infty}^{\infty} \lim_{T \to \infty} \mathbb{E}\left[\frac{|X_T(f)|^2}{T}\right] df \qquad (4.75)$$

from which it can be seen that the variance of the process can be expressed as

$$\sigma_x^2 = \int_{-\infty}^{\infty} S_{xx}(f) df \qquad (4.76)$$

where

$$S_{xx}(f) = \lim_{T \to \infty} \frac{\mathbb{E}\left[|X_T(f)|^2\right]}{T} \qquad (4.77)$$

Consequently, the average power of the process (the variance) is decomposed over frequency in terms of the function $S_{xx}(f)$. This function is the *power-spectral density function* of the process. This is a useful function which has a direct physical interpretation. It can also be related directly to the autocovariance function in the following way.

$$S_{xx}(f) = \int_{-\infty}^{\infty} R_{xx}(\tau) e^{-2\pi i f \tau} d\tau \qquad (4.78)$$

and

$$R_{xx}(\tau) = \int_{-\infty}^{\infty} S_{xx}(f)e^{2\pi i f \tau} df \tag{4.79}$$

These relations are sometimes called the Wiener–Khinchin theorem.
Note: If ω is used, the equivalent result is

$$S_{xx}(\omega) = \int_{-\infty}^{\infty} R_{xx}(\tau)e^{-i\omega\tau} d\tau \tag{4.80}$$

and

$$R_{xx}(\tau) = \frac{1}{2\pi} \int_{-\infty}^{\infty} S_{xx}(\omega)e^{i\omega\tau} d\omega \tag{4.81}$$

4.5.6.1 Comments on the power-spectral density function

1. If $x(t)$ is in volts, $S_{xx}(f)$ has units of volts squared per hertz.
2. $S_{xx}(f)$ is an even function of frequency.
3. A one-sided power-spectral density can be defined, called

$$\begin{aligned} G_{xx}(f) &= 2S_{xx}(f) & f > 0 \\ &= S_{xx}(f) & f = 0 \\ &= 0 & f < 0 \end{aligned} \tag{4.82}$$

4.5.6.2 Examples of power spectra

1. If $R_{xx}(\tau) = \sigma^2 e^{-\lambda|\tau|}$, then

$$S_{xx}(f) = \int_{-\infty}^{\infty} R_{xx}(\tau)e^{-2\pi i f \tau} d\tau = \frac{2\lambda\sigma^2}{\lambda^2 + 4\pi^2 f^2}$$

that is, an exponentially decaying correlation function results in a 'low-frequency' power-spectral density function.

2. If $R_{xx}(\tau) = k\delta(\tau)$, then $S_{xx}(f) = k$; that is, a 'narrow' autocorrelation function results in a broad spectrum.

3. If $R_{xx}(\tau) = (A^2/2)\cos 2\pi f_0 \tau$, then

$$S_{xx}(f) = \frac{A^2}{4}\delta(f+f_0) + \frac{A^2}{4}\delta(f-f_0)$$

that is, an oscillatory correlation corresponds to spikes in the spectral density.

4.5.6.3 Cross-spectral density function

This may be defined (generalising the Wiener–Khinchin theorem) as

$$S_{xy}(f) = \int_{-\infty}^{\infty} R_{xy}(\tau) e^{-i2\pi f \tau} d\tau \tag{4.83}$$

with the inverse

$$R_{xy}(\tau) = \int_{-\infty}^{\infty} S_{xy}(f) e^{i2\pi f \tau} df \tag{4.84}$$

Alternatively, $S_{xy}(f)$ may be defined as

$$S_{xy}(f) = \lim_{T \to \infty} \frac{\mathbb{E}\left[X_T^*(f) Y_T(f)\right]}{T} \tag{4.85}$$

where $X_T(f)$, $Y_T(f)$ are Fourier transforms of $x(t)$, $y(t)$ for $-T/2 \le t \le T/2$.

Note: The cross-spectral density is, in general, complex and has the interpretation of being the frequency-domain equivalent of the cross-covariance function, that is, $S_{xy}(f) = |S_{xy}(f)| e^{i \arg(S_{xy}(f))}$, where $|S_{xy}(f)|$ is the *cross-amplitude spectrum*, which shows whether frequency components in one time series are associated with large or small amplitudes at the same frequency in the other series. $\arg(S_{xy}(f))$ is the *phase spectrum* and this shows whether frequency components in one series lag or lead the components at the same frequency in the other series.

4.5.6.4 Properties and associated definitions

1. $S_{xy}(f) = S_{yx}^*(f)$. \hfill (4.86)

This may be proved using the fact that $R_{xy}(\tau) = R_{yx}(-\tau)$

2. A one-sided cross-spectral density $G_{xy}(f)$ is defined as

$$G_{xy}(f) = 2S_{xy}(f) \quad f > 0$$
$$= S_{xy}(f) \quad f = 0$$
$$= 0 \quad f < 0 \quad (4.87)$$

3. The coincident spectral density (cospectrum) and quadrature spectral density (quadspectrum) are defined as

$$S_{xy}(f) = C_{xy}(f) - iQ_{xy}(f), \quad f \geq 0 \quad (4.88)$$

C_{xy}, Q_{xy} are the so-called co- and quadspectra, respectively.

4. An inequality satisfied by the cross-spectral density is

$$|S_{xy}(f)|^2 \leq S_{xx}(f)S_{yy}(f) \quad (4.89)$$

4.6 INPUT–OUTPUT RELATIONSHIPS AND SYSTEM IDENTIFICATION

We shall now develop input–output relationships for linear systems operating with random inputs. We shall consider time-invariant systems and stationary excitations.

4.6.1 Single input–single output systems

Figure 4.5 depicts a linear, time-invariant system with input $x(t)$ and output $y(t)$. Consider Equations 4.13 and 4.14, restated here for convenience:

$$y(t) = \int_{-\infty}^{t} h(t - t_1)x(t_1)dt_1 \quad (4.13)$$

$$y(t) = \int_{0}^{\infty} h(\tau)x(t - \tau)d\tau \quad (4.14)$$

Assuming the input is stationary and random, then one may consider the autocovariance function $R_{xx}(\tau)$. The output autocovariance function $R_{yy}(\tau)$, assuming $\mu_x = 0$, is

$$R_{yy}(\tau) = \mathbb{E}[y(t)y(t+\tau)] = \mathbb{E}\left[\int_0^\infty \int_0^\infty h(\tau_1)x(t-\tau_1)h(\tau_2)x(t+\tau-\tau_1)d\tau_1 d\tau_2\right]$$

that is to say,

$$R_{yy}(\tau) = \int_0^\infty \int_0^\infty h(\tau_1)h(\tau_2)R_{xx}(t+\tau_1-\tau_2)d\tau_1 d\tau_2 \quad (4.90)$$

The frequency-domain equivalent is more compact. Fourier transforming this equation gives

$$S_{yy}(f) = |H(f)|^2 S_{xx}(f) \quad (4.91)$$

where $H(f) = \int_0^\infty h(\tau)e^{-i2\pi f\tau}dt$ is the *frequency-response function* of the system.

The frequency-domain form is clearly simpler than the time-domain version and indicates how the power-spectral density of the input is 'shaped' by the frequency-response characteristic of the system. The output variance is represented by the area under the $S_{yy}(f)$ curve. The expression above is purely real (phase plays no part). This is in contrast to the cross-covariance function in the following where phase is important.

The input–output cross-covariance function $\mathbb{E}[x(t)y(t+\tau)]$ is

$$R_{xy}(\tau) = \mathbb{E}[x(t)y(t+\tau)] = \mathbb{E}\left[\int_0^\infty h(\tau_1)x(t)x(t+\tau-\tau_1)d\tau_1\right]$$

$$R_{xy}(\tau) = \int_0^\infty h(\tau_1)R_{xx}(\tau-\tau_1)d\tau_1 \quad (4.92)$$

The frequency-domain equivalent of this is obtained by transforming this to give

$$S_{xy}(f) = H(f)S_{xx}(f) \quad (4.93)$$

Note that this expression contains the phase of the frequency-response function, that is, $\arg(S_{xy}(f)) = \arg(H(f))$. Equation 4.93 is often the basis of system identification schemes because of this property.

4.6.2 Coherence function

The coherence between input $x(t)$ and output $y(t)$ is defined as

$$\gamma_{xy}^2(f) = \frac{|S_{xy}(f)|^2}{S_{xx}(f)S_{yy}(f)} \tag{4.94}$$

and from the inequality (Equation 4.89),

$$0 \leq \gamma_{xy}^2(f) \leq 1 \tag{4.95}$$

For the single input–single output system in Figure 4.5, the coherence function may be shown (using Equations 4.70 and 4.72) to be

$$\gamma_{xy}^2(f) = \frac{|H(f)|^2 S_{xx}^2(f)}{S_{xx}(f)|H(f)|^2 S_{xx}(f)} = 1 \tag{4.96}$$

So, if two processes are *linearly* related, the coherence is unity, and if $x(t)$ and $y(t)$ are completely unrelated (linearly), the coherence function is zero. If the coherence function is greater than zero but less than one, then $x(t)$ and $y(t)$ are in part linearly related but, in addition, at least one of the following holds:

1. The measurements are 'noisy'.
2. $x(t)$ and $y(t)$ are not *purely* linearly related.
3. $y(t)$ is an output due to input $x(t)$ *and other inputs*.

4.6.2.1 Example

Consider the effect of measurement noise (see Figure 4.28). $y_m(t)$ is a measurement. The coherence between $x(t)$ and $y_m(t)$ is

$$\gamma_{xy_m}^2(f) = \frac{|S_{xy_m}(f)|^2}{S_{xx}(f)S_{y_my_m}(f)}$$

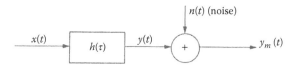

Figure 4.28 Effect of measurement noise on the output.

Assuming x and n are uncorrelated, with power spectra $S_{xx}(f)$ and $S_{nn}(f)$, and using the standard input–output relationships, then

$$\gamma^2_{xy_m}(f) = \frac{1}{1 + \dfrac{S_{nn}(f)}{S_{y_m y_m}(f)}} = \frac{S_{y_m y_m}(f)}{S_{y_m y_m}(f) + S_{nn}(f)}$$

that is, $\gamma^2_{xy_m}(f)$ is the fractional portion of the power of the output $y_m(t)$ which is contributed by input $x(t)$, and $1 - \gamma^2_{xy_m}(f)$ is the fractional portion of the power of the output *not* accounted for by input $x(t)$.

The term *coherent output power* describes the power of that part of $y_m(t)$ that is due to $x(t)$; that is, $S_{yy}(f) = \gamma^2_{xy_m}(f) S_{y_m y_m}(f)$. Noise power (or incoherent output power) is that part of $y_m(t)$ that is due to noise $n(t)$, that is, $S_{nn}(f) = (1 - \gamma^2_{xy_m}(f)) S_{y_m y_m}(f)$.

4.6.3 System identification

This section is concerned with the identification of linear time-invariant systems using random input–output data. As a preliminary, we consider a problem in least-squares estimation.

Let us assume that we make measurements of two quantities, not necessarily time series, x and y, and that we intend to find a linear relationship between them of the form $y = ax$. If, for each x_j, the associated value of y is y_j, then we might plot the results as a scatter diagram (Figure 4.29).

Now consider three situations:

1. Variable x_j is known exactly, but y_j may be 'noisy'. In this case, we attempt to find the linear relation by forming an error $e_j = y_j - \hat{a}x_j$ and find the $\hat{a}$ that minimises $\sum e_j^2$. This is

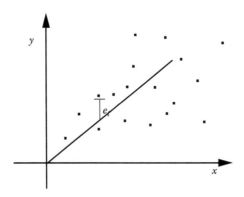

Figure 4.29 Scatter diagram relating variables x (input) and y (output); errors on output.

$$\hat{a} = \frac{\sum x_i y_i}{\sum x_i^2} \quad (4.97)$$

(see Figure 4.29).
2. Variable y_i is known exactly but x_i may be 'noisy'. Now, we form an error, $e_i = (x_i - (y_i/\hat{a}))$, and minimising $\sum e_i^2$ yields

$$\hat{a} = \frac{\sum y_i^2}{\sum x_i y_i} \quad (4.98)$$

(see Figure 4.30).
3. Both variables x_i and y_i are noisy. In this case, the perpendicular distance to the line $y = \hat{a}x$ is minimised (see Figure 4.31). This is referred to as the total least-squares approach and the error is

$$e_i = \frac{(y_i - \hat{a}x_i)}{\sqrt{1 + \hat{a}^2}} \quad (4.99)$$

and $\hat{a}$ is found by minimising $\sum e_i^2$ with respect to $\hat{a}$.

Following these preliminary observations, we return to linear system identification.

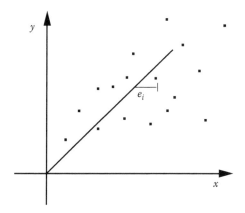

Figure 4.30 Scatter diagram; errors on input.

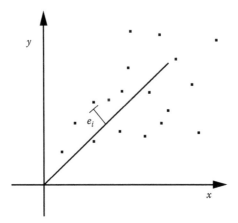

Figure 4.31 Scatter diagram; errors on input and output.

4.6.3.1 Frequency-response identification

For a linear system, we know that

$$Y(f) = H(f)X(f) \qquad (4.100)$$

Let us suppose that for each f we have a series of results $X_j(f)$ and $Y_j(f)$. (We shall assume appropriate use of estimation theory, as in Section 4.7.) Then, by using the equations in Section 4.6.3 we seek $H(f)$ (analogous to parameter a) relating X and Y for each f. The only difference here is that the quantities are complex, and so the analysis requires appropriate modification. For Case 1, namely that of output noise only, we obtain

$$\hat{H}(f) = \frac{\sum X_j^*(f)Y_j(f)}{\sum |X_j(f)|^2} \qquad (4.101)$$

which, with suitable averaging and limiting operations, is $S_{xy}(f)/S_{xx}(f)$. This is sometimes referred to as estimator $H_1(f)$, which is unbiased with respect to the presence of output noise.

For Case 2, namely that of input noise only, we see that

$$\hat{H}(f) = \frac{\sum |Y_j(f)|^2}{\sum X_j(f)Y_j^*(f)} \qquad (4.102)$$

which leads to $S_{yy}(f)/S_{xy}^*(f)$. This is sometimes referred to as estimator $H_2(f)$, which is unbiased with respect to input noise.

For Case 3, the results are somewhat more complicated and lead to the so-called H_v estimator (Leuridan et al. 1985), sometimes called the H_s estimator (Wicks and Vold 1986) and H_w (White et al. 2006). It is noted that a full text with comprehensive coverage of this topic can be found in Shin and Hammond (2008).

4.6.3.2 Effects of feedback

It is often the case that feedback is present in a dynamical system (see, for example, Figure 4.32). Here signal r is the (controlled) input and n is an extraneous input. $H(f)$ is the system to be determined.

Forming $H_1(f) = S_{xy}(f)/S_{xx}(f)$ leads to $H(f) S_{rr}(f) + G^*(f) S_{nn}(f)/S_{rr}(f) + |G(f)|^2 S_{nn}(f)$, which is certainly not the required $H(f)$. Indeed, if $S_{nn}(f)/S_{rr}(f)$ is large, then the ratio $S_{xy}(f)/S_{xx}(f)$ is $1/G(f)$. It can be easily verified, however, that using the three signals $r(t)$, $x(t)$ and $y(t)$ and forming the ratio of two cross-spectral densities $S_{ry}(f)/S_{rx}(f)$ (sometimes referred to as estimator $H_3(f)$) yields the required $H(f)$.

4.7 RANDOM PROCESSES AND ESTIMATION

In Section 4.5, random processes were described in terms of probability functions, correlation functions and spectral densities. The definition of these 'theoretical' concepts serves as the basis on which measured data may be interpreted. In this section, we introduce and discuss estimation methods for random signals, based on a single recording of the process.

4.7.1 Estimator errors and accuracy

Suppose the mean value of a random process is μ_x (unknown to us), and we only have a record length T of the signal. A reasonable estimate $\bar{x}$ of μ_x is

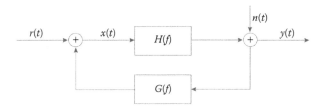

Figure 4.32 System with feedback.

$$\bar{x} = \frac{1}{T}\int_0^T x(t)\,dt \qquad (4.103)$$

The value obtained is a sample value of a random variable $\bar{X}$, having its own probability distribution (called a sample distribution), and $\bar{x}$ is a single realisation. Each $\bar{x}$ computed from a (different) record of length T will, in general, be different. If the estimation procedure is to be useful, we would hope that (i) the scatter of values of $\bar{x}$ is not 'too great' and should lie 'close' to μ_x; (ii) the more data we take (the larger T is), the 'better' the estimate. We can formalise these ideas as follows.

Let ϕ be a parameter we wish to estimate (in the introductory example, ϕ is μ_x), and let $\hat{\phi}$ be an estimator (above $\bar{x}$) for ϕ; then $\hat{\phi}$ is a random variable with the probability distribution, for example, as in Figure 4.33, where it can be seen that estimates $\hat{\phi}$ of random variable $\hat{\phi}$ would predominantly take values near ϕ.

The sampling distribution $p(\hat{\phi})$ is often difficult to obtain, and so we shall settle for a few summarising properties.

4.7.1.1 Bias error

The bias of an estimator is defined:

$$\mathbb{B}\!\left[\hat{\phi}\right] = \mathbb{E}\!\left[\hat{\phi}\right] - \phi \qquad (4.104)$$

This is the difference between the average of the estimator and the true value. If $\mathbb{B}\!\left[\hat{\phi}\right]=0$, then $\hat{\phi}$ is said to be 'unbiased'. While it might seem desirable to use an unbiased estimator, it is sometimes prudent to tolerate some bias if it means that the variability of the estimate is reduced (relative to that of an unbiased estimator).

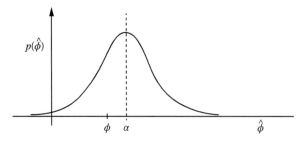

Figure 4.33 Probability-density function, $p(\hat{\phi})$, of the estimates $\hat{\phi}$ for the parameter ϕ, the true value of which is indicated in the figure.

4.7.1.2 Variance

The variance of an estimator is a measure of the spread of about its own mean value

$$\mathbb{V}\left[\hat{\phi}\right] = \mathbb{E}\left[\left(\hat{\phi} - \mathbb{E}\left[\hat{\phi}\right]\right)^2\right] \quad (4.105)$$

The square root of the variance is the standard deviation ($\sigma(\hat{\phi})$) of the estimator. We would hope that this would be 'small'; that is, the probability-density function should be narrow, but this requirement often increases the bias. A measure of the relative importance of bias and variance is the mean-square error.

4.7.1.3 Mean-square error

The mean-square error (mse) of an estimator is a measure of the spread of values about the true value, that is,

$$\mathrm{mse}\left[\hat{\phi}\right] = \mathbb{E}\left[\left(\hat{\phi} - \phi\right)^2\right] \quad (4.106)$$

The right-hand side of Equation 4.106 can be rewritten as

$$\mathrm{mse}\left[\hat{\phi}\right] = \mathbb{V}\left[\hat{\phi}\right] + \mathbb{B}^2\left[\hat{\phi}\right] \quad (4.107)$$

showing that the mse reflects both variance and bias. If the mse reduces as the sample size (amount of data used) increases, then the estimator is said to be *consistent*.

4.7.1.4 Confidence intervals

The estimates $\hat{\phi}$ referred to so far are *point estimates*, that is, single values. It is frequently desirable to know that a parameter is likely to fall within a certain *interval of values*. For example, if we estimate a mean value $\bar{x}$ as 5, perhaps it is 'likely' that μ_x lies in the interval 4.5–5.5. These estimates are *interval estimates*, and are called *confidence intervals* when we attach a number describing the likelihood of the parameter falling within the interval.

If, for example, we state that 'a 95% confidence interval' for μ_x is (4.5, 5.5), then this means we are 95% confident that μ_x lies in the range (4.5, 5.5). Note that this does *not* mean that the probability that μ_x lies in the interval (4.5, 5.5) is 0.95 (μ_x is not a random variable and so we do not assign probabilities to it). We mean that if we could realise a large number of samples and find a confidence interval of μ_x for *each* sample, then approximately 95% of such intervals would contain μ_x. To calculate confidence intervals, we need to know the sampling distribution of the estimator. We shall return to this problem for a specific estimator.

4.7.2 Estimators for stochastic processes

The problem of the study of estimators for stochastic processes is a wide one, and some idea of the extent of the field can be obtained from Jenkins and Watts (1968). Several other texts are not as detailed and give useful summaries (Beauchamp 1973; Bendat and Piersol 1980, 2010; Otnes and Enochson 1978).

We shall pick out the autocorrelation function and power-spectral density as examples, and try to indicate some of the features that should be considered when analysing a sample time history. Following this we briefly discuss cross-spectra and system identification. Unless otherwise stated, we shall assume we are dealing with realisations of a continuous, stationary random process. In practice, data are sampled so that the computation may be done and is therefore subject to the attendant 'aliasing' problems. It is assumed throughout that the sampling rates are 'sufficiently high'.

4.7.3 Autocorrelation function

4.7.3.1 Definition

If we have a signal defined for $0 \le t \le T$, then there are two (Jenkins and Watts 1968) commonly used estimators of the function $R_{xx}(\tau)$. Following Jenkins and Watts, we write them as $\hat{R}_{xx}(\tau)$ and $\hat{R}'_{xx}(\tau)$, where (assuming a zero-mean process)

$$\hat{R}_{xx}(\tau) = \frac{1}{T} \int_0^{T-|\tau|} x(t)x(t+|\tau|)dt \quad 0 \le |\tau| \le T$$

$$= 0 \quad |\tau| > T \qquad (4.108)$$

and

$$\hat{R}'_{xx}(\tau) = \frac{1}{T-|\tau|} \int_0^{T-|\tau|} x(t)x(t+|\tau|)dt \quad 0 \le |\tau| \le T$$

$$= 0 \quad |\tau| > T \qquad (4.109)$$

Often, Equation 4.109 is used since it is unbiased, but it is the mse that should be the criterion for a choice between the two.

4.7.3.2 Statistical considerations

For τ in the range $0 \leq |\tau| \leq T$, $\hat{R}'_{xx}(\tau)$ is an unbiased estimator, whereas $\mathbb{E}[\hat{R}_{xx}(\tau)] = R_{xx}(\tau)(1 - (|\tau|/T)$, showing Equation 4.109 to be only 'asymptotically unbiased'.

An approximate expression for the variance $\hat{R}_{xx}(\tau)$ is

$$\mathbb{V}\left[\hat{R}_{xx}(\tau)\right] \approx \frac{1}{T}\int_{-\infty}^{\infty}\left[R^2_{xx}(v) + R_{xx}(v+\tau)R_{xx}(v-\tau)\right]dv$$

showing this to be a consistent estimator. When $\tau \ll T$, there is little to choose between the two estimators, but as τ approaches T, the variance of the unbiased estimator $\hat{R}'_{xx}(\tau)$ diverges (hardly surprising, considering the divisor in Equation 4.109).

Another important feature of the estimators is that values computed for neighbouring values of τ are samples of (in general) strongly correlated random variables, meaning that $\hat{R}_{xx}(\tau)$ is more strongly correlated than the original time series $x(t)$, so the estimate may not decay as rapidly as one might expect (see section 4.3 of Jenkins and Watts 1968).

4.7.4 Power-spectral density

Estimation methods for spectra considered here will relate to the traditional, nonparametric methods rather than the alternative parametric approaches. The standard methods are (1) the Fourier transform of the autocorrelation function (indirect method); (2) the direct method of averaging periodograms; and (3) filtering (either band-pass or complex demodulation (heterodyning and low-pass filtering). We shall outline (1) and (2). An interesting paper (Bingham et al. 1967) considers various aspects (computational and statistical) of spectral estimation.

4.7.4.1 Fourier transform of the autocorrelation function

An estimate of the power-spectral density from a length of data T (here assumed to be defined for $|\tau| < T/2$) is

$$\hat{S}_{xx}(f) = \frac{|X_T(f)|^2}{T} = \int_{-T}^{T}\hat{R}_{xx}(\tau)e^{-i2\pi f\tau}d\tau \qquad (4.110)$$

This is termed the 'raw' spectral density, since it turns out (see Section 4.7.4.2) that the variability of this estimator is independent of the data length T.

To reduce the sampling fluctuations, the estimate must be smoothed in the frequency domain, and this can be achieved by multiplying the autocorrection function estimate by a 'lag window' $w(\tau)$ which has a Fourier transform $W(f)$ (a 'spectral window'), and Fourier transforming the product. This defines the estimator $\tilde{S}_{xx}(f)$, where

$$\tilde{S}_{xx}(f) = \int_{-T}^{T} \hat{R}_{xx}(\tau) w(\tau) e^{-i2\pi f \tau} d\tau \qquad (4.111)$$

which by Section 4.3.3 is the convolution of the 'raw' spectral density with $W(f)$, that is, $\hat{S}_{xx}(f)*W(f)$. This estimation procedure may be viewed as a smoothing operation in the frequency domain. In the time domain, the lag window can be regarded as reducing the 'importance' of values of $\hat{R}_{xx}(\tau)$ as τ increases (since the variability of the estimator increases as τ increases).

4.7.4.2 Statistical properties

To discuss the properties of estimators $\hat{S}_{xx}(f)$ and $\tilde{S}_{xx}(f)$, it is necessary to first outline a few preliminary ideas in statistics (Jenkins and Watts 1968).

4.7.4.2.1 Chi-squared distribution

If $X_1, X_2, \ldots, X_n$ are n independent random variables each having a normal distribution with zero mean and unit standard deviation, and if

$$\chi_n^2 = X_1^2 + X_2^2 + \cdots + X_n^2 \qquad (4.112)$$

then the distribution of χ_n^2 is called the chi-squared distribution with n degrees of freedom (the number of degrees of freedom representing the number of independent random variables X_i entering the expression). The first two moments of χ_n^2 are

$$\mathbb{E}\left[\chi_n^2\right] = n \quad \text{and} \quad \mathbb{V}\left[\chi_n^2\right] = 2n \qquad (4.113)$$

If a set of independent χ_n^2 variables are added together, that is,

$$\chi_v^2 = \chi_{n_1}^2 + \chi_{n_2}^2 + \cdots + \chi_{n_k}^2 \qquad (4.114)$$

then the sum is χ_v^2 (as indicated), and $v = n_1 + n_2 + \cdots + n_k$. These properties of the chi-squared distribution are used in the development of some of the following results (see Jenkins and Watts 1968).

4.7.4.2.2 Bias

$\hat{S}_{xx}(f)$ is an asymptotically unbiased estimator, that is, $\mathbb{E}[\hat{S}_{xx}(f)] = S(f)$ as $T \to \infty$.

4.7.4.2.3 Variance

Since $\tilde{S}_{xx}(f)$ is a squared quantity (and if we assume $x(t)$ is Gaussian), then it can be argued that $2\hat{S}_{xx}(f)/S_{xx}(f)$ is distributed as a chi-squared random variable with two degrees of freedom, that is, χ_2^2 (for all values of data length T). From Equation 4.113 it follows, therefore, that the variance of $\hat{S}_{xx}(f)$ is $S_{xx}^2(f)$. So, $\hat{S}_{xx}(f)$ is an inconsistent estimator of $S_{xx}(f)$, and the standard deviation of the estimate is as large as the quantity being estimated.

4.7.4.3 Smoothed spectral estimators

Since $\tilde{S}_{xx}(f)$ is a convolution of $\hat{S}_{xx}(f)$ with $W(f)$, it turns out that $\tilde{S}_{xx}(f)$ is biased. In fact (for large T)

$$\mathbb{E}\left[\tilde{S}_{xx}(f)\right] \approx \int_{-\infty}^{\infty} S_{xx}(g) W(f-g) dg \qquad (4.115)$$

and

$$\mathbb{B}\left[\tilde{S}_{xx}(f)\right] \approx \int_{-\infty}^{\infty} [w(\tau) - 1] R_{xx}(\tau) e^{-i2\pi f \tau} d\tau \qquad (4.116)$$

Equation 4.115 indicates how the estimate is distorted by smearing and leakage of the spectral density owing to the spectral window width and shape (side lobes), and the window discussions earlier relate directly to this situation. Owing to the complicated shape of the $W(f)$ functions, we can only comment broadly that the general effect is to reduce the dynamic range of the spectra; that is, peaks are underestimated and troughs are overestimated, and the effect is reduced as the spectral window gets narrower, that is, as the data window T gets larger. However, the windows cannot be made too narrow since then there is little smoothing and the random errors increase. Consequently, there is always the problem of trading bias errors against random errors. (The bias problem is sometimes referred to under 'bandwidth considerations'.)

Since $\tilde{S}_{xx}(f)$ is a weighted sum of values of $\hat{S}_{xx}(f)$, it is argued (Jenkins and Watts 1968) that $n\tilde{S}_{xx}(f)/S_{xx}(f)$ is approximately distributed as χ_n^2 where n (the number of degrees of freedom) = $2BT$ and B is the resolution bandwidth. It follows that

$$\mathbb{V}\left[\tilde{S}_{XX}(f)\right] = \frac{S_{xx}^2(f)}{n/2} = \frac{S_{xx}^2(f)}{BT} \tag{4.117}$$

and in Jenkins and Watts(1968), estimator accuracies are tabulated for various lag windows.

The choice of window should depend on whether the concern is for statistical *stability* (low variance) or small bias. If the bias is small for all f, then we have high *fidelity*. Of course, in a general situation one must be careful with both. For example, if the spectral function has narrow peaks of importance, we may willingly tolerate some loss of stability to resolve the peaks properly, while if the spectral function is smooth, then bias errors are not likely to be so important.

4.7.4.4 Confidence intervals

We now give 'interval estimates' based on the 'point estimates' for the smoothed spectral density. Since $n\tilde{S}_{xx}(f)/S_{xx}(f)$ is distributed as a χ_n^2 random variable, then the probability-density function has the appearance as in Figure 4.34.

If we choose a number α $(0 < \alpha < 1)$ such that sections of area $\alpha/2$ are marked off as shown by the points $x_{n,\alpha/2}$ and $x_{n,1-\alpha/2}$, then

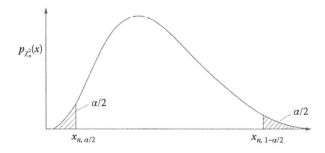

Figure 4.34 Creation of confidence intervals from the sample distribution.

$$P\left[x_{n,\alpha/2} \le \frac{n\tilde{S}_{xx}(f)}{S_{xx}(f)} \le x_{n,1-\alpha/2}\right] = 1 - \alpha \qquad (4.118)$$

The points $x_{n,\alpha/2}$ and $x_{n,1-\alpha/2}$ can be obtained from tables of χ_n^2 for different α and the inequality 'solved' for the true spectral density $S_{xx}(f)$ to yield the equivalent inequality

$$\frac{n\tilde{S}_{xx}(f)}{x_{n,1-\alpha/2}} \le S_{xx}(f) \le \frac{n\tilde{S}_{xx}(f)}{x_{n,\alpha/2}} \qquad (4.119)$$

So, for a particular point estimate $\tilde{S}_{xx}(f)$, the $100(1 - \alpha)\%$ confidence limits for $S_{xx}(f)$ are

$$\frac{n}{x_{n,1-\alpha/2}}\tilde{S}_{xx}(f), \frac{n}{x_{n,\alpha/2}}\tilde{S}_{xx}(f) \qquad (4.120)$$

The confidence *interval* is the difference between them. Note that on a linear scale, the confidence interval depends on the estimate, but on a log scale the width of the interval is $\log(n/x_{n,\alpha/2}) - \log(n/x_{n,1-\alpha/2})$, which is a constant independent of $\tilde{S}_{xx}(f)$.

4.7.4.5 Segment averaging

A method of spectral estimation that has become very popular (owing mainly to its speed of computation) is that of segment averaging, discussed in Jenkins and Watts (1968) (Section 4.3.4) as Bartlett's procedure and in some detail by Welch (1967) and Yuen (1979). The books by Bendat and Piersol (2010), Otnes and Enochson (1978) and Beauchamp (1973) also discuss it in detail. The basic procedure is outlined with reference to Figure 4.35.

A data length T is split up into q segments of length T_r. In practice, these segments are usually overlapped. The raw periodogram is formed for each segment, that is,

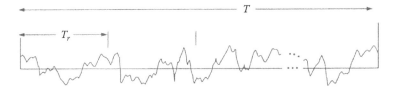

Figure 4.35 Basis of segment averaging.

$$\hat{S}_{xx_i}(f) = \frac{1}{T_r}|X_{T_{r_i}}(f)|^2 \quad \text{for} \quad i = 1,\ldots,q \tag{4.121}$$

and to reduce the fluctuations, the average

$$\hat{S}_{xx}(f) = \frac{1}{q}\sum_{j=1}^{q}\hat{S}_{xx_j}(f) \tag{4.122}$$

is computed. It can be argued that, since each $2\hat{S}_{xx_j}(f)/S_{xx}(f)$ is a χ_2^2 random variable, then, assuming these variables are independent, $2\tilde{S}_{xx_q}(f)/S_{xx}(f)$ is χ_{2q}^2, from which it follows that

$$\mathbb{V}\left[\tilde{S}_{xx}(f)\right] \approx \frac{S_{xx}^2(f)}{q} \tag{4.123}$$

Note that the independence of the raw periodogram estimates is rarely satisfied; overlapping serves to undermine the assumption of independence. Consequently, this analysis only holds in an approximate fashion.

Since the resolution bandwidth B is approximately $1/T_r$, then

$$\mathbb{V}\left[\tilde{S}_{xx}(f)\right] \approx \frac{S_{xx}^2(f)}{BT} \quad \text{(as before)} \tag{4.124}$$

While this summarises the essential features of the method, Welch (1967) and Bingham et al. (1967) give more elaborate procedures and insight. An important aspect is the use of data windows (or linear tapering as it is sometimes called) that are assigned to each segment (it is important to realise that the word linear here does not refer to the window shape, but to the fact that it operates on the data directly and not on the autocorrelation function, where it is sometimes called quadratic tapering). Linear windowing, in effect, ignores some data because of the tapered shape of the window. Intuitively, the overlapping of segments compensates for this in some way, and Welch (1967) has results for this case, though it is not easy to relate these results to the indirect method of Fourier transforming the autocorrelation function.

The use of a data window on a segment before transformation reduces leakage but has the disadvantage of also reducing the total energy of that segment, and this must be compensated. This results in the calculation of modified periodograms for each segment of the form

$$\hat{S}_{xx_i}(f) = \frac{\dfrac{1}{T_r}\left| \displaystyle\int_{j^{th}\ \text{interval}} x(t)w(t)e^{-i2\pi ft}dt \right|^2}{\dfrac{1}{T_r}\displaystyle\int_{-T_r/2}^{T_r/2} w^2(t)dt} \quad (4.125)$$

where the denominator compensates for the 'energy reduction'.

4.7.5 Cross-spectra, coherence and transfer functions

4.7.5.1 Cross-spectra

These basic considerations also relate to cross-spectral estimation, together with some additional features. Detailed results can be found in chapter 9 of Jenkins and Watts (1968). If the smoothed cross-spectral density $\tilde{S}_{xy}(f)$ is written as $|\tilde{S}_{xy}(f)| \exp(i \arg(\tilde{S}_{xy}(f)))$, then the variances of the amplitude $(|\tilde{S}_{xy}(f)|)$ and phase $(\arg(\tilde{S}_{xy}))$ estimators are proportional to $1/BT$ but are also strongly dependent on the coherence $\tilde{\gamma}^2_{xy}(f)$. As Jenkins and Watts emphasise, the sampling properties of the cross-amplitude and phase estimators may be dominated by the (uncontrollable) influence of the coherency spectrum rather than by the (controllable) influence of the smoothing factor $1/BT$.

4.7.5.2 Coherence function

The estimate for the coherence function is made up from estimates of smoothed power and cross-spectra, that is,

$$\tilde{\gamma}^2_{xy}(f) = \frac{|\tilde{S}_{xy}(f)|^2}{\tilde{S}_{xx}(f)\tilde{S}_{yy}(f)} \quad (4.126)$$

It should be noted that if 'raw' spectra are used in this estimate, it is a simple matter to verify that the sample coherence function is unity for all frequencies for any signals $x(t)$ and $y(t)$. Statistical errors and confidence limits are given in Bendat and Piersol (1980, 2010) and Otnes and Enochson (1978), and Carter (1980) discusses the bias of the estimator. Results derived by Jenkins and Watts (1968) show that the bias is proportional to the square of the derivative of the phase spectrum with respect to frequency between x and y. This means that the estimator is sensitive to delays between x and y, and can also be expected to be inaccurate when, for example, $x(t)$ and $y(t)$ are input and output for a lightly damped oscillator. The bias error can

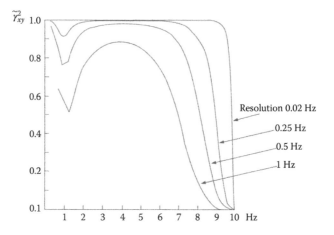

Figure 4.36 Effect of bias on the estimated coherence function.

be reduced by improving the resolution of the estimate. In Figure 4.36, the coherence function is estimated for simulated input–output results for an oscillator with a centre frequency of $f_0 = 1$ Hz. The theoretical value of coherence is unity, and the resolutions used are shown in the figure.

4.7.5.3 Frequency-response functions

These are estimated using smoothed spectra from the expression $\tilde{H}_1(f) = \tilde{S}_{xy}(f)/\tilde{S}_{xx}(f)$, and results for errors and confidence limits can be found in Bendat and Piersol (1980, 2010) and Otnes and Enochson (1978). A few results from Bendat and Piersol (1980, 2010) are quoted.

The normalised random error is defined as the standard deviation of the error divided by the quantity being estimated, that is, $\varepsilon_r = (\sigma(\hat{\phi}))/\phi$ assuming $\phi \neq 0$. The factor n_d refers to the number of averages used and assumes the data segments used are mutually uncorrelated (Table 4.2).

4.8 CONCLUDING REMARKS

This chapter is an attempt to highlight and summarise some of the essential aspects of the fundamentals of signal processing that are commonly used for data analysis. Practitioners should have a good understanding of the underlying assumptions and limitations of the validity of the many procedures that are now so accessible on special-purpose and personal computer (PC)-based analysers. Inevitably, 'real' data do not conform neatly to the various categories described and so the analyst must draw on a variety of methods to elicit the information contained in a signal.

Fundamentals of signal processing 207

Table 4.2 Normalized random error of various quantities

Estimator	Random error
$\tilde{S}_{xx}(f)$	$\dfrac{1}{\sqrt{n_d}}$
$\left\|\tilde{S}_{xy}(f)\right\|$	$\dfrac{1}{\left\|\gamma_{xy}\right\|/\sqrt{n_d}}$
$\arg\left\{\tilde{S}_{xy}(f)\right\}$	$\dfrac{\left(1-\gamma_{xy}^2\right)^{1/2}}{\left\|\gamma_{xy}\right\|\sqrt{2n_d}}$
$\tilde{\gamma}_{xy}^2(f)$	$\dfrac{\sqrt{2}\left[1-\gamma_{xy}^2\right]}{\left\|\gamma_{xy}\right\|\sqrt{n_d}}$
$\left\|\tilde{H}_1(f)\right\|$	$\dfrac{\left(1-\gamma_{xy}^2\right)^{1/2}}{\left\|\gamma_{xy}\right\|\sqrt{2n_d}}$

The development of signal processing throughout the years has led to a variety of excellent texts providing more detail than can be conveyed in a single chapter such as this. The interested reader can find a detailed treatment of the fundamental material around Fourier analysis and systems theory in Oppenheim and Schafer (1975), whereas Bendat and Piersol (2010) provide further insight surrounding the processing of random signals, and Shin and Hammond (2008) set all of this in the specific context of sound and vibration problems.

4.9 QUESTIONS

1. With reference to the $x(t)$ and $h(t)$ shown in the figure below, explain what is meant by the convolution of two signals, $y(t) = h(t) * x(t)$. Calculate and sketch $y(t)$.

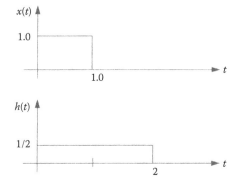

2. For the system shown below, the processes $x_1(t)$, $y_1(t)$, $n_1(t)$ and $n_2(t)$ are stationary random processes, and n_1, n_2 and x_1 are mutually uncorrelated.

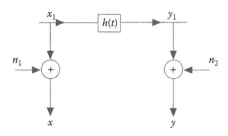

(i) Obtain expressions for

$$H_1(f) = \frac{S_{xy}(f)}{S_{xx}(f)} \text{ and } H_2(f) = \frac{S_{yy}(f)}{S_{yx}(f)}$$

(ii) If $S_{n_1 n_1}(f)/S_{x_1 x_1}(f) = S_{n_2 n_2}(f)/S_{x_1 x_1}(f) = \rho$, explain how you would use expressions H_1 and H_2 as estimators for $H(f)$ if $H(f)$ has both resonances and antiresonances.

3. Two signals $x(t)$ and $h(t)$ are defined below:

$$x(t) = e^{-at} \sin bt \quad t \geq 0$$
$$= 0 \quad t < 0$$

and $h(t) = \delta(t) + 0.1\delta(t - t_0)$ ($\delta(t)$ denotes a delta function).
With reference to these signals, illustrate carefully what is meant by the convolution of two signals, that is, $y(t) = h(t)*x(t)$, where * denotes convolution. Sketch and compare the Fourier transforms of $x(t)$ and $y(t)$.

4. If the aim is to 'identify' a system using a random test signal, and it is expected that the system behaves like a second-order single degree of freedom system with resonant frequency about 100 Hz and damping ratio less than 0.05, then select, with reasons, the following:
   (a) The bandwidth of the excitation signal
   (b) The sample rate for digital analysis of the data
   (c) The amount of data to be captured (assume that FFT sizes are limited to 1024 points)
   For the parameters you have selected, describe the errors that may be expected in the estimates of $S_{yy}(f)$, $S_{xy}(f)$ and the coherence function $\gamma^2_{xy}(f)$.

5. (a) A signal is shown below:

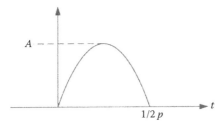

It is described by

$$x(t) = A\sin 2\pi pt \quad \text{for } 0 < t < \frac{1}{2p}$$

$$= 0 \quad \text{elsewhere}$$

Calculate the Fourier transform $X(f)$ of $x(t)$ and sketch the modulus and phase of this transform.

(b) A second signal, $x_1(t)$, is formed from $x(t)$ and is described by

$$x_1(t) = x(t) + 0.5x(t - \Delta)$$

How does the Fourier transform of $x_1(t)$ differ from that of $x(t)$?

(c) A third signal, $x_1(t)$, is formed from $x(t)$ and is described by

$$x_2(t) = \sum_{n=-\infty}^{\infty} x\left(t - \frac{n}{2p}\right)$$

What is the Fourier representation of $x_2(t)$?

REFERENCES

Ables, J.G. (1974) Maximum entropy spectral analysis. *Astronomy and Astrophysics Supplement Series*, **15**, 383–393.

Beauchamp, K.G. (1973) *Signal Processing: Using Analog and Digital Techniques*. George Allen & Unwin, London.

Bendat, J.S. and Piersol, A.G. (1980) *Engineering Applications of Correlation and Spectral Analysis*. Wiley Interscience, New York.

Bendat, J.S. and Piersol, A.G. (2010) *Random Data: Analysis and Measurement Procedures*, 4th edn, John Wiley & Sons, Hoboken, NJ.

Bingham, C., Godfrey, M., and Tukey, J.W. (1967) Modern techniques of power spectrum estimation. *IEEE Transactions Audio and Electroacoustics*, **AU-15** (2), 56–66.

Boashash, B. (1987) Nonstationary signals and the Wigner-Ville distribution. *Proceedings of the ASTED International Symposium on Signal Processing and Its Applications*, 261–268, Brisbane, Australia.

Bogert, B., Healey, M., and Tukey, J. (1963) The frequency analysis of time series for echoes. In M. Rosenblatt (ed.), *Proceedings of Symposium on Time Series Analysis*, 209–243, John Wiley & Sons, New York.

Brigham, E.O. (1974) *The Fast Fourier Transform*. Prentice Hall, Englewood Cliffs, NJ.

Carter, G.C. (1980) Bias in magnitude-squared coherence estimation due to misalignment. *IEEE Transactions on Acoustics, Speech and Signal Processing*, **ASSP-28** (1), 97–99.

Childers, D. and Durling, A. (1975) *Digital Filtering and Signal Processing*. West Publishing, St Paul.

Cohen, L. (1989) Time frequency distributions: A review. *Proceedings of the IEEE*, **77** (7), 941–981.

Davies, P. (1983) A recursive approach to Prony parameter estimation. *Journal of Sound and Vibration*, **89** (4), 571–583.

Hammond, J.K. and White, P.R. (1996) The analysis of nonstationary signals using time-frequency methods. *Journal of Sound and Vibration*, **190** (3), 419–447.

Harris, F.J. (1978) On the use of windows for harmonic analysis with the discrete Fourier transform. *Proceedings of the IEEE*, **66** (1), 51–83.

Hsu, H.P. (1970) *Fourier Analysis*. Simon & Schuster, New York.

Jenkins, G.M. and Watts, D.G. (1968) *Spectral Analysis and Its Applications*. Holden Day, San Francisco.

Leuridan, J., De Vis, D., Van der Auweraer, H., and Lembregts, F. (1985) *Proceedings of 4th International Modal Analysis Conference*, 908–918, Los Angeles.

Oppenheim, A.V. and Schafer, R.W. (1975) *Digital Signal Processing*. Prentice Hall, Englewood Cliffs, NJ.

Otnes, R.K. and Enochson, L. (1978) *Applied Time Series Analysis. Vol. 1. Basic Techniques*. John Wiley & Sons, New York.

Priestley, M.B. (1967) Power spectral analysis of nonstationary processes. *Journal of Sound and Vibration*, **6**, 86–97.

Rabiner, L.R. and Gold, B. (1975) *Theory and Application of Digital Signal Processing*. Prentice Hall, Englewood Cliffs, NJ.

Randall, R.B. (1977) *Frequency Analysis*. Brüel and Kjær Publications, Naerum, Denmark.

Shin, K. and Hammond, J.K. (2008) *Fundamentals of Signal Processing for Sound and Vibration Engineers*. John Wiley & Sons, Chichester.

Spitznogle, F.R. and Quazi, A.H. (1970) Representation and analysis of time-limited signals using a complex exponential algorithm. *Journal of the Acoustical Society of America*, **47** (5, part 1), 1150–1155.

Welch, P.D. (1967) The use of the fast Fourier transform for the estimation of power spectra: A method based on time averaging over short, modified periodograms. *IEEE Transaction Audio and Electroacoustics*, **AU-15** (2), 70–73.

White, P.R. Tan, M.H., and Hammond, J.K. (2006) Analysis of the maximum likelihood, total least squares and principal component approaches for frequency response function estimation. *Journal of Sound and Vibration,* **290**, 676–689.

Wicks, A.L. and Vold, H. (1986) The H frequency response function estimator. *Proceedings of 4th International Modal Analysis Conference*, 897–899, Los Angeles, CA.

Yuen, C.K. (1979) Comments on modern methods for spectrum estimators. *IEEE Transactions Acoustics, Speech and Signal Processing*, **ASSP** 27 (3), 298–299.

Zadeh, L. and Desoer, C.A. (1963) *Linear System Theory.* McGraw-Hill, New York.

Chapter 5

Noise control

David Thompson

5.1 INTRODUCTION

Noise is unwanted, undesirable or harmful sound. It is present in many situations and may have a number of adverse effects on people, as will be described in Chapter 6. It is clearly desirable to be able to control noise. In practice, there are two main motivating factors to address noise issues.

First, pressure to control noise comes from legislation. For example, many products have their noise output controlled by European Union (EU) regulations. Limits on the drive-by noise from road vehicles have existed in Europe since the early 1970s (EU 2014); more recently, EU directives have been introduced covering machinery noise (EU 2006) and noise from outdoor equipment (EU 2000).

Second, a major incentive to control noise is related to customer demand. For many types of product, quiet design may be used by manufacturers to gain competitive advantage. In contrast to legal limits, this may lead to a different strategy. Legal limits are well defined and must be met, albeit with a margin of safety, but there is no incentive to do better than this. Customer expectations, on the other hand, are continually developing and require continuous improvement of products. Moreover, customer requirements are subjective, so that meeting a simple overall noise level is insufficient. For example, if a car sounds too quiet it may be perceived as lacking character or power.

The introduction of new products, particularly where these involve a step change in the design or operation, may bring undesirable changes to the noise that is produced. It is essential, therefore, that noise is considered early in the design process, as many of the fundamental design decisions affect the noise in a way that cannot be remedied later by simple countermeasures. In practice, this is not always done and the acoustician may be asked to solve noise problems at a later stage in the process with little remaining scope for fundamental changes to the design. It must not be forgotten that there are many constraints to effective noise control, including financial expense, safety, weight, volume, robustness and durability, efficient operation or ease of maintenance.

This chapter aims to show how targets can be set for noise control and how noise sources can be better understood. The various principles behind noise control treatments are described, followed by more detailed descriptions of the operation of various noise control systems and their application in practice.

Throughout the chapter, only passive noise control techniques will be considered. Active control, in which secondary sound sources are used to cancel the primary sound (or vibration) field, can be considered as an alternative or as a complement to these techniques, particularly for low-frequency sound in enclosed spaces such as inside vehicles and aircraft (see Nelson and Elliott 1991; Fuller et al. 1996; Elliott and Nelson 2004). Active vibration control is also quite commonly applied in ships.

5.2 SOUND SOURCES: CATEGORIES AND CHARACTERISATION

5.2.1 What is a sound source?

What we consider to be a sound source depends on our perspective. Consider, for example, a car travelling along a road. For the local residents, the road as a whole with its passing traffic forms the noise source, although some vehicles may stand out more than others. However, each individual vehicle is considered as a separate source of noise when it is assessed against EU legislation using a drive-by test EU (2014). Furthermore, a noise, vibration and harshness (NVH) engineer may be interested, for instance, in the engine (or the exhaust, intake, tyres or aerodynamic noise) and would consider this as a source of noise within the vehicle. To the powertrain engineer, the combustion process within the engine itself or the out-of-balance forces due to rotating components may be the actual source to be studied. In this chapter, we are concerned with the engineering aspects of the control of noise sources and, therefore, a sound source is considered to be a component or mechanism which generates sound directly by acting on the surrounding air.

Sound sources are extremely diverse in their physical forms, mechanisms and characteristics. Any single noise-generating system, such as a vacuum cleaner, machine tool, car engine or rolling tyre, may produce noise by a number of distinct mechanisms. Examples include jet emission, fans, structural vibration, friction and impact. Hence, in technical terms, each system can be considered to consist of a number of separate sources. The remainder

Noise control   215

of this section will consider different ways in which sound sources can be categorised or characterised.

5.2.2 Measurement quantities

5.2.2.1 Amplitude

Before discussing sound pressure, it is worthwhile recalling the quantities used to measure sound. The sound pressure level in decibels is defined (see Section 2.1.3) in terms of the mean-square pressure, $\overline{p^2}$ (see Equation 2.4), where the overbar denotes a time-averaged quantity. Because of their relation to measured quantities, mean-square values will be used extensively in this chapter, although (complex) amplitudes are also used where it is necessary to account for phase information (see also Chapters 2 and 3). The square root of the mean-square value, the r.m.s. or *root mean square*, is also convenient. For a harmonic signal, $\overline{p^2} = p_{rms}^2 = \frac{1}{2}|p|^2$.

5.2.2.2 Frequency spectra

The other main dimension of acoustic signals is their frequency content. The frequency content of a signal can be obtained using the Fourier-transform techniques described in Chapter 4. In practical analysis, however, it is often convenient to work with spectra that are expressed in terms of broader frequency bands. The human auditory system resolves frequencies according to a logarithmic scale; for example, a doubling of frequency corresponds to a fixed musical interval which is termed an octave. Early frequency analysis consisted of a set of contiguous band-pass filters, each with a bandwidth of one octave. Ten such octaves are required to span the audible range (20 Hz–20 kHz). This was later refined into one-third octave bands, in which each octave is divided into three equal bands (on a logarithmic scale). There are 10 one-third octaves in a decade (factor of 10) of frequency. The octave and one-third octave bands are described by their nominal centre frequency, as listed in Table 5.1. The lower (f_L) and upper (f_U) limits of the bands (the *3 dB down* points of the corresponding filter characteristics) are more correctly defined as

$$f_L = 10^{(n-\frac{1}{2})/10}, \quad f_U = 10^{(n+\frac{1}{2})/10} \tag{5.1}$$

where n is the frequency band number, as listed in Table 5.1. The centre frequencies can be obtained from $f_c = 10^{n/10}$. To evaluate frequency spectra in one-third octave or octave bands, the pressure signal is filtered before determining the mean-square value in each band. The measurement of sound, including the practical use of one-third octave bands, is described in Section 8.6.1.

Table 5.1 Standard octave and one-third octave frequency bands and A-weighting corrections

Band number	Standard frequency bands (Hz)				A-weighting corrections	
	Octave-band centre frequency	One-third-octave band centre frequency	Band limits		Octave-band weightings (dB)	One-third-octave band weightings (dB)
			Lower	Upper		
13	–	20	18	22	–	−50.5
14	–	25	22	28	–	−44.7
15	31.5	31.5	28	35	−39.4	−39.4
16	–	40	35	44	–	−34.6
17	–	50	44	57	–	−30.2
18	63	63	57	71	−26.2	−26.2
19	–	80	71	88	–	−22.5
20	–	100	88	113	–	−19.1
21	125	125	113	141	−16.1	−16.1
22	–	160	141	176	–	−13.4
23	–	200	176	225	–	−10.9
24	250	250	225	283	−8.6	−8.6
25	–	315	283	353	–	−6.6
26	–	400	353	440	–	−4.8
27	500	500	440	565	−3.2	−3.2
28	–	630	565	707	–	−1.9
29	–	800	707	880	–	−0.8
30	1,000	1,000	880	1,130	0.0	0.0
31	–	1,250	1,130	1,414	–	0.6
32	–	1,600	1,414	1,760	–	1.0
33	2,000	2,000	1,760	2,250	1.2	1.2
34	–	2,500	2,250	2,825	–	1.3
35	–	3,150	2,825	3,530	–	1.2
36	4,000	4,000	3,530	4,400	1.0	1.0
37	–	5,000	4,400	5,650	–	0.5
38	–	6,300	5,650	7,070	–	−0.1
39	8,000	8,000	7,070	8,800	−1.1	−1.1

Table 5.1 (Continued) Standard octave and one-third octave frequency bands and A-weighting corrections

Band number	Standard frequency bands (Hz)				A-weighting corrections	
	Octave-band centre frequency	One-third-octave band centre frequency	Band limits		Octave-band weightings (dB)	One-third-octave band weightings (dB)
			Lower	Upper		
40	–	10,000	8,800	11,300	–	–2.5
41	–	12,500	11,300	14,140	–	–4.3
42	16,000	16,000	14,140	17,600	–6.6	–6.6
43	–	20,000	17,600	22,500	–	–9.3

5.2.2.3 Weighted spectra

The sensitivity of the human auditory system is frequency dependent, as described in more detail in Chapter 6. It is, therefore, appropriate to 'weight' sound spectra to improve the correlation between objective measurements and subjective effects. The most widely used weighting curve is the A-weighting (see Figure 6.6). This is described by a filter characteristic that is applied to the time signal (see IEC 2013); it can also be approximated by applying the weighting corrections listed in Table 5.1 to one-third-octave or octave band spectra. The overall A-weighted sound level can be obtained by adding the mean-square pressure from each band (see Equation 2.10). However, it should be noted that the use of such correction terms can give misleading results if a strong tonal component present in a band is not close to its centre frequency.

5.2.3 Categories of sound source

We can divide noise sources into two main types. The first is mechanical, comprising vibrating solid structures. The radiation of sound from such vibration is discussed in detail in Section 5.5. The second is produced by fluctuating air motion without any accompanying structural vibration; such sources are classed as aeroacoustic. For example, clapping one's hands or blowing over a rigid surface (such as a finger or the edge of a credit card) produces considerable noise without any attendant vibration of the surface itself. Jet noise and rotating fan blades are important examples of aeroacoustic sources, and many musical instruments, particularly woodwind and brass, rely on aeroacoustic mechanisms of sound production.

Aeroacoustic sources, in particular, can be divided into a number of categories, which have differing efficiencies:

1. The most efficient sources are monopole in nature; these are sources which produce an oscillating displacement of fluid volume or mass. Examples include the unsteady component of the gas flow through the exhaust system of an internal combustion engine, and a compressed air siren. A loudspeaker mounted in a cabinet, although not an aeroacoustic source, also has the form of a monopole source (at low frequencies), in this case due to vibration of the loudspeaker cone.
2. Dipole sources are those which apply fluctuating or moving forces to the fluid. Rotating fan blades and singing telephone wires are common examples. Airflow over parts of vehicles, particularly protruding mirrors and aerials, also produces dipole-type aeroacoustic noise. The deployed landing gear of aircraft on approach forms a particularly important example. A freely suspended, unboxed/unbaffled loudspeaker also forms a dipole source which is much less efficient than when the same speaker is mounted in a cabinet, due to partial cancellation of the sound radiated from the front and back of the speaker.
3. Unsteady flow, involving only fluctuating stresses on the fluid in the absence of solid surfaces, forms quadrupole sources. Jet noise is a common example (Lighthill 1952), as well as the turbulent boundary layer formed by vehicle motion, although this also has a dipole component.

The basic mathematical models of acoustic monopoles, dipoles and quadrupoles are described in Sections 2.4 and 2.5. Figure 5.1 shows some examples of these various source categories.

5.2.4 Acoustic efficiency

It is well known that when a machine operates, only a part of its mechanical power is converted into the desired operation; much is lost in heat. However, a small fraction is converted into mechanical vibration; of this vibrational energy, only a very small fraction is converted into sound. For many engineering noise sources, the fraction of the mechanical power used to drive a machine that is radiated as sound power is extremely small, often of the order of one-millionth.

The ratio of radiated sound power to the mechanical power of the machine or system can be defined as the *acoustic efficiency*, η_{ac}. While this should ideally be as large as possible for sound-generating systems, such as loudspeakers or musical instruments, it should be as small as possible for engineering noise sources. It varies considerably with the type of source, the mechanical design, the damping present and the form and material of

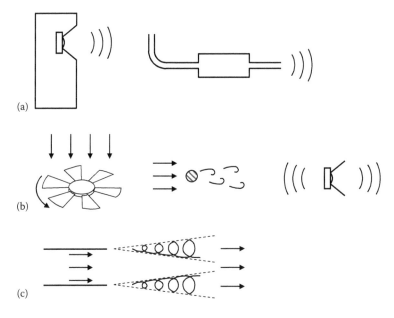

Figure 5.1 Examples of noise source categories. (a) Monopole: fluctuating volume or mass injection; (b) dipole: rotating or fluctuating force on fluid; (c) quadrupole: fluctuating fluid shear stress.

the structure. Some typical values are listed in Table 5.2 (see Heckl and Müller 1994).

For aeroacoustic sources, the three categories identified in Section 5.2.3 have differing acoustic efficiencies. Monopole sources have an efficiency (sound power relative to the rate of flow of kinetic energy of the fluid) that is proportional to the flow Mach number, $M = U/c_0$, where U is the flow speed and c_0 is the speed of sound in the ambient external fluid. Their high efficiency is exemplified by the results for unsilenced internal combustion engines and sirens (see Table 5.2). The noise emitted from the silencer of an internal combustion engine is also monopole in nature but with a much attenuated amplitude.

The efficiency of aeroacoustic dipole-type sources varies with flow speed at a rate of M^3 and that of quadrupole sources at a rate of M^5. Thus, although they are less efficient, they are much more sensitive to changes in the flow speed. The high speed dependence of jet noise, in particular, underlies the success in reducing the noise from civil jet aircraft by introducing high-bypass-ratio engines and thereby reducing the effective flow speed. This has enabled modern aircraft to be around 20–30 dB quieter than the first generation jet aircraft.

The acoustic efficiency is only used as a rough rule-of-thumb indicator. However, it reveals that the noise creation process is often extremely

Table 5.2 Some typical examples of acoustic efficiency, η_{ac}

Source type		η_{ac}
Electric motor		$2\times 10^{-8} - 2\times 10^{-7}$
Diesel engine	3000 rev min^{-1} (from engine block)	5×10^{-6}
	Outlet noise (unsilenced)	10^{-4}
Fan in optimal working situation	Pressure drop, $\Delta p < 250$ Pa	10^{-6}
	Pressure drop, $\Delta p > 250$ Pa	$(\Delta p/250)\times 10^{-6}$
Road vehicle	Tyre noise	10^{-6}
Excavators, wheeled loaders etc.		$10^{-7} - 10^{-6}$
Propeller aircraft		5×10^{-3}
Human voice		5×10^{-4}
Pipe organ		$10^{-3} - 10^{-2}$
Oboe, flute	Fortissimo	1×10^{-3}
Trumpet, saxophone	Fortissimo	1×10^{-2}
Loudspeaker		5×10^{-2}
Horn loudspeaker		4×10^{-1}
Siren		$1-7\times 10^{-1}$

inefficient. It is this inherent inefficiency that often makes it extremely difficult to reduce noise at the source. Moreover, for a given type of source, although the size of the machine may vary considerably, the acoustic efficiency will vary much less, subject to differences between good and bad practice. Often, it is only possible to reduce noise substantially by making significant design or operational changes, for example by reducing speeds; or by changing the principle of operation, for example by using an electric motor instead of a combustion engine.

5.2.5 Operational speed

The dependence of sound power (or pressure) on operational speed can be useful in identifying the types of source that are dominant in a particular situation, as well as estimating the potential effects of changing the speed.

5.2.5.1 Aeroacoustic sources

For aeroacoustic sources, to find the dependence of the sound power on the flow speed, it is necessary to combine the acoustic efficiency with the mechanical power associated with the flow. The mechanical power

Table 5.3 Speed dependence of aeroacoustic sources

	Acoustic efficiency	Sound power	Sound power level (dB)	
Monopole-type	$\eta_{ac} \sim U$	$W \sim U^4$	$L_W \sim 40 \log_{10} U$	12 dB per doubling of speed
Dipole-type	$\eta_{ac} \sim U^3$	$W \sim U^6$	$L_W \sim 60 \log_{10} U$	18 dB per doubling of speed
Quadrupole-type	$\eta_{ac} \sim U^5$	$W \sim U^8$	$L_W \sim 80 \log_{10} U$	24 dB per doubling of speed

Note: ~ indicates 'varies approximately as'.

transmitted per unit area by a fluid flow is proportional to U^3. As a result, the speed dependence of the acoustical power output of aeroacoustic sources is as listed in various forms in Table 5.3. The sound pressure level will tend to increase with the same speed dependence as the sound power level (apart from directivity effects). Note that the frequency dependence also changes as the speed increases, so the A-weighted sound level may have a greater speed dependence than that shown in the table.

5.2.5.2 Mechanical sources

Mechanical systems generate noise by various different physical mechanisms (see also Section 5.4.2). It is, therefore, not always possible to attribute a unique index to the speed dependence of the sound power. Empirical data are available for many systems, however, examples of which are listed in Table 5.4.

Figure 5.2 shows an example of the dependence on speed of the noise from a passing train. At speeds up to around 300 km h^{-1}, the mechanical (rolling) noise dominates and the speed dependence follows a $30 \log_{10} V$ line. Aerodynamic noise has a greater speed dependence, appearing here as around

Table 5.4 Typical speed dependence of mechanical sources

Train rolling noise	$W \sim V^3$	$L_W \sim 30 \log_{10} V$
Tyre/road noise	$W \sim V^3$ to V^4	$L_W \sim (30–40) \log_{10} V$
IC petrol engines – at low speed	$W \sim N^2$	$L_W \sim 20 \log_{10} N$
– at high speed	$W \sim N^5$	$L_W \sim 50 \log_{10} N$
Naturally aspirated direct injection diesel	$W \sim N^3$ to N^4	$L_W \sim (30–40) \log_{10} N$
Turbocharged DI diesel engines	$W \sim N^2$ to N^3	$L_W \sim (20–30) \log_{10} N$
Electric motors	$W \sim N^{1.5}$	$L_W \sim 15 \log_{10} N$
Gears	$W \sim N^2$ to N^3	$L_W \sim (20–30) \log_{10} N$
Propeller fans	$W \sim N^5$	$L_W \sim 50 \log_{10} N$

Note: V = vehicle speed (km h^{-1}), N = rotational speed (rev min^{-1})

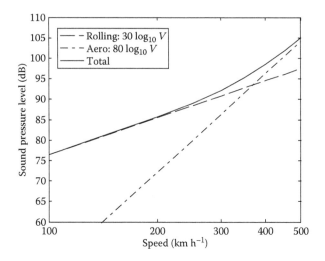

Figure 5.2 Typical speed dependence of train noise.

$80 \log_{10} V$ in terms of the A-weighted level. Consequently, this becomes dominant at higher speeds, typically above about 350 km h^{-1} (Thompson 2009).

5.2.5.3 Consequence of speed dependence

Generally, sound power increases with increasing mechanical power; this is a consequence of the fact that the acoustic efficiency is roughly constant. However, 'high-power' machines are not necessarily noisier than 'low-power' machines and may even be quieter in some situations.

A given power output can be achieved by a combination of low torque and high speed or by a higher torque and lower speed. The former can be satisfied by a small, compact machine which is light, efficient and cheap, but will generally be noisier than a larger machine running more slowly. The larger machine may be overdimensioned for the operational task, but this may allow it to operate more quietly. Consequently, a compromise has to be found between the requirements of the noise control engineer and the production engineer.

Example: Consider two diesel engines rated at 100 and 200 kW at 3000 rev min^{-1}. Estimate the difference in noise produced if they are each required to produce 100 kW. We make the following assumptions:

1. Assume a constant acoustic efficiency of, say, $\eta_{ac} = 5 \times 10^{-6}$ at 3000 rev min^{-1} (see Table 5.2).
2. Assume for simplicity a constant torque versus speed (this is not generally true), so the mechanical power will be proportional to the rotational speed N.

3. Assume a speed dependence of diesel engine noise of about N^3 (see Table 5.4).

At 3000 rev min^{-1}, the smaller (100 kW) engine will produce $10^5 \times \eta_{ac} = 0.5$ W of acoustic power and the larger engine will similarly produce 1.0 W, that is, 3 dB more than the smaller engine. However, from assumption (2), to produce a power of 100 kW the smaller engine must run at 3000 rev min^{-1} but the larger engine can run at a reduced speed of 1500 rev min^{-1}. At this reduced speed, the sound power will be $1.0 \times (1500/3000)^3 = 0.125$ W, or 6 dB less than the smaller engine.

5.2.6 Frequency spectra

Any sound-pressure time history (even isolated transients) can be mathematically decomposed into an infinite number of sinusoids of infinite duration using the Fourier transform (see Chapter 4). The amplitudes and phases of these sinusoids form the spectrum of the signal. Periodic sounds, such as compressor whines, or cyclic events, such as regularly repeated impacts, have spectra with nonzero components only at discrete frequencies which are multiples (known as harmonics) of the fundamental frequency. Aperiodic sounds, such as subsonic jet noise or isolated impacts, have continuous frequency spectra. Some sources contain both, such as fans where tonal components at the blade-passing frequency and its harmonics are superimposed on a broadband noise.

Correctly resolved spectra can show whether periodic or aperiodic effects dominate. Moreover, the relative importance of the various harmonics in a periodic signal can be used for diagnostic purposes; for example, by relating the fundamental frequency or the main harmonics to the operation of the machine. The strength of different harmonics can be related to the pulse shape of a periodic event. The presence of non-integer harmonic or subharmonic components could be related to other events, for example through gearing.

It can also be useful to analyse the frequency content of acoustic signals as a function of operational speed. Figure 5.3 shows a series of spectra from a car engine measured inside the vehicle, a large family hatchback, at different engine rotational speeds. By arranging these spectra behind one another, trends with increasing speed can be seen. Figure 5.4 shows the same measured data in a different way: here, the magnitude at various engine 'orders' has been extracted and is plotted as a function of engine speed. An engine *order* is a multiple of the engine rotational speed. In the example shown, a four-cylinder four-stroke engine, the dominant component at most speeds is the second order of engine rotation, referred to as 2E, which corresponds to the frequency of firing events. In this example, a large peak occurs in the signal in the 2E component

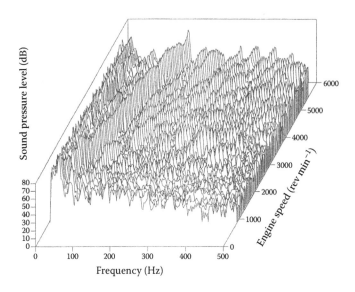

Figure 5.3 'Waterfall' plot of noise spectra from an automotive engine measured inside the vehicle during a gradual increase of the engine speed from idle to maximum.

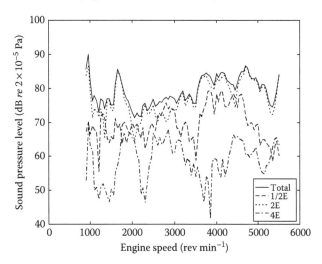

Figure 5.4 Same data as Figure 5.3 plotted as 'order analysis' showing overall sound pressure level and main engine orders as a function of engine speed.

at about 1650 rev min^{-1} which constitutes an annoying 'boom' in the vehicle. Ideally, the sound in the vehicle should increase gradually with increasing speed. Often, boom corresponds to the coincidence of an excitation frequency with the mechanical or acoustic resonance frequencies of the vehicle. In the present case, the 2E component at 1650 rev min^{-1}

Figure 5.5 Basic noise transmission model.

corresponds to a frequency of 55 Hz, which is the first acoustic resonance of the interior of this vehicle; it is also likely that it coincides with a mechanical resonance.

5.2.7 Directivity and geometric attenuation

The sound emitted by a source can be quantified by its sound power. However, the noise received at a particular location also depends on the transmission of sound to that location. This is shown in general terms in Figure 5.5. For outdoor sound transmission, for example, the sound pressure depends on the distance from source to receiver, the physical extent of the source, the heights of source and receiver above the ground, the ground shape and absorption, intervening objects, meteorological conditions and the directivity of the source (ISO 1996a).

5.2.7.1 Dependence of sound pressure on distance from source (geometric spreading)

A simple model of geometric spreading describes the dependence on distance for a source in a free field. For simplicity, we neglect the effects of ground, reflective or absorptive objects and any air absorption. The sound at sufficient distance from the source in relation to its physical size (i.e. in the far field) can be approximated locally as a plane wave. Hence, the mean-square pressure (in a frequency band) is given by

$$\overline{p^2} = \rho_0 c_0 I \tag{5.2}$$

where:

I is the time-averaged intensity (see Section 2.3.2 for the definition)
$\rho_0 c_0$ is the characteristic specific* acoustic impedance

---

* The word 'specific' refers to the ratio of pressure to particle velocity, to distinguish this from an acoustic impedance of the form pressure over volume velocity (see Fahy 2000 and Morfey 2001). In the remainder of this chapter, it is referred to as the *acoustic impedance* unless there is any ambiguity.

ρ_0 is the density of air
c_0 is the speed of sound (see Section 2.2.8)

For an omnidirectional point source emitting sound power W into a free field, the power spreads out over a sphere so that the mean-square pressure at distance r is given by

$$\overline{p^2} = \frac{\rho_0 c_0 W}{4\pi r^2} \quad \text{or} \quad L_p = L_W - 20\log_{10} r - 11 \tag{5.3}$$

for the usual reference values of 2×10^{-5} Pa for the sound pressure level and 10^{-12} W (watts) for the sound power level. Thus, the sound-pressure level decreases by 6 dB per doubling of distance.

If the source is located on a rigid ground surface, these formulae are modified to allow for spreading over a hemisphere. Thus, for an omnidirectional point source,

$$\overline{p^2} = \frac{\rho_0 c_0 W}{2\pi r^2} \quad \text{or} \quad L_p = L_W - 20\log_{10} r - 8 \tag{5.4}$$

For a line source, the sound spreads out over an infinitely long cylinder and the dependence on distance is -3 dB per doubling of distance.

5.2.7.2 Far-field directivity

In the far field, the sound field can be approximated as having a certain dependence on distance, as discussed, and a separate dependence on direction, denoted as D_θ. At a distance r from a point source in a free field, the intensity in the radial direction averaged over a spherical surface is

$$\langle I \rangle = \frac{W}{4\pi r^2} \tag{5.5}$$

The directivity factor D_θ is then defined as the ratio of the radial intensity in a particular direction to the mean intensity. In decibels, this is expressed as the directivity index (DI), written as $DI = 10\log_{10} D_\theta$. So, Equation 5.3 is modified to give

$$L_p = L_W - 20\log_{10} r - 11 + DI \tag{5.6}$$

At low frequencies, most compact sources radiate sound with a fairly simple directivity. At high frequencies, however, the sound radiation from many sources can become strongly directional.

5.3 NOISE SOURCE QUANTIFICATION

5.3.1 Purpose of source quantification

Noise control strategies can be based on reductions at source or reductions of transmission from source to receiver (see Figure 5.5). In both cases, it is useful to quantify the source output, for which a suitable measurement quantity is required to represent the source strength. This should ideally be

- A property of the source alone, independent of its location
- Representative of the sound from the whole source
- Directly related to the receiver quantity (usually sound pressure level)

In practice, sound sources are usually measured in terms of their sound power. However, it should be recognised that this quantity does not completely satisfy these criteria. In particular, the sound power can vary with source location. For example, when located close to a rigid surface, a compact monopole source can produce twice as much sound power as when it is located in a free field* (see Section 2.5.3). Nevertheless, provided that suitable precautions are taken to ensure that the measurement is representative of normal operation, sound power can be, and is, used to quantify sound sources. Decoupling of vibrating machines from support/connected structures is also important in noise control. This is explained in Section 5.10.

Sound sources are quantified for a variety of reasons:

1. To compare different machines to allow user selection
2. For a manufacturer to check the acceptability of components from subsuppliers against targets
3. To check that a machine complies with regulatory or legal requirements
4. For source labelling according to regulations
5. To predict the sound pressure at an operator position (for assessment of the hearing hazard) or in the neighbourhood (the environmental impact)
6. To identify source mechanisms (diagnostics)
7. To understand the physics of a source to develop models for the purpose of improving the design
8. As input to models of transmission paths for assessing noise control by reduced transmission

The requirements in these situations differ: in some, the use of a standard method is essential, whereas in others, bespoke measurements may be more

---

* This applies for a source that is very small compared with the wavelength, for which the volume velocity is unaffected by the presence of the rigid surface, located closer than a quarter of a wavelength from the rigid surface.

appropriate. The level of precision required will also vary depending on the application.

Sound power cannot be measured directly, so it is necessary to measure some other quantity (e.g. pressure, intensity) and infer the power from a model. The main methods used with sound pressure measurements are based on either a free field or a reverberant field model. Intensity measurements do not require these assumptions. The various international measurement standards are summarised in Table 5.5. Although the standards are summarised in the following sections, it is important that the reader checks the details of the current standards before applying them.

5.3.2 Free-field sound power measurement methods

To approximate free-field conditions, an anechoic (or semi-anechoic) chamber is required, at least to perform precision measurements (ISO 2012). Such a chamber minimises reflections from the room walls and also the background noise. A fictitious measurement surface is defined around the source and divided into n segments; for a precision-grade measurement, this surface is a sphere or, where the measurement is carried out over a rigid ground, a hemisphere (see Figure 5.6). It is implicitly assumed that, in the far field, the sound field is locally a plane wave and, therefore, the time-averaged intensity normal to the surface is given by $p^2/\rho_0 c_0$.

Table 5.5 Summary of sound power measurement methods for different grades of precision (numbers refer to ISO standards)

	Precision	Engineering	Survey
Direct			
Pressure-based methods			
Anechoic environment	3745		
Semi-anechoic environment	3745	3744	
Reverberant environment	3741	3743	
Large rooms		3744	
In situ (indoors or outdoors)		3744	3746
Intensity-based methods			
Single points		9614-1	9614-1
Surface scanning	9614-3	9614-2	9614-2
Indirect			
Source substitution method			3747
Sound pressure at standard location		Various	

Noise control

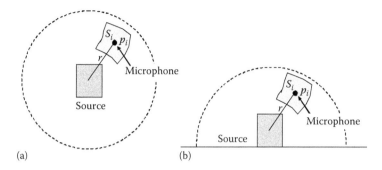

Figure 5.6 Fictitious measurement surfaces for 'free-field' measurement methods. (a) Spherical surface; (b) hemispherical surface for source located on rigid ground.

The pressure is sampled at a number of points i on the measurement surface, each associated with an area S_i, and the power is derived from

$$W \approx \sum_{i=1}^{n} \left(\frac{\overline{p_i^2}}{\rho_0 c_0} \right) S_i \tag{5.7}$$

Various conditions applying to the measurement surfaces, the distance from the walls and the background noise are described in the appropriate standards (ISO 2010e, 2010f, 2012).

5.3.3 Reverberant-field sound power measurement methods

Measurements based on sound pressure are also possible in a reverberant environment (ISO 2009, 2010c, 2010d). The model in this case assumes a diffuse sound field, which means that the sound pressure level is approximately independent of the location within the room (away from the source and the walls). This approximation is usually valid only for broadband sources, so that for tonal sources additional precautions are necessary. This requires a room with low absorption, that is, with hard reflecting walls. The absorption can be measured by determining the reverberation time T_{60}, which is defined as the time taken for the sound level to reduce by 60 dB after switching off a source (see Section 2.7.3). T_{60} is usually expressed in one-third octave frequency bands. The standards also warn against having too little absorption, particularly at low frequencies; the absorption should be sufficient to minimise the effect of individual room modes.

For the sound field to be diffuse, the frequency should be above the Schroeder frequency (Equation 2.188) which implies that a large room is needed. For valid measurements in the 100 Hz one-third octave band, a volume of at least 200 m³ is required (ISO 2010c). A large room also means that a large number of acoustic modes of the room have their natural frequencies within each frequency band of interest. Moreover, to ensure that the acoustic modes are spread evenly within a given frequency band, the room dimensions should be different from each other and not close to an integer ratio. Reverberation rooms are often non-rectangular, sometimes five-sided, and diffusing panels or rotating diffusers are also often deployed.

In an ideal diffuse field, the intensity at all points and in all directions is equal. There is, therefore, no net intensity in any direction. However, we can define a one-sided intensity (see Section 2.7.2) which is the intensity passing through a surface in one direction. In a diffuse field, this is given by

$$I_{1-s} = \langle \overline{p^2} \rangle \frac{1}{4\rho_0 c_0} \tag{5.8}$$

where $\langle \overline{p^2} \rangle$ is the spatially averaged mean-square pressure.

If the room boundaries have total surface area S and average diffuse-field absorption coefficient α (the proportion of incident intensity that is absorbed, see Section 5.9), the power absorbed is $I_{1-s} S\alpha$. In a steady state, this can be equated to the power input by the source to give

$$W = \langle \overline{p^2} \rangle \frac{S\alpha}{4\rho_0 c_0} \tag{5.9}$$

The random-incidence (Sabine) absorption area $S\alpha$ of a room of volume V containing air at 20°C is related to the reverberation time T_{60} by (see Section 2.7.3)

$$S\alpha = \frac{55.3 V}{c_0 T_{60}} = \frac{0.161 V}{T_{60}} \tag{5.10}$$

so that the sound power is given by

$$W = \langle \overline{p^2} \rangle \frac{13.8 V}{\rho_0 c_0^2 T_{60}} \tag{5.11}$$

The spatially averaged mean-square pressure can be measured by averaging over a number of microphone positions or using a microphone which is rotated slowly on a boom. At high frequencies, air absorption becomes significant and $S\alpha$ in Equations 5.9 and 5.10 should be replaced by $S\alpha + 4m_0 V$, where m_0 is the air absorption coefficient.

The reverberation-room method is much quicker and simpler to carry out than the free-field method, but it does not give any information about the directivity of the source. Moreover, it faces difficulties with tonal noise unless rotating diffusers are employed.

5.3.4 Measurement methods based on sound intensity

Intensity methods employ the direct measurement of the component of sound intensity normal to a closed measurement surface. The Gauss integral theorem is exploited to suppress the influence of extraneous noise sources located outside the measurement surface. Note that this suppression only applies to the overall sound power estimate, not to the intensity at individual points. It also only applies to time-stationary sources and extraneous sources. The integral of the normal intensity over the surface gives the sound power of the sources within the surface, provided that there is little or no absorption present within the volume enclosed by the surface. The measurement surface is divided into segments and the normal intensity I_n is measured either at a fixed point at the centre of each segment (ISO 1993), or by scanning the intensity probe over each segment (ISO 1996b, 2002).

$$W \approx \sum_{i=1}^{n} I_{ni} S_i \quad \text{or} \quad W \approx \sum_{i=1}^{n} \langle I_n \rangle_i S_i \qquad (5.12)$$

The intensity probe measures the pressure and particle velocity simultaneously (see Section 8.7). In the case of probes based on two microphones, the pressure gradient, and hence the particle velocity, is obtained approximately from the difference between the sound pressures at the two microphone locations (Fahy 1995).

The intensity-based method allows measurements to be conducted in the near field of a source, in the presence of reflecting surfaces, and in proximity to other noise sources. The method is thus well suited to use *in situ*. The standards also specify a number of indicators that allow the reliability of the result to be quantified.

5.3.5 Source substitution method

In the source substitution method (ISO 2010g), a special-purpose reference source, calibrated by a standards laboratory in terms of its radiated sound-power spectrum, $L_{W,ref}$, is used. The actual source is switched off and is replaced by the substitution source; if the actual source is moveable it can be removed. Measurements have to be made in a reasonably reverberant environment containing no extraneous sources. The spatially averaged mean-square sound-pressure spectrum is measured only with the source under test in operation $\langle L_p \rangle$ and then only with the reference source in operation $\langle L_{p,\text{ref}} \rangle$. The sound-power spectrum of the actual source under investigation is derived from

$$L_W = L_{W,\text{ref}} + \langle L_p \rangle - \langle L_{p,\text{ref}} \rangle \tag{5.13}$$

5.3.6 Sound pressure standards

To quantify a sound source, it can sometimes be sufficient to measure a sound pressure level (or spectrum) at a standardised location. Sound pressure measurements are generally simpler than sound power determination. Examples include the following:

1. The EU Machinery Directive (EU 2006) requires a sound pressure level to be measured at the operator position for all machines, and additionally a sound power level to be measured where the sound pressure levels are higher than 85 dB(A), or for certain types of machine.
2. Automotive engine noise is often quantified by means of a measurement in a semi-anechoic test cell, with microphones placed at a distance of 1 m from each face of a rectangular fictitious surface enclosing the engine. As these positions are standardised, the results are not converted to sound power but are used in the form of an average sound pressure level over the six positions. However, the results can be affected by the acoustic treatment in the test cell.
3. Road-vehicle noise is quantified at standard positions at a distance of 7.5 m either side either side of the test track centre line and 1.2 m above the ground during a standard test (ISO 2007; EU 2014).
4. Environmental noise sources are often quantified in terms of sound-pressure level at a standard position. For example, railway noise is measured at positions 7.5 or 25 m from the track centre line (ISO 2013).

5.4 PRINCIPLES OF NOISE CONTROL

5.4.1 Noise-path models

Although sound is transmitted from a source to a receiver by many different paths, these can be principally grouped into airborne and structure-borne paths, both of which are operative in many practical cases.

- Airborne paths are those for which the source radiates sound into the surrounding air. The sound is then transmitted mainly through the air, although it may also pass through solid partitions (walls, floors, etc.) from the air on one side to that on the other.
- Structure-borne paths are those for which the source transmits mechanical vibration, for example through mounting points, into a receiving structure. This vibration is transmitted through the structure and eventually radiates (airborne) sound to the receiver location.

As a first step to systematic noise control, separate paths should be identified and, where possible, their contribution quantified. Once the relative importance of each path is known, all paths making a significant contribution should be treated. It is often helpful to construct a *noise-path model* in the form of a block diagram, an example of which is shown in Figure 5.7. Such a noise-path model can be used to give insight and can also be used as a basis for quantifying the contributions from various sources and paths. For this, methods such as transfer path analysis (van der Auweraer et al. 1995) and substitution source methods (Verheij 1997a, b; Janssens et al. 1999) can be used.

If noise control is sought via changes to transmission paths, it is important to note that when noise is reduced at one location it may be increased at another. The reduction of sound by direct control at source is, therefore, generally the most effective method, although it is often much more difficult to achieve.

5.4.2 Control at source

To suppress the noise generation at source presents a significant technical challenge. The physics of the complete noise-generating system(s) must be understood while modifications must not adversely affect the production costs or time, safety, the operational cost-effectiveness or the efficiency of the system. This requires effective collaboration between the noise control

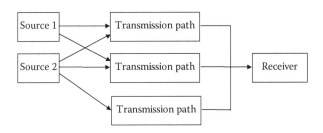

Figure 5.7 Example of noise transmission model with multiple sources and multiple paths.

engineers and those responsible for other aspects of the design; compromises often have to be found between differing requirements.

5.4.2.1 General principles

Some general principles for the control of noise sources are listed in ISO (1995) and ISO (2000). However, there is such a diversity of noise sources that a comprehensive review cannot be accommodated here.

Noise-generating mechanisms include

- Impacts: as well as direct sound from the impact, these may lead to vibration of the impacting bodies and subsequent sound radiation (see Section 5.4.2.2).
- Imbalance of rotating components: rotating parts should be balanced, accelerating masses minimised and steadiness of motion improved.
- Friction and self-excitation: frictional excitation is often caused by a difference between static and dynamic friction coefficients, leading to stick-slip excitation. Control measures include lubrication to minimise friction levels, selection of appropriate materials to reduce the difference in friction coefficients and damping treatments (see Section 5.10.2) to prevent self-excited vibration occurring at lightly damped resonances.
- Rolling components including bearings: it should be ensured that contacting surfaces are smooth, effective lubrication is applied, and tolerances are minimised to prevent rattling.
- Gear noise: noise is caused by time-varying forces transmitted by the gears. To mitigate this, the contact time between gear teeth should be increased, the number of teeth on gear wheels should not be too small, helical gears should be used where possible to smooth out the forces, the quality of tooth geometry should be improved and the rigidity of the housing should be sufficient.
- Turbulence: flow speeds should be minimised, pressure drops reduced, obstacles or sharp corners in the flow avoided, the inlet flows to fans should be as spatially uniform as possible and the tip speed of rotors in fans minimised.

Examples of the application of some of these principles are shown in Figure 5.8. Further discussion of various mechanical sources of noise is given by Brennan and Ferguson (2004).

5.4.2.2 Impacts

Impacts are particularly noisy and can excite high-frequency vibration. The noise produced can be substantially reduced by interposing a resilient

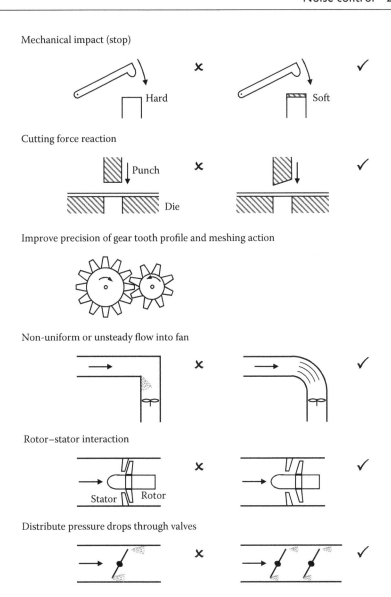

Figure 5.8 Examples of noise control at source.

element between the impacting bodies. The effect may be understood in terms of the change in the spectrum of the force, see Figure 5.9. The impulse (the time integral of the force, which equals the change of momentum) remains the same but, by lengthening the duration of the pulse, the peak force is reduced. Moreover, the mean-square force is reduced and hence also the average noise level. Finally and importantly, the frequency content

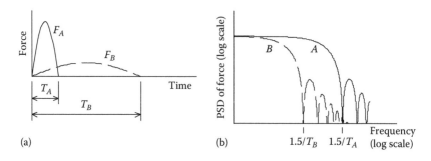

Figure 5.9 Illustration of increasing the duration of an impulsive force to reduce the frequency content of the power spectrum. (a) Force time history; (b) force spectrum.

is limited to frequencies below about $1.5/T$, where T is the pulse duration. This means that the A-weighted noise level may be reduced further as T is increased.

5.4.3 Control of airborne and structure-borne paths

Two distinct physical principles are used to control sound transmission by both airborne and structure-borne paths: these are the principles of *reflection* and *dissipation*.

To control airborne transmission, the principle of reflection is used in partitions, enclosures and barriers, as described in Sections 5.6 to 5.8. This reduces the sound energy reaching the receiver, but this may be at the expense of increasing it elsewhere. Dissipation is used to convert sound energy to heat through absorptive treatments (see Section 5.9). Many effective noise control systems use a combination of these two principles; for example, enclosures with absorptive linings, barriers with an absorptive face and double-skin partitions with an absorptive treatment between the two skins.

Structure-borne transmission can similarly be controlled, either by reflection, through impedance mismatching or vibration isolation, or by dissipation through damping treatments. These techniques are described in more detail in Section 5.10. In addition, single-frequency problems can often be solved by detuning resonances or applying dynamic neutralisers (see Brennan and Ferguson 2004). It is difficult and costly to suppress structure-borne transmission in large structures such as buildings or ships, because load-bearing structures cannot easily accommodate resilient inserts. Moreover, there are often many parallel transmission paths to be taken into account. Therefore, it is important to suppress structure-borne sound transmission as close as possible to the source or the receiver.

Table 5.6 Summary of noise control principles

	Structure-borne sources	Airborne sources
At source	Reduce forces or velocities	Reduce forces or velocities
	Out of balance	Radiating area
		Radiation ratio
Reflection	**Isolation**	**Insulation**
	Impedance mismatch	Partitions
	Resilient mounting	Double skins/double glazing
	Reduced radiation ratio	Enclosures
		Barriers
Dissipation	**Damping**	**Absorption**
	Constrained layer	Absorptive linings
	Unconstrained layer	Seats, trim etc.
	Tuned absorbers	
Discrete frequencies	Detuning	

Table 5.6 summarises the various principles that can be used to control noise. The remainder of the chapter discusses the practical implementation of each of these noise control principles.

5.5 SOUND RADIATION FROM VIBRATING STRUCTURES

5.5.1 Radiation ratio

As has been noted in Section 5.2.3, many sound sources take the form of vibrating solid objects that generate sound by exciting the air around them into vibration. This section introduces an engineering method for estimating the sound radiation from vibrating structures.

Consider first the idealised case of a plane surface generating one-dimensional plane waves propagating in the x-direction, as shown in Figure 5.10. This may be a vibrating piston radiating sound into a tube or an infinite flat surface vibrating in phase. For harmonic motion at a circular frequency ω, it is assumed that the surface is vibrating with a normal velocity, $v(t) = v_0 \cos(\omega t)$. As described in Section 2.3.3, this generates plane propagating waves with particle velocity:

$$u(x,t) = v_0 \cos(\omega t - kx) \tag{5.14}$$

where $k = \omega/c_0$ is the acoustic wavenumber (see Section 2.2.7). For plane progressive waves, the time-averaged intensity is related to the mean-square pressure $\overline{p^2}$ or the mean-square particle velocity $\overline{u^2}$ by the acoustic impedance $\rho_0 c_0$ (see Section 2.3.3):

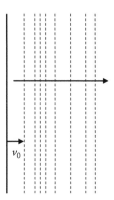

Figure 5.10 Vibrating surface producing plane waves in the fluid. Dashed lines indicate instantaneous fluid displacement.

$$I = \frac{\overline{p^2}}{\rho_0 c_0} = \rho_0 c_0 \overline{u^2} \tag{5.15}$$

Hence, the radiated power per unit area is given by

$$\frac{W}{S} = I = \rho_0 c_0 \overline{v^2} \tag{5.16}$$

where $\overline{v^2}$ $(= v_0^2/2)$ is the mean-square velocity of the vibrating surface.

The vibration of most mechanical systems exhibits complex frequency-dependent spatial distributions of amplitude and phase. Nevertheless, it is useful to compare the sound radiation from actual structures with this simple case. We can write the radiated sound power of an arbitrary structure as

$$W_{\text{rad}} = \rho_0 c_0 S \langle \overline{v^2} \rangle \sigma \tag{5.17}$$

where:

- S is the total surface area
- $\rho_0 c_0$ is the acoustic impedance of air
- $\langle \overline{v^2} \rangle$ is the surface-averaged mean-square normal velocity
- σ is called the radiation ratio or radiation efficiency

W_{rad}, σ and $\langle \overline{v^2} \rangle$ are all functions of frequency; Equation 5.17 is typically expressed in one-third octave bands. This can be rearranged to give

$$\sigma = \frac{W_{\text{rad}}}{\rho_0 c_0 S \langle \overline{v^2} \rangle} \tag{5.18}$$

which is the definition of σ: the ratio of the radiated sound power to that which would be produced by a flat surface vibrating in phase with the same average velocity and producing plane waves. The radiation ratio is not an efficiency in the usual sense and may have values greater than unity. Generally, σ is much smaller than unity at low frequencies, where the dimensions of the vibrating object are small compared with the acoustic wavelength, and close to unity at high frequencies, where the size of the object becomes large compared with the acoustic wavelength. In decibel form, $L_\sigma = 10 \log_{10} \sigma$ is called the radiation index.

5.5.2 Radiation from simple sources

The sound power radiated by a *pulsating* sphere of radius a is described in Section 2.4.4 (see Equation 2.107) and may be expressed in the form of a radiation ratio, giving

$$\sigma = \frac{(ka)^2}{1+(ka)^2} \tag{5.19}$$

where $k = \omega/c_0$ is the acoustic wavenumber. At low frequencies ($ka \ll 1$), the frequency dependence is given by $\sigma \approx (ka)^2 \propto f^2$, whereas at high frequencies ($ka \gg 1$), σ tends to unity. Similarly, the radiation ratio of an *oscillating* rigid sphere is given by

$$\sigma = \frac{(ka)^4}{4+(ka)^4} \tag{5.20}$$

At low frequencies, the frequency dependence is given by $\sigma \approx (ka)^4/4 \propto f^4$ but, again, at high frequencies σ tends to unity. These two results are plotted in Figure 5.11 for the example of a sphere of radius 0.1 m. The transition frequency, at which $ka = 1$, occurs at 550 Hz in this example. These results can also be used more generally for vibrating objects which vibrate as rigid bodies, not only for spheres.

Note that if the sphere is oscillating with r.m.s. amplitude v_{rms}, the surface-averaged mean-square normal velocity, used in the derivation of Equation 5.20, is $v_{\mathrm{rms}}^2/3$. In some texts, a different normalisation is used to allow for this factor.

5.5.3 Rayleigh integral

The sound radiation from more complex vibrating structures can be determined using numerical techniques. The simplest of these is the Rayleigh integral (Rayleigh 1896), which can be used to calculate the

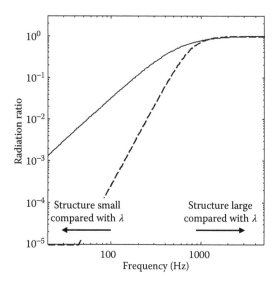

Figure 5.11 Radiation ratio of a sphere of radius 0.1 m. (—): pulsating sphere; (– – –): oscillating sphere. Transition frequency $ka=1$ corresponds to 550 Hz.

radiated sound field in the idealised case due to a planar source located in an infinite flat rigid surface (baffle) (see Section 2.5.4). Despite its simplifying assumptions, the Rayleigh integral is widely used. For example, Figure 5.12 (from Xie et al. 2005) shows the radiation ratio of vibration in each of the modes of a 3 mm thick aluminium baffled rectangular plate of dimensions 0.6×0.5 m and the overall result for arbitrary (spatially averaged) forcing. In the region between about 100 Hz and 4 kHz (for this plate geometry), the average radiation ratio is well below unity, rising to a peak at about 4 kHz. The significance of this will be explained in the following section.

Numerical methods, such as acoustic finite elements and boundary elements, can also be used to predict sound radiation for more complex geometries, although they are usually limited to relatively low frequencies (see Petyt and Jones 2004).

5.5.4 Bending waves and critical frequency

Although stiff or compact sources can often be represented by the simple sources considered in Section 5.5.2, many practical structures consist of plates and beams which vibrate in flexure (or bending). Flexure is governed by a fourth-order differential equation (see Section 3.7.2), which leads to the following equation for the bending wavenumber for a plate:

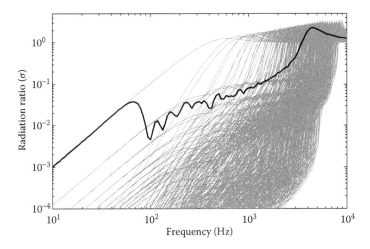

Figure 5.12 Radiation ratio of a rectangular simply supported 3 mm thick aluminium plate 0.6 × 0.5 m mounted in a rigid baffle. Thin lines: result for each mode; thick line: overall result for a point force. (Reprinted from Xie, G., Thompson, D.J. and Jones, C.J.C., *Journal of Sound and Vibration* 280, 181–209, 2005. With permission.)

$$k_B = \left(\frac{\mu\omega^2}{D}\right)^{1/4} = \omega^{1/2}\left(\frac{12(1-\nu^2)\rho_m}{Eh^2}\right)^{1/4} \quad (5.21)$$

where:
$D = Eh^3/(12(1-\nu^2))$ is the bending stiffness per unit width
E is Young's modulus
ν is Poisson's ratio
$\mu = \rho_m h$ is the mass per unit area
ρ_m is the material density
h is the thickness

The bending wavenumber can be written as $k_B = 2\pi/\lambda$, with λ the bending wavelength; its dependence on ω is known as the dispersion relation. This wavenumber governs the behaviour of waves propagating in an unbounded plate or, equivalently, the modes of vibration of a finite plate.

Bending waves are dispersive, since their wavespeed, $c_B = \omega/k_B$, depends on their frequency. For acoustic waves in a fluid (e.g. air), $k = \omega/c_0$, where c_0, the speed of sound, is independent of frequency and these waves are non-dispersive. The curves of k_B and k cross at a frequency known as the critical frequency, as shown in Figure 5.13. For bending waves, the wavenumber and wavelength also depend on the plate thickness, as indicated in

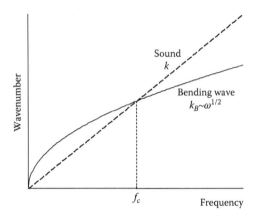

Figure 5.13 Dispersion curves for acoustic waves and bending waves. The frequency at which they cross is the critical frequency, f_c.

Figure 5.14, which shows these curves in the form of wavelength plotted against frequency f (in Hz) on a logarithmic scale for various thicknesses of the plate. To find the critical frequency $f_c = \omega_c/2\pi$, the bending wavenumber can be equated to the wavenumber in air:

$$k = \frac{\omega_c}{c_0} = \left(\frac{\mu \omega_c^2}{D}\right)^{1/4} = k_B \tag{5.22}$$

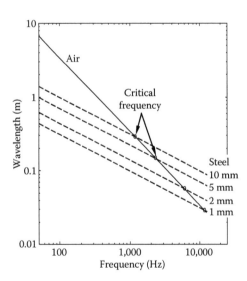

Figure 5.14 Wavelengths of acoustic waves in air and bending waves in steel plates of different thicknesses.

For a homogeneous panel of thickness h this leads to

$$f_c = \frac{\omega_c}{2\pi} = \frac{c_0^2}{2\pi h}\sqrt{\frac{12\rho_m(1-\nu^2)}{E}} \qquad (5.23)$$

The critical frequency is important both for the radiation of sound from panels and the transmission of sound through panels (see Section 5.6). Table 5.7 lists some typical material properties and the corresponding critical frequency f_c. For example, for a steel plate of thickness 1 mm, $f_c = (12.4/0.001)$ Hz = 12.4 kHz. Note that the product hf_c is similar for steel, aluminium and glass, as the ratio E/ρ_m is similar in each case.

5.5.5 Radiation from a plate in bending: engineering approach

In this section, asymptotic formulae for the frequency band average radiation ratio of a plate set in an infinite baffle are presented, based on results originally derived by Maidanik (1962). It is based on the assumption that there are many resonant modes with natural frequencies within a frequency band.

Figure 5.15 shows the radiation ratio of a baffled plate schematically. The frequency range can be divided into three main regions. At high frequencies, above the critical frequency f_c, the radiation ratio tends to unity with a peak occurring at f_c. Below the plate's fundamental natural frequency $f_{1,1}$, the radiation is similar to that from a monopole. Between these two regions, *acoustic short-circuiting* occurs, leading to a reduction in the radiation compared with a rigid vibrating surface of the same area. These three frequency regions are considered separately in the next sections.

Table 5.7 Properties of different materials and corresponding critical frequency at 20°C in air ($c_0 = 343$ m s^{-1})

Material	Young's modulus E (N m^{-2})	Poisson's ratio ν	Density ρ_m (kg m^{-3})	hf_c m s^{-1}
Steel	2.0×10^{11}	0.28	7800	12.4
Aluminium	7.1×10^{10}	0.33	2700	12.0
Plastic	1.5×10^{10}	0.35	1410	19.1
Magnesium	4.5×10^{10}	0.30	1800	13.5
Copper	1.25×10^{11}	0.35	8900	16.3
Glass	6.0×10^{10}	0.24	2400	12.7
Perspex	5.6×10^{9}	0.4	1200	27.7

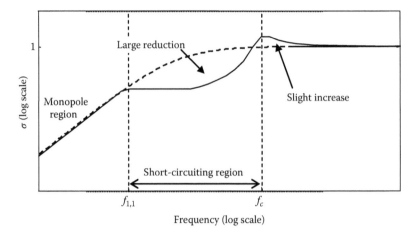

Figure 5.15 Schematic diagram of radiation ratio of a finite plate. Acoustic short-circuiting occurs between first plate natural frequency $f_{1,1}$ and critical frequency f_c.

5.5.5.1 High frequencies

Considering first the high-frequency region, the bending wavelength λ_B is larger than the acoustic wavelength λ (Figure 5.14), so that some angle φ exists such that

$$\lambda = \lambda_B \sin\varphi \quad \text{or} \quad k_B = k \sin\varphi \quad \text{for } f \geq f_c \tag{5.24}$$

As shown in Figure 5.16, a wave in the plate radiates sound at this angle φ to the plate normal. This matching of the wavenumbers of the plate and the fluid is known as *coincidence* and occurs at some angle for any frequency above the critical frequency. At the critical frequency, $\sin\varphi = 1$, and sound is radiated parallel to the plate. Below the critical frequency, no such angle φ exists, as $k_B > k$.

The radiation ratio for a bending wave in this high-frequency region is given by Maidanik (1962)

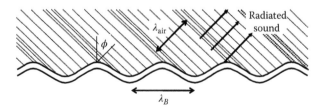

Figure 5.16 Illustration of coincidence at angle φ to the normal.

$$\sigma = \frac{1}{\left(1 - f_c/f\right)^{1/2}} \quad \text{for } f \geq f_c \qquad (5.25)$$

This gives a result greater than unity. In practice, the value of σ is limited in the region around f_c by effects of the finite size of the plate and by fluid loading. A limiting value is given by Vér and Beranek (2005)

$$\sigma = 0.45 \sqrt{\frac{P f_c}{c_0}} \left(\frac{b}{a}\right)^{1/4} \qquad (5.26)$$

where $P = 2(a+b)$ is the perimeter length and $b < a$. This formula gives lower results than those originally given by Maidanik (1962) and gives similar results to the more complex formulae of Leppington et al. (1982).

5.5.5.2 Low frequencies

Below the fundamental natural frequency of the plate, the whole surface moves in phase and the motion is predominantly that of the fundamental mode shape. The fundamental natural frequency of a simply supported rectangular plate of dimensions $a \times b$ occurs at

$$f_{1,1} = \frac{\pi}{2} \left(\frac{1}{a^2} + \frac{1}{b^2}\right) \sqrt{\frac{D}{\mu}} \qquad (5.27)$$

Assuming that the plate is set in an infinite baffle, the plate motion radiates sound as a monopole. This has a radiation ratio of (Wallace 1972; see also Xie et al. 2005)

$$\sigma \approx \frac{4 f^2 S}{c_0^2} \quad \text{for } f < f_{1,1} \qquad (5.28)$$

where S is its surface area. Note that the factor 4 is slightly different from the case of a pulsating hemisphere or a circular piston in a baffle, where the factor is 2π, due to the distribution of vibration in the fundamental mode of the plate.

5.5.5.3 Acoustic short-circuiting region

For a free bending wave in an infinite homogeneous plate, there is no sound radiation below the critical frequency. According to Equation 5.25, the radiated power is imaginary; an imaginary value of radiated power corresponds to a situation in which the pressure and particle velocity are in quadrature and can be interpreted in terms of a nonradiating near field.

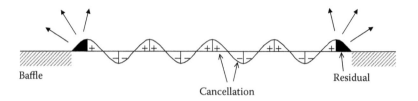

Figure 5.17 Acoustic short-circuiting below the critical frequency. Net sound radiation occurs only from the corners (and edges).

The fact that there is no sound radiation below the critical frequency can be explained in terms of acoustic short-circuiting between the radiation from the maxima and minima of the vibration pattern, which are close together compared with the acoustic wavelength. For a finite plate, however, the edges and corners disturb this cancellation, resulting in a net radiation of sound, as indicated in Figure 5.17. Non-uniform mass or stiffness distribution also reduces the cancellation and increases the radiation ratio.

For a simply supported plate vibrating in a natural mode (m, n), the components of the wavenumber vector in the x and y directions are $k_x = m\pi/a$, $k_y = n\pi/b$, which satisfy $k_B^2 = k_x^2 + k_y^2$ at the natural frequency. Below the critical frequency, where $k_B > k$, it is possible to distinguish two cases:

1. At lower frequencies, $k_B \gg k$ and both trace wavenumbers k_x and k_y are greater than k for all modes. This is called the *corner-mode* region. For each mode in this region, acoustic short-circuiting occurs over the whole area of the plate except at the corners. There is then only a small net sound radiation, leading to a radiation ratio of (Beranek 1971)*

$$\sigma = \frac{4\pi^2}{c_0^2 S} \frac{D}{\mu} \tag{5.29}$$

2. At intermediate frequencies, it is still the case that $k_B > k$, but *some* modes exist with either k_x or k_y smaller than k. This is called the *edge-mode* region. For these edge modes, net radiation occurs along two opposite edges. The average radiation efficiency in this frequency region can be expressed as (Maidanik 1962)

$$\sigma = \frac{Pc_0}{4\pi^2 S f_c} \times \frac{\left(1-\beta^2\right)\ln\left(\dfrac{1+\beta}{1-\beta}\right) + 2\beta}{\left(1-\beta^2\right)^{3/2}} \tag{5.30}$$

where $\beta = \sqrt{f/f_c}$ and $P = 2(a+b)$ is the perimeter length.

---

* The corner mode region is not given separately by Maidanik (1962) or Leppington et al. (1982). This formula is given in a design chart in Beranek (1971), where it is stated that it will provide an overestimation in this region.

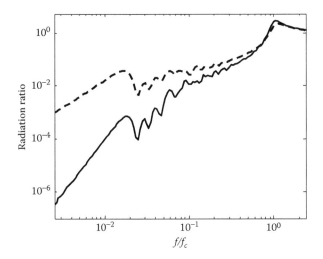

Figure 5.18 Radiation ratio of 3 mm thick aluminium rectangular plate 0.65 × 0.5 m. (—): unbaffled; (– – –): in rigid baffle. (Reprinted from Putra, A. and Thompson, D.J., *Applied Acoustics* 71, 1113–1125, 2010. With permission.)

Comparing the result of this model in Figure 5.15 with the result for an actual plate geometry, shown in Figure 5.12, similar trends can be seen, although the results differ in detail, especially in the corner-mode region where the number of modes in a frequency band is low and where the plate is no longer large compared with the wavelength.

For an unbaffled source, this formulation does not hold. The Rayleigh integral is no longer useful, as neither the pressure nor the normal velocity outside the boundaries of the plate are known *a priori*. Several calculation methods have been proposed for an unbaffled plate. For example, Figure 5.18 shows results from Putra and Thompson (2010a) using the model of Laulagnet (1998) for a rectangular plate. Compared with the baffled case, the radiation is much reduced below the fundamental natural frequency, where it has a dipole instead of monopole nature, and also in the corner-mode region. However, at high frequencies the results are similar.

5.5.6 Reducing sound radiation

It can be seen from Equation 5.17 that the sound radiated from a structure can be reduced by reducing the vibration amplitude, the surface area or the radiation ratio. In fact, reducing the radiating area usually also leads to a reduction in the radiation ratio at some frequencies. To illustrate this, Figure 5.19 shows the radiated sound power for an oscillating sphere when the radius is halved from 0.1 to 0.05 m for a constant r.m.s. vibration

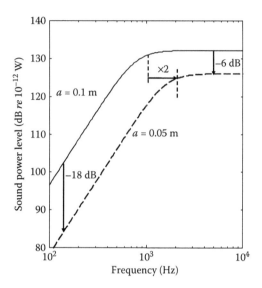

Figure 5.19 Radiated sound power from an oscillating sphere with r.m.s. velocity 1 m s^{-1}. (—): radius 0.1 m; (— — —): radius 0.05 m.

amplitude of 1 ms^{-1}. Similar effects can be expected for oscillating bodies of other shapes.

The radiation ratio for an oscillating sphere is given by Equation 5.20, while the vibrating surface area $S = 4\pi a^2$. By reducing the radius a, both S and σ are affected. At high frequencies, $\sigma \to 1$, so the reduction in sound power is due to S alone, and is proportional to a^2; thus, a halving of the radius leads to a 6 dB reduction. At low frequencies, however, $\sigma \approx (ka)^4/4$, so the overall result is proportional to a^6, giving a reduction of 18 dB for a halving of the radius.

The situation for thin-plate structures is more complicated. As discussed in the previous section, acoustic short-circuiting occurs in the frequency region between the fundamental natural frequency and the critical frequency. Changing the thickness of the plate will affect both of these limits – a thinner plate will have a lower fundamental natural frequency and a higher critical frequency (see Equations 5.23 and 5.27). Consequently, the radiation ratio will be reduced in the acoustic short-circuiting region. This is illustrated in Figure 5.20 with one-third octave band results for plates of different thicknesses. It should be remembered, however, that changing the thickness of a plate may also affect its vibration amplitude (see Section 5.10).

Changing the area of a plate will not affect its critical frequency, but it will affect the fundamental natural frequency. As shown in Figure 5.21, a smaller plate will have a smaller radiation ratio at very low frequencies and a higher radiation ratio in the short-circuiting region. In the corner-mode

Noise control 249

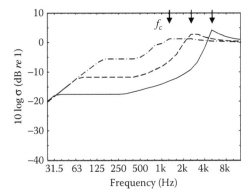

Figure 5.20 Effect of plate thickness on radiation ratios of rectangular plates 1 m×0.3 m. (—): 2.5 mm; (– – –): 5 mm; (·—·—·—·): 10 mm. The critical frequency is indicated by the arrow in each case.

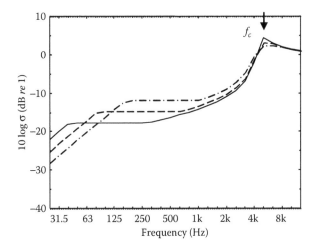

Figure 5.21 Effect of plate dimensions on radiation ratios of 2.5 mm thick rectangular plates. (—): 1 m×0.3 m; (– – –): 0.5 m×0.3 m; (·—·—·—·): 0.25 m×0.3 m. The critical frequency is indicated by the arrow.

region, $\sigma \propto 1/S$ (Equation 5.29), so the radiated power, which is proportional to $S\sigma$, is independent of the plate dimensions.

Adding stiffening beams to a plate structure has the effect of dividing the plate field into a number of separate regions. At higher frequencies, where the bending wavelength on the plate is small compared with the separation between the stiffeners, the radiation ratio can be approximated in terms of that of the individual plate regions. Thus, the addition of stiffeners increases the radiation ratio in the short-circuiting region as the plate regions become

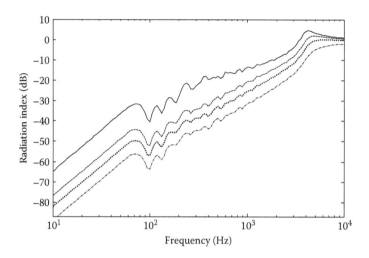

Figure 5.22 Radiation index for a rectangular 3 mm thick aluminium plate, 0.65 × 0.5 m, with 10 mm diameter holes. (—): solid plate; (– – –): 10% perforated area; (·····): 20% perforated area; (·–·–·–·): 40% perforated area. (Reprinted from Putra, A. and Thompson, D.J., *Journal of Sound and Vibration* 329, 4227–4250, 2010. With permission.)

smaller. At lower frequencies, the stiffeners increase the bending stiffness of the stiffened plate, leading to a similar effect to an increase in thickness, again giving an increase in σ. Thus, to minimise noise radiation, it is actually preferable to remove stiffeners from a plate, provided that the vibration amplitude is not thereby increased (see Section 5.10).

Finally, if it is practicable, the radiation ratio can be considerably reduced by the use of perforated structures. Figure 5.22 shows the radiation ratio predicted for a rectangular 3 mm thick aluminium plate with 10 mm diameter holes (Putra and Thompson 2010b). The curves show results for different degrees of perforation, which indicate a 10–20 dB reduction at most frequencies. At high frequencies, particularly where the wavelength becomes short compared with the hole separation, the effect will be negligible. Again, a perforated structure may vibrate more than a solid one but the net reduction in sound radiation is still likely to be substantial.

5.6 SOUND TRANSMISSION THROUGH PARTITIONS

5.6.1 Introduction

The problem of airborne sound transmission through partitions has been studied extensively for buildings – walls, windows, floors and ceilings

Noise control 251

(Hopkins 2008). More recently, the commercial demand to reduce noise inside vehicles, such as cars, aircraft and trains, as well as the need to reduce noise inside factories and that produced by industrial plant, has led to studies of a wide range of geometries and materials.

In practice, partitions do not have simple geometry and construction: they may be inhomogeneous and non-uniform; they may comprise combinations of different materials and have indeterminate boundary conditions; they are also often subject to complex sound fields generated in rooms, enclosures or ventilation ducts. Surprisingly, in many cases the sound transmission can be described quite well by a relatively simple model which ignores the details of the boundaries or the incident sound fields because forced wave transmission is dominant below the critical frequency.

In this section, a mathematical model of sound transmission through a single-leaf, uniform, homogeneous partition is introduced. The importance of the coincidence phenomenon, introduced in Section 5.5.3, for the sound transmission is then explained. The implications of multiple-layer partitions are described and finally the methods of measuring sound transmission are briefly covered.

5.6.2 Sound transmission under normal incidence

Consider an infinite, plane panel which is thin and uniform, characterised by its mass per unit area μ. Fluid is assumed to be present on both sides with density ρ_0 and sound speed c_0. The model may be extended for the case of different fluids on each of the sides (e.g. for application to marine structures), but this is not covered here (see Fahy 2000; Fahy and Gardonio 2006). Harmonic motion at circular frequency ω is assumed.

As indicated in Figure 5.23, a plane wave with complex pressure amplitude A is assumed to be incident in the normal direction from the left-hand side. The complex pressure amplitude is given by

$$p_i(x) = A\,e^{-ikx} \qquad (5.31)$$

where $k = \omega/c_0$ is the acoustic wavenumber. The panel vibrates with velocity amplitude v_0. A reflected wave of amplitude B and a transmitted wave of amplitude C are generated:

$$p_r(x) = B\,e^{ikx}, \quad p_t(x) = C\,e^{-ikx} \qquad (5.32)$$

Taking A as known, three equations are required to find the three unknowns B, C and v_0, of which C is required to find the sound transmission. The first two of these equations come from the requirement

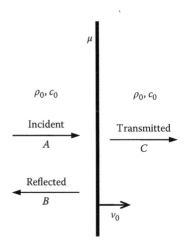

Figure 5.23 Sound transmission through a partition of mass per unit area μ.

that the particle velocity in the fluid at the panel boundary must be equal to the panel velocity. The particle velocity in a plane wave in the direction of propagation is given by the pressure divided by the impedance, $\rho_0 c_0$:

$$\frac{A-B}{\rho_0 c_0} = v_0 = \frac{C}{\rho_0 c_0} \qquad (5.33)$$

where the reflected wave contains a minus sign due to its opposite direction of propagation (see Section 2.2.8). The third equation comes from the equation of motion of the panel, where the pressure difference causes acceleration of the mass per unit area of the panel (for harmonic motion the acceleration amplitude is $i\omega v_0$):

$$i\omega \mu v_0 = A + B - C \qquad (5.34)$$

Eliminating B and v_0 leads to

$$\frac{C}{A} = \frac{1}{1 + \dfrac{i\omega \mu}{2\rho_0 c_0}} \qquad (5.35)$$

We now introduce the sound power transmission coefficient, which is defined as the ratio of the transmitted sound power to the incident power. From Equation 5.35, as the fluid is the same on both sides, this is found as

$$\tau = \left|\frac{C}{A}\right|^2 = \frac{1}{1+\left(\dfrac{\omega\mu}{2\rho_0 c_0}\right)^2} \tag{5.36}$$

It should be noted that this contains the ratio of the impedances of the panel and the fluid. For typical partitions in air, this ratio, $\omega\mu/\rho_0 c_0$, known as the fluid-loading parameter, is usually much greater than unity, so that Equation 5.36 can be approximated to

$$\tau \approx \left(\frac{2\rho_0 c_0}{\omega\mu}\right)^2 \tag{5.37}$$

This simplification is not valid at very low frequencies or for very light-weight partitions such as plastic film sheets.

It is conventional to use a decibel form, known as the sound reduction index (or transmission loss), given by $R = 10\log_{10}(1/\tau)$. From the approximate expression in Equation 5.37, this can be further simplified, for the case of panels in air, to

$$R(0) \approx 20\log_{10}(f\mu) - 42 \tag{5.38}$$

in which (0) is used to signify normal incidence and f is frequency in hertz. This equation expresses the normal incidence *mass law*. Note that $R(0)$ increases by 6 dB for a doubling of either the mass or the frequency.

At low frequencies, at and below the fundamental bending natural frequency of a finite panel, this formulation can be modified by replacing the panel impedance $i\omega\mu$ by $i\omega\mu - is(1+i\eta)/\omega$, where $s = \omega_{1,1}^2\mu$ is the stiffness per unit area associated with the fundamental mode, $\omega_{1,1} = 2\pi f_{1,1}$ is the natural frequency of this mode and η is the damping loss factor associated with the mode (see Section 3.3.4). This leads to

$$\tau = \frac{1}{\left(1+\dfrac{s\eta/\omega}{2\rho_0 c_0}\right)^2 + \dfrac{1}{4}\left(\dfrac{\omega\mu}{\rho_0 c_0} - \dfrac{s}{\omega \rho_0 c_0}\right)^2} \tag{5.39}$$

For a simply supported plate, the fundamental panel natural frequency is given by Equation 5.27. Figure 5.24 (solid line) shows typical predicted behaviour for a 2 mm thick steel plate with dimensions 0.5×0.4 m. At the panel resonance, the sound reduction index has a minimum which is controlled by the panel damping. Below this frequency, the sound-reduction

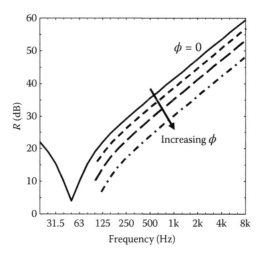

Figure 5.24 Sound reduction index of a 2 mm thick steel plate with dimensions 0.5×0.4 m at various angles of incidence.

index is *stiffness controlled* and increases towards lower frequencies with a frequency dependence of −20 dB per decade.

5.6.3 Oblique and diffuse incidence

The case of oblique incidence is much more complicated than that of normal incidence, because the incident pressure induces bending in the partition. Consequently, the bending stiffness and damping of the panel have an influence. For more details, see Fahy (2000) and Fahy and Gardonio (2006).

Consider a plane wave incident at an angle φ to the normal on an infinite, flexible partition, as shown in Figure 5.25a. The incident pressure is given by

$$p_i(x,y) = A e^{-ik_x x - i k_y y} \tag{5.40}$$

where k_x and k_y are the components of the acoustic wavenumber vector (see Section 2.6.3) in the x and y directions, with x representing the direction normal to the panel. These satisfy

$$k_x^2 + k_y^2 = \frac{\omega^2}{c_0^2} = k^2 \tag{5.41}$$

The propagation direction φ is given by $\tan \varphi = k_y/k_x$. Equivalently, we can write $k_x = k \cos \varphi$ and $k_y = k \sin \varphi$. As shown in Figure 5.25b, the

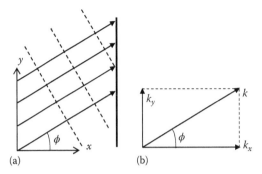

Figure 5.25 (a) Incident wavefronts (dashed lines) propagating in the direction φ to the normal; (b) wavenumber components in x and y directions.

component k_x acts in the direction normal to the partition while k_y is the trace wavenumber in the direction parallel to it. The incident acoustic field forces the panel to vibrate with this wavenumber k_y.

At a given frequency, free-bending waves (or modes) in a plate can be characterised by their bending wavenumber, as given by Equation 5.21. Figure 5.26 shows the bending wavenumber k_B and the acoustic trace wavenumber $k_y = k \sin \varphi$ schematically for various angles of incidence φ. As seen in Section 5.5, below the critical frequency $k_B > k$, while above the critical frequency $k_B < k$.

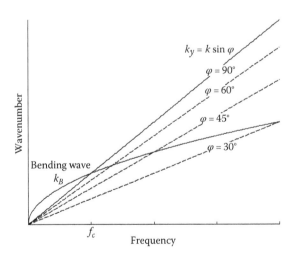

Figure 5.26 Comparison of bending wavenumber with acoustic trace wavenumber k_y for different angles of incidence φ. The critical frequency f_c is the lowest coincidence frequency.

5.6.3.1 Below the critical frequency

As seen in Section 5.5.4, below the critical frequency there is no angle φ that satisfies $k_B = k \sin \varphi = k_y$. Hence, for all angles of incidence, the partition is forced to vibrate in bending with the wavenumber k_y, which is different from the free-bending wavenumber k_B. Consequently, the sound reduction index is *mass controlled*, similar to that found for normal incidence, and is given by

$$R(\varphi) \approx 20 \log_{10} \left(\frac{\omega \mu \cos \varphi}{2 \rho_0 c_0} \right) = R(0) + 20 \log_{10}(\cos \varphi) \leq R(0) \quad (5.42)$$

This is shown in Figure 5.24 for an example case.

In many practical cases, for example in buildings, the incident sound field can be approximated by an ideal *diffuse field* in which plane waves are assumed to propagate in all directions with equal probability (see Section 2.7 and Section 5.3.3). Integrating this expression over all angles of incidence (provided that $R(0) \gg 0$) leads to

$$R_d = R(0) - 20 \log_{10}(0.23 R(0)) \quad (5.43)$$

In practice, it is found that better results are obtained if the angles of incidence close to grazing are excluded. An empirical formula, corresponding to a limit of 78°, is

$$R_f = R(0) - 5 \quad (5.44)$$

This is termed the *field incidence sound reduction index*. Note, however, that the limit angle of 78° has no theoretical justification.

5.6.3.2 Above the critical frequency

As shown in Figure 5.26, above the critical frequency $k_B < k$, and there is an angle φ for which $k_B = k_y$. At this angle of incidence, free waves can be excited in the panel. Thus, for all frequencies above f_c, coincidence can occur for some angle $\varphi = \varphi_c$. Similarly, for a given angle of incidence, there is a coincidence frequency at which the sound transmission index has a minimum (the transmitted power has a maximum). The depth of this minimum is controlled by the panel damping. The critical frequency is the lowest possible coincidence frequency. The sound transmission at various angles of incidence is illustrated in Figure 5.27.

For diffuse incidence at any frequency above the critical frequency, the component incident at the coincident angle φ_c will be transmitted most

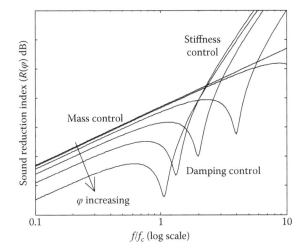

Figure 5.27 Variation of sound reduction index with frequency for various angles of incidence (0° to 75° in 15° steps).

strongly. This transmission is *damping controlled*. For smaller angles the transmission will be mass controlled whereas at larger angles it will be stiffness controlled.

The average sound reduction index in the frequency region above the critical frequency is given approximately by Cremer (1942):

$$R_d \approx R(0) + 10\log_{10}\left(\frac{f}{f_c}\right) + 10\log_{10}\eta - 2 \qquad (5.45)$$

where η is the damping loss factor of the panel. This relation is shown in Figure 5.28. It has a frequency dependence of 30 dB per decade (9 dB per octave). Although it is designated 'damping controlled', it is, of course, still influenced by the panel mass through $R(0)$ and both the mass and the bending stiffness through f_c. Between $f_c/2$ and f_c, it is possible to use a linear interpolation against $\log_{10} f$ as shown in Figure 5.28. More detailed formulations for this region are found, for example, in Bies and Hansen (2009). Drawing these results together, Figure 5.29 shows the general form of the sound reduction index of a single panel as a function of frequency.

5.6.3.3 Bounded partitions

The fact that real partitions are finite modifies the results below the critical frequency and can lead to an increase in the low-frequency values of R (see Fahy and Gardonio 2006). This effect was quantified by Sewell (1970),

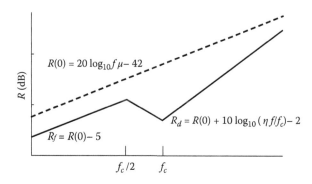

Figure 5.28 Theoretical diffuse-field sound reduction index below and above critical frequency.

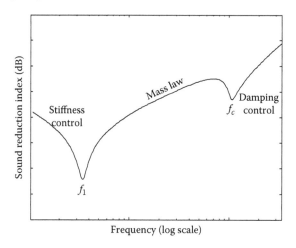

Figure 5.29 General form of diffuse field sound reduction index.

who gave the sound reduction index of a panel of area S below the critical frequency as

$$R \approx R(0) - 10\log_{10}\left(\ln(kS^{1/2})\right) + 20\log_{10}\left(1 - (f/f_c)^2\right) \quad (5.46)$$

where k is the acoustic wavenumber. More recent models for this effect include Villot et al. (2001).

5.6.4 Double-leaf partitions

As has been seen, the panel mass is the only parameter affecting the sound reduction of a single-leaf partition for much of the frequency region. At a given frequency, the sound reduction index of a panel increases by 6 dB per doubling of mass, provided that no significant flanking transmission or coincidence-controlled transmission occurs. In practice, however, it is often necessary for structures to have low weight and yet to provide high sound reduction. Examples include aircraft fuselages, automotive structures, partition walls in tall buildings and moveable walls in studios. Since increasing the mass is neither desirable nor particularly effective in reducing the sound transmission, a common solution is to employ constructions comprising two separate leaves separated by an air space or cavity.

Ideally one might expect the sound reduction index to be the sum of those of the two partitions. In practice, however, the limited width of the cavity means that the two partitions are dynamically coupled by the air between them, with the result that the sound reduction index of the combination may fall below this ideal, sometimes by a large amount. In fact, it can actually become worse at some frequencies.

Theoretical analysis of the sound transmission behaviour of double-leaf partitions is less well developed than for the single-leaf case. In particular, it is difficult to include the effects of mechanical connections between leaves in mathematical models.

5.6.4.1 Theoretical performance

Consider two plane partitions separated by an air cavity of thickness d, as shown in Figure 5.30. At low frequencies, the air cavity can be represented as a spring connecting the two panels. The stiffness per unit area of this air cavity at low frequencies is

$$s_a = \frac{\rho_0 c_0^2}{d} \tag{5.47}$$

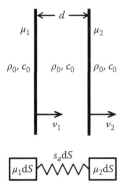

Figure 5.30 Double-leaf partition represented as mass-spring-mass system.

For normal incidence, neglecting the pressure at the receiver side, the *specific transfer impedance* (i.e. the velocity response at Side 2 to a pressure at Side 1) of this mass-spring-mass system can be determined as

$$z = \frac{p}{v_2} = i\omega(\mu_1 + \mu_2) - i\omega^3 \frac{\mu_1 \mu_2}{s_a} \qquad (5.48)$$

The sound reduction index can be found by replacing $i\omega\mu$ in Equation 5.35 by this impedance z. This gives the following expression for R:

$$R(0) \approx 10 \log_{10} \left(\left(\frac{\omega(\mu_1 + \mu_2)}{2\rho_0 c_0} - \frac{\omega^3 \mu_1 \mu_2 d}{2\rho_0^2 c_0^3} \right)^2 \right) \qquad (5.49)$$

The double-leaf system exhibits a *mass-air-mass* resonance at a frequency given by

$$f_{MAM} = \frac{c_0}{2\pi} \sqrt{\frac{\rho_0 (\mu_1 + \mu_2)}{d \mu_1 \mu_2}} \approx 60 \sqrt{\frac{(\mu_1 + \mu_2)}{d \mu_1 \mu_2}} \qquad (5.50)$$

where the final form applies to air. At this frequency, $R(0)$ in Equation 5.49 becomes small, dropping below the mass-law result. In practice, this dip is limited by damping.

Below f_{MAM}, the stiffness of the air layer causes the two leaves to move together as a single partition of mass $\mu_1 + \mu_2$, as seen in the first term in Equation 5.49. Above f_{MAM}, R increases at 60 dB per decade:

$$R \approx 20 \log_{10} \left(\frac{\omega^3 \mu_1 \mu_2 d}{2\rho_0^2 c_0^3} \right) = R_{\mu_1 + \mu_2} + 40 \log_{10} \left(\frac{f}{f_{MAM}} \right) \qquad (5.51)$$

where $R_{\mu_1 + \mu_2}$ is the mass-law result for the combined mass of the double-panel system. This expression holds as long as the air gap is small compared with the wavelength. At higher frequencies, standing waves occur in the air gap – when $d = n\lambda/2$ for integer values of n, the air gap 'locks up' and the two partitions are effectively rigidly connected. For a practical system, and for diffuse incidence, the sound transmission behaviour does not exhibit such clear peaks and dips and a good approximation at high frequencies is

$$R \approx R_{\mu 1} + R_{\mu 2} + 6 \qquad (5.52)$$

where $R_{\mu 1}$ and $R_{\mu 2}$ are the (field incidence) mass-law results for the separate panels (Sharp 1978).

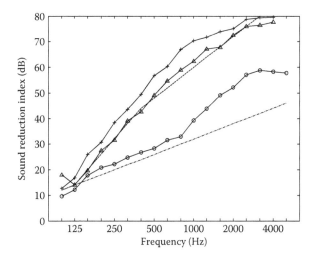

Figure 5.31 Performance of a mechanically isolated double-leaf partition. (o–o): measured for 6 mm hardboard, air gap of 160 mm and 3 mm hardboard; (Δ–Δ): measured with 50 mm of fibreglass in air cavity; (+–+): measured with 100 mm of fibreglass in air cavity; (·—·—·—·): predicted using mass law with mass of both panels; (– – –): predicted using Equations 5.51 and 5.52. (Redrawn from Sharp, B., *Noise Control Engineering* 11, 53–63, 1978. With permission.)

5.6.4.2 Practical considerations

Internal cavity resonances involving standing waves in the lateral direction can degrade the performance. Figure 5.31 shows the measured results of a mechanically isolated double-leaf partition constructed of 6 mm and 3 mm hardboard with an air gap of 160 mm, from Sharp (1978). The measured results with no absorptive material fall well below the curve predicted using Equations 5.51 and 5.52. These resonances of the cavity are suppressed when a layer of fibreglass is included in the air cavity.

For good low-frequency sound insulation, f_{MAM} should be as low as possible. From Equation 5.50, it can be seen that this requires heavy 'leaves' (also needed for high $R(0)$). A wide air gap is also required, although this in turn lowers the frequencies of standing waves in the cavity. If possible, reduced values of c_0 and ρ_0 in the cavity can be beneficial, for example by introducing a partial vacuum or a gas other than air.

Example 5.1

Consider two layers of glass 3 mm thick, which have $\mu = 7.2$ kg m^{-2}. For a cavity depth of 10 mm, the mass-air-mass resonance is found to be 320 Hz. This can be reduced by using thicker panes of glass and a wider air gap.

Example 5.2

For a typical high-performance lightweight partition, panels of 25 mm plasterboard (two layers of 12.5 mm thickness) give a mass per unit area of 16 kg m^{-2}. For a cavity depth of 100 mm, the mass-air-mass resonance is 67 Hz. Such a partition would normally include at least 50 mm of mineral-wool absorbent and would have staggered timber studs or flexible metal frames to avoid flanking transmission. Increasing the air gap to 200 mm lowers the resonance to 47 Hz.

It is generally beneficial to avoid having two partitions with the same thickness, because they would have the same critical frequency. Mechanical connections between the leaves, such as studding in a wall, can short-circuit the air spring and degrade the performance of the partition unless care is taken. Figure 5.32 shows some example measured results based on data from Halliwell et al. (1998). The results from the mass law and from Equation 5.51 are also shown for comparison. The mass-air-mass resonance in this case is around 85 Hz. There is a large dip at the critical frequency of 2.5 kHz. Steel studs can be seen to give an enhanced performance compared with wooden studs. Wooden studs that are staggered avoid a direct path between the two leaves and can give some improvement. The inclusion of mineral fibre in the cavity can also be seen to give significant benefit for both metal or staggered wooden studs.

5.6.5 Sound transmission measurements

To measure the sound reduction index of a partition, it should normally be placed between two reverberation rooms that are isolated both mechanically and acoustically from each other and from the surrounding building, as shown in Figure 5.33 (see ISO 2010a).

Broadband sound sources are located in the source room (Room 1), and produce an approximately diffuse field. The spatially averaged mean-square sound pressure is measured in both rooms, using either a fixed array of microphones or a spatial sweep on a rotating microphone boom. Measured values are usually stated in one-third octave bands.

From the assumption of a diffuse field, the sound power incident on a partition of area S is (cf. Equation 5.9)

$$W_i = \frac{\langle p_1^2 \rangle}{4\rho_0 c_0} S \qquad (5.53)$$

Of this, a proportion $W_t = \tau W_i$ is transmitted into the receiving room (Room 2). This leads to a mean-square pressure in Room 2 which is related to the transmitted power by

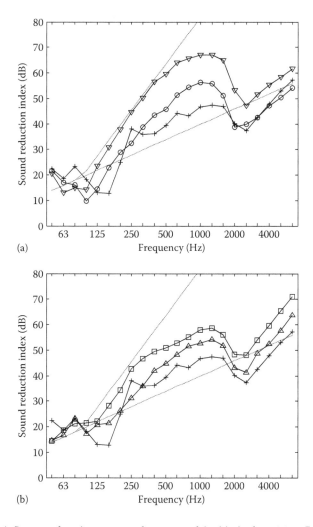

Figure 5.32 Influence of stud type on performance of double-leaf partition. Each has two leaves of 16 mm plasterboard with 90 mm studs. (a) o: steel studs, cavity not filled; ▽: steel studs, cavity filled with mineral fibre; +: wooden studs, cavity filled with mineral fibre; ······: Equation 5.51 and combined mass law (b) △: staggered wooden studs, cavity 108 mm not filled; □: staggered wooden studs, cavity 108 mm filled with mineral fibre; +: wooden studs, cavity filled with mineral fibre; ······: Equation 5.51 and combined mass law. (Redrawn from Halliwell, R.E., Nightingale, T.R.T., Warnock, A.C.C. and Birta, J.A., Gypsum board walls: transmission loss data, National Research Council, Canada, Internal Report IRC-IR-761, 1998. With permission.)

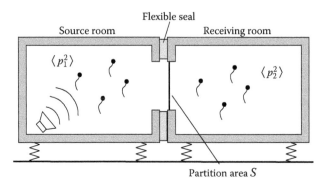

Figure 5.33 Schematic view of airborne sound transmission test suite. In practice, multiple source positions should be used.

$$W_t = \frac{\langle \overline{p_2^2} \rangle}{4\rho_0 c_0} A_2 \qquad (5.54)$$

where $A_2 = \alpha_2 S_2$ is the Sabine absorption area of Room 2 (see also Equation 5.9). It can be measured from the reverberation time, using the Sabine formula (Equation 5.10). Note that there is no need to measure the reverberation time of the source room. The transmission coefficient is thus given by

$$\tau = \frac{W_t}{W_i} = \frac{\langle \overline{p_2^2} \rangle}{\langle \overline{p_1^2} \rangle} \frac{A_2}{S} \qquad (5.55)$$

and the sound reduction index is found as

$$R = 10\log_{10}(1/\tau) = L_{p1} - L_{p2} + 10\log_{10}(S/A_2) \qquad (5.56)$$

Note that $D = L_{p1} - L_{p2}$ is the *noise reduction* or sound level difference of the partition, but this depends on the absorption in Room 2 as well as on R.

Measurements of sound insulation *in situ* (ISO 2014) follow the same principles but inevitably include flanking transmission. Here, the sound level difference D is measured. This can be normalised to eliminate the effects of the reverberation time T of the receiving room ($D_{n,T} = D + 10 \log_{10}(T/T_0)$, for a reference value T_0, usually 0.5 s). Alternatively, it can be expressed as an apparent sound reduction index, $R' = D + 10 \log_{10}(S/A_2)$.

An alternative laboratory measurement method is to use a single reverberant room as the source room and to use a sound-intensity scan over the receiver side of the partition to determine the transmitted power (ISO 2003a). This is also very useful for identifying the principal surfaces involved in flanking transmission. A similar method is available for field measurements (ISO 2010b).

5.7 NOISE CONTROL ENCLOSURES

Free-standing enclosures are widely used to control noise from machinery and plant. Often, absorbent material is added on the inside of the enclosure walls to prevent the build-up of reverberant sound within the enclosure. The following simple analysis of an enclosure is based on the assumption of a diffuse field within the enclosure. This is rarely a valid assumption, of course. In particular, a diffuse field is less likely as the absorption within the enclosure increases; it is also less reliable at low frequencies where the volume is too small. Nevertheless, it is instructive in showing the basic behaviour of an enclosure.

Figure 5.34 shows a free-standing enclosure constructed from an impermeable outer sheet lined with absorptive material. A source of acoustic power W is assumed to be located within the enclosure. The wall area S_W is assumed to have a transmission coefficient τ and an average absorption coefficient α (which is assumed to be unaffected by vibration of the walls). An aperture of area S_A is also included, which for simplicity is assumed to have a transmission coefficient of unity (in practice it is much smaller at low frequencies). Note that for a floor-mounted enclosure, the total surface area $S_W + S_A$ is equal to that of the four walls plus the top but excluding the bottom.

From the assumption of a diffuse field, the one-sided intensity I_{1-s} is given by Equation 5.8. In the steady state, the power emitted by the source is transmitted through the aperture (W_A), through the wall (W_W) or absorbed ($W\alpha$). These three components of power can be written in terms of the one-sided intensity as $W_A = I_{1-s} S_A$, $W_W = I_{1-s}\tau S_W$ and $W_\alpha = I_{1-s}\alpha S_W$. Thus, the power balance is given by

$$W = W_A + W_W + W_\alpha = I_{1-s}\left[S_W\left(\alpha + \tau\right) + S_A\right] \qquad (5.57)$$

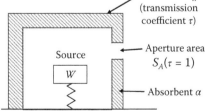

Figure 5.34 Free-standing noise control enclosure.

Therefore, the total power transmitted to the exterior per unit source power is

$$\frac{W_T}{W} = \frac{(W_A + W_W)}{W} = \frac{(S_A + \tau S_W)}{S_W(a+\tau) + S_A} \tag{5.58}$$

This can be written in decibel form as the sound power insertion loss, which is the difference in the exterior level without and with the enclosure: $IL = 10 \log_{10}(W/W_T)$.

In particular, for the case where there is no absorbent or damping, $W_T/W = 1$, that is, $IL = 0$ dB. Usually, however, $a \gg \tau$ (except at low frequencies) and $S_W \gg S_A$. Then, this formulation can be approximated as

$$IL \approx 10 \log_{10}\left(\frac{a}{(S_A/S_W) + \tau}\right) \tag{5.59}$$

This shows clearly that the absorption coefficient, a, should be maximised, while both S_A/S_W and τ should be minimised. For the particular case where there is no aperture, $S_A = 0$:

$$IL = 10 \log_{10}\left(\frac{S_W(a+\tau)}{\tau S_W}\right) \approx 10 \log_{10}\left(\frac{a}{\tau}\right) \tag{5.60}$$

From both Equations 5.59 and 5.60, it can be seen that $IL < 10 \log_{10}(1/\tau) = R$. Thus, it can also be seen that the insertion loss is always less than the sound reduction index of the walls, a consequence of the build-up of reverberant sound in the enclosure.

The limitations of this analysis should be noted. It has been assumed that the enclosure is large compared with the wavelength of sound, which may not be true at low frequencies. Moreover, it has been assumed that the source is small compared with the volume inside the enclosure, where in practice the enclosure may be closely fitting. Finally, it has also been assumed that there is a diffuse field, which may not be the case; for example, if the absorption is large. Other models for enclosures, in particular where they are close fitting, are given, for example, in Vér and Beranek (2005) and Fahy (2000).

5.8 NOISE BARRIERS

Noise barriers, or screens, are employed outdoors to reduce the levels of noise at the facades of buildings due to road or rail traffic or industrial operations. They are also used within offices and industrial workspaces to attenuate the propagation of sound within large rooms. This section gives a brief introduction to the acoustic performance of barriers.

The phenomenon of diffraction underlies the behaviour of a barrier. Sound waves can 'bend around' corners so that, in the 'shadow' region behind a barrier, the sound is attenuated but not eliminated. The extent of this diffraction effect depends on a non-dimensional parameter known as the *Fresnel number*.

Consider a point source S located in front of a semi-infinite barrier, as shown in Figure 5.35, and a receiver P behind the barrier. The path length difference δ is defined as the difference between the direct path S–P that would exist in the absence of the barrier and the path S–O–P via the *diffraction edge* at the top of the barrier:

$$\delta = A + B - d \tag{5.61}$$

The Fresnel number is then defined as $N = 2\delta/\lambda$, where λ is the wavelength of sound. For a given geometry, N is small at low frequencies and large at high frequencies. The insertion loss of a barrier was predicted by Maekawa (1968) as a function of N. It was found that the attenuation (insertion loss) is positive even when the line S–P passes above the top of the barrier; an insertion loss of 5 dB is found when S–P just passes over the barrier.

A good approximation to the barrier performance for a point source is provided by the Kurze–Anderson formula (Kurze and Anderson 1971):

$$\text{IL} = 5 + 20\log_{10}\left(\frac{\sqrt{2\pi N}}{\tanh\sqrt{2\pi N}}\right) \tag{5.62}$$

For an incoherent line source, representative of a stream of cars passing along a road or of a train, the barrier gives a reduced performance, which can be approximated by (Anderson and Kurze 2005)

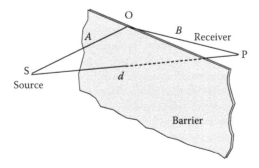

Figure 5.35 Barrier geometry showing path S–P in absence of barrier and path S–O–P via the top of the barrier.

$$\text{IL} = 5 + 15\log_{10}\left(\frac{\sqrt{2\pi N}}{\tanh\sqrt{2\pi N}}\right) \tag{5.63}$$

This tends to overestimate the insertion loss at small values of N but gives reasonable results at larger values. The results of the two formulae are compared in Figure 5.36. An alternative calculation method is provided in ISO (1996a).

The acoustic performance depends on many practical factors, including the following:

- Barrier length. Diffraction can occur around the ends of a barrier as well as over its top.
- Presence of nearby reflecting objects (buildings, far-side barrier, the vehicle itself). These may lead to reflected sound that increases the effective source height and reduces the barrier attenuation. This can be ameliorated by the use of absorbing material on the side of the barrier facing the source or inclined barrier surfaces.
- Use of absorbing material on the barrier top. This can reduce diffracted sound by 2–4 dB, depending on the width of the barrier.
- Any apertures in the barrier or beneath it will compromise the performance, especially at high frequencies.
- Details of the source (geometry, frequency content) can be important. Clearly, high-frequency sources are shielded more effectively than low-frequency sources. Elevated sources, such as raised exhausts or

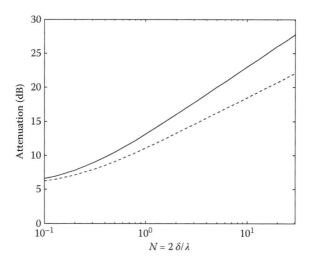

Figure 5.36 Barrier attenuation predicted by Kurze–Anderson formulae. (—): point source; (– – –): line source.

air inlets on trucks or pantographs and roof-mounted equipment on trains, are less well shielded by a barrier than those located close to the ground, such as wheels.
- Topography and acoustic impedance of ground. On soft ground, the barrier may destroy the 'ground dip' and thereby cause an increase in sound level at some frequencies. A full barrier calculation will include the contributions of image sources due to reflections from the ground on both the source and receiver sides of the barrier.
- Wind speed and direction. Turbulence, particularly at the top of the barrier, can lead to increased diffraction into the shadow zone.

It is generally not practical to expect an overall attenuation for an outdoor barrier to exceed about 15–20 dB(A) without a prohibitively high cost.

5.9 SOUND ABSORPTION

Sound-absorbent materials such as plastic foams, mineral wool or glass fibre are widely used in noise control. They are applied to the surfaces of rooms (ceilings, walls, floors) to control reflections; in factories and offices they limit the sound level and facilitate speech communication; and in auditoria they are installed to control reverberation and facilitate speech intelligibility or musical appreciation. Their application in enclosures has already been seen in Section 5.7, and incorporation in noise barriers has been mentioned in Section 5.8. Other applications include installation within flow ducts to attenuate sound from fans, within lightweight double-panel walls and ceilings, and in vehicles. In very hot conditions, such as gas-turbine exhaust ducts, alternative materials such as sintered steel sheets may be employed.

In this section, the absorption coefficient of absorptive treatments is first derived in terms of their impedance. Then, various physical parameters describing porous absorbing materials are introduced to give a 'physical' model.

5.9.1 Local reaction

Consider a plane boundary between an absorbent material ($x > 0$) and a non-absorbent medium, for example air ($x < 0$), as shown in Figure 5.37. The absorbent material itself is not modelled in this section; the boundary is simply described by its surface normal acoustic impedance:

$$z_n = \frac{p}{u_n} \tag{5.64}$$

where p is the complex pressure amplitude at the point and u_n is the complex amplitude of the normal component of acoustic particle velocity

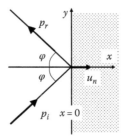

Figure 5.37 Sound incident at angle φ to the normal at a plane boundary between absorbent material ($x>0$) and air ($x<0$).

at that point; z_n is usually frequency dependent. The real part of z_n is called the specific acoustic *resistance*, r_n, and is associated with "damping" (energy loss to the fluid either by dissipation or transmission). The imaginary part of z_n is called the specific acoustic *reactance*, x_n, and is associated with the stiffness of the material (if it is negative) or its mass (if positive).

In general, the normal velocity at a point on a surface depends on the pressure at all points. Systems in which waves can travel freely along a surface, for example sheets of glass, metal or wood, exhibit *extended reaction* and z_n cannot be uniquely defined. However, for many common forms of absorbent material, wave motion parallel to the surface within the medium is strongly attenuated by viscous dissipation. This promotes a condition known as *local reaction*. Then, the normal velocity at a point **x** depends only on the pressure at that point **x**, and we can write

$$u_n(\mathbf{x}) = z_n p(\mathbf{x}) \tag{5.65}$$

For a system with local reaction, the relation between pressure and normal velocity is independent of the form of the incident field. Consequently, z_n is effectively independent of the incident angle φ.

5.9.2 Reflection and absorption at an impedance interface

Assume local reaction and consider a plane wave which is incident on the plane boundary at $x=0$ at an angle φ. This has components of the wavenumber vector (see Section 2.6.3) $k_x = k \cos \varphi$ in the x-direction and $k_y = k \sin \varphi$ in the y-direction:

$$p_i(x,y) = A\, e^{-ik_x x} e^{-ik_y y} \tag{5.66}$$

The reflected wave will have the form

$$p_r(x,y) = B\, e^{ik_x x} e^{-ik_y y} \tag{5.67}$$

At $x=0$, the pressure is

$$p(0,y) = (A+B)\, e^{-ik_y y} \tag{5.68}$$

and the normal (x) component of particle velocity is

$$u_n(0) = -\frac{1}{i\omega\rho_0}(-ik_x A + ik_x B) e^{-ik_y y} \tag{5.69}$$

Taking the ratio of p to u_n and noting that $k_x = k\cos\varphi$, the impedance can be written as

$$z_n = \frac{p(0)}{u_n(0)} = \frac{(A+B)\rho_0 c_0}{(A-B)\cos\varphi} \tag{5.70}$$

Rearranging, the ratio of reflected to incident wave amplitudes, $R_p = B/A$ is

$$R_p = \frac{B}{A} = \frac{(z'_n \cos\varphi - 1)}{(z'_n \cos\varphi + 1)} \tag{5.71}$$

where $z'_n = z_n/\rho_0 c_0$ is the non-dimensionalised normal specific acoustic impedance.

The sound power absorption coefficient is defined as the ratio of the absorbed power to the incident power. The sound power incident on the surface per unit area is $W'_i = |A|^2 \cos\varphi/(2\rho_0 c_0)$ while the reflected power per unit area is $W'_r = |B|^2 \cos\varphi/(2\rho_0 c_0)$. Hence, the sound power absorption coefficient for an angle of incidence φ is

$$\alpha(\varphi) = 1 - \left|\frac{B}{A}\right|^2 = 1 - \left|\frac{(z'_n \cos\varphi - 1)}{(z'_n \cos\varphi + 1)}\right|^2 \tag{5.72}$$

This can also be written in terms of the resistance and reactance:

$$\alpha(\varphi) = \frac{4 r'_n \cos\varphi}{(r'_n \cos\varphi + 1)^2 + (x'_n \cos\varphi)^2} \tag{5.73}$$

From this, it is clear that $0 \leq \alpha(\varphi) \leq 1$, as is required from physical argument. The sound is perfectly absorbed if $z'_n \cos \varphi = 1$, which requires z_n to be real. If the impedance is complex, $x'_n \neq 0$ and $\alpha(\varphi) < 1$. For either infinite or zero impedance, the sound is perfectly reflected. For grazing incidence ($\varphi = \pi/2$), the absorption coefficient tends to zero. Ideally, a large absorption coefficient requires z'_n to be real and close to 1 – in fact a value slightly larger than unity is most efficient in allowing it to be effective for a range of angles.

For a diffuse sound field (see Section 2.7.1), the sound is incident from all directions with equal probability. The total incident sound power per unit area W'_i is found by integrating over all possible incident directions on a hemisphere:

$$W'_i = \frac{1}{2\pi} \int_0^{\pi/2} \left\{ \int_0^{2\pi} (I_i \cos \varphi) \sin \varphi d\theta \right\} d\varphi = I_i \int_0^{\pi/2} \sin \varphi \cos \varphi d\varphi = I_i / 2 \qquad (5.74)$$

where I_i is the incident intensity in each direction. Similarly, the absorbed power per unit area is

$$W'_a = I_i \int_0^{\pi/2} \alpha(\varphi) \sin \varphi \cos \varphi d\varphi \qquad (5.75)$$

Hence, the diffuse field absorption coefficient is given by

$$\alpha_d = \frac{W'_a}{W'_i} = \int_0^{\pi/2} \alpha(\varphi) \sin 2\varphi d\varphi \qquad (5.76)$$

which is known as Paris's formula.

5.9.3 Parameters describing porous materials

Many sound absorbers are based on porous materials. Sound absorption (the conversion of sound energy into heat) in porous materials occurs by a variety of mechanisms:

1. Viscous losses due to oscillatory flow in the pores of porous materials (e.g. mineral wool, fibreglass, open-cell plastic foam, porous plaster, wood wool)
2. Heat conduction through porous materials
3. Vibration of the solid skeleton of porous materials (e.g. closed-cell plastic foam)

4. Vibration of impermeable panels by incident sound (wood, glass, metal, etc.) – internal and support friction causes energy loss
5. Viscous and heat conduction losses in acoustic (Helmholtz) resonators

Porous absorbing materials have very complex geometrical forms, and so are not amenable to deterministic models. Instead, phenomenological models are used with spatially 'smeared' properties. Comprehensive models considering these effects require many parameters (Allard and Atalla 2009). Empirical models also exist, such as in Delany and Bazley (1970), but these do not give physical insight. Instead, a simplified model is presented here for those porous materials for which mechanism 1 is dominant and the others can be neglected. The most important parameters are:

1. Porosity h: this is defined as the ratio of the volume of voids to the total volume occupied by the porous structure. It is often between 0.9 and 0.95 for effective absorptive materials.
2. Flow resistivity r: to define this, consider a steady flow of air through a layer of porous material. The viscous resistance to flow gives rise to a static pressure gradient. The flow resistivity is defined as this pressure gradient divided by the mean flow velocity (volume flow rate per unit cross-sectional area). For porous absorbing materials, typical values are in the range 10^3–10^5 kg s^{-1} m^{-3}. It is found that, except at extremely high incident sound levels, the acoustic flow resistivity is very close to the static flow resistivity.
3. Tortuosity s: this is a non-dimensional quantity that expresses the influence of the geometric form of the structure on the effective density of the fluid; $s \geq 1$. The vibrating fluid is forced to undergo accelerations in various directions by the tortuous and irregular forms of the pores/channels within a porous material. Generally, this parameter cannot be readily estimated theoretically.

5.9.4 Equations for fluid in a porous material

Introducing these parameters into the description of an equivalent fluid, the modified one-dimensional wave equation for a porous material is found as (Fahy 2000)

$$\frac{\partial^2 p}{\partial x^2} = \frac{s}{c_0^2}\frac{\partial^2 p}{\partial t^2} + \frac{rh}{\rho_0 c_0^2}\frac{\partial p}{\partial t} \tag{5.77}$$

Compared with the usual wave equation (see Equation 2.29), the first term on the right-hand side is modified by the factor s (increasing the effective mass), while the second term introduces damping. Seeking harmonic wave-type solutions of the form $p(x,t) = A e^{i(\omega t - \bar{k}x)}$, where $\bar{k}$ is a complex wavenumber, results in a relation between ω and $\bar{k}$:

$$\bar{k}^2 = \frac{s}{c_0^2}\omega^2 - \frac{rh}{\rho_0 c_0^2}i\omega \tag{5.78}$$

The wavenumber has both real and imaginary parts. The real part is the phase constant and the imaginary part is the attenuation constant. Rearranging Equation 5.78 and expressing the complex wavenumber in non-dimensional form gives

$$k' = \frac{\bar{k}}{k} = s^{1/2}\left(1 - \frac{i}{\Omega}\right)^{1/2} \tag{5.79}$$

where $\Omega = \omega \rho_0 s / rh$ is a non-dimensional frequency and k is the wavenumber in air. At the frequency $\Omega = 1$, or $\omega = rh/\rho_0 s$, the viscous effects are equal to the inertial effects in the pores. Typically, this occurs within the audible frequency range. Below this frequency, viscous effects are dominant (second term in Equation 5.78), while at higher frequencies, inertial effects are dominant (first term in the equation).

At low frequencies,

$$k' \approx \left(\frac{-is}{\Omega}\right)^{1/2} = \left(\frac{s}{2\Omega}\right)^{1/2}(1-i) \quad \text{for } \Omega \ll 1 \tag{5.80}$$

which has equal real and (negative) imaginary parts, whereas at high frequencies,

$$k' \approx s^{1/2}\left(1 - \frac{i}{2\Omega}\right) \approx s^{1/2} \quad \text{for } \Omega \gg 1 \tag{5.81}$$

which is mainly real with a wavespeed slightly smaller than in air.

The non-dimensional characteristic specific acoustic impedance of the porous medium is given by

$$z'_c = \frac{z_c}{\rho_0 c_0} = \frac{k'}{h} \tag{5.82}$$

which allows it to be written as

$$z'_c = \frac{s^{1/2}}{h}\left(1 - \frac{i}{\Omega}\right)^{1/2} \tag{5.83}$$

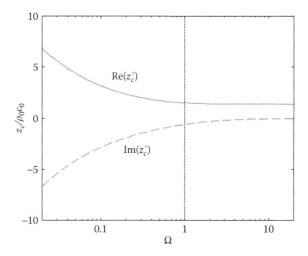

Figure 5.38 Predicted non-dimensional characteristic impedance of a porous material as a function of non-dimensional frequency Ω.

This impedance corresponds to an infinitely thick layer of the porous material. It is plotted in Figure 5.38 against non-dimensional frequency for example values of $h=0.9$ and $s=1.5$. At low frequencies, similarly to Equation 5.80, it can be approximated as

$$z'_c \approx \frac{1}{h}\left(\frac{-is}{\Omega}\right)^{1/2} = \frac{1-i}{h}\left(\frac{s}{2\Omega}\right)^{1/2} \quad \text{for } \Omega \ll 1 \tag{5.84}$$

which has equal real and (negative) imaginary parts. As frequency reduces, both the real and the imaginary parts of z'_c become large. At high frequencies, similarly to Equation 5.81,

$$z'_c \approx \frac{s^{1/2}}{h} \quad \text{for } \Omega \gg 1 \tag{5.85}$$

which is almost real and slightly greater than 1, which makes it ideal for an absorbing material.

Equation 5.73 can be used to find the absorption coefficient from this impedance. For simplicity, only the result for normal incidence is given here. This is plotted in Figure 5.39 for various values of the flow resistivity. The low- and high-frequency asymptotes are found as

$$a(0) \approx 2h\left(\frac{2\Omega}{s}\right)^{1/2} \quad \text{for } \Omega \ll 1 \tag{5.86}$$

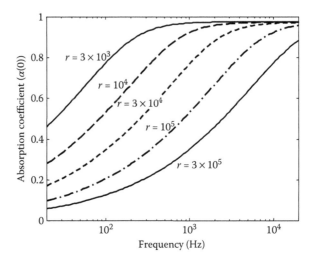

Figure 5.39 Predicted normal incidence absorption coefficient for infinitely thick layers of porous material of different flow resistivities r (Rayls m^{-1}).

which becomes small at low frequencies in proportion to $\omega^{1/2}$, and

$$\alpha(0) \approx \frac{4s^{1/2}/h}{(s^{1/2}/h + 1)^2} \quad \text{for } \Omega \gg 1 \tag{5.87}$$

which tends to a constant value at high frequencies.

5.9.5 Absorption due to a layer of porous material

Usually, sound absorption is provided by a porous layer of finite depth mounted on a rigid backing. The surface normal specific acoustic impedance of such a layer of thickness l is

$$z'_n = -i z'_c \cot \bar{k} l \tag{5.88}$$

This is shown in Figure 5.40 for the same example parameters and $l = 0.03$ m. After some algebra, the following asymptotes are obtained. At low frequencies $\left(\Omega \ll 1, |\bar{k} l| \ll 1\right)$,

$$z'_n \approx \frac{-ic_0}{\omega h l} + \frac{rl}{2\rho_0 c_0} \quad \text{for } \Omega \ll 1 \tag{5.89}$$

which becomes large and mainly imaginary (stiffness controlled) with a constant real part. This leads to

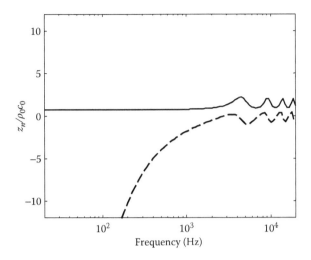

Figure 5.40 Predicted non-dimensional impedance of a porous material of thickness 0.03 m, flow resistivity (in Rayls m^{-1}) $r=3\times10^4$ Rayls m^{-1}, porosity $h=0.9$ and tortuosity $s=1.5$. (—): real part; (– – –): imaginary part.

$$\alpha(0) \approx \frac{2r\omega^2 h^2 l^3}{\rho_0 c_0^3} \quad \text{for } \Omega \ll 1 \tag{5.90}$$

which is small and proportional to ω^2; it was proportional to $\omega^{1/2}$ for the infinite layer. At high frequencies $(\Omega \gg 1, |\bar{k}l| \gg 1)$,

$$z'_n \approx \frac{s^{1/2}}{h}\left(1 + 2\,e^{-i2\bar{k}l}\right) \quad \text{for } \Omega \gg 1 \tag{5.91}$$

This oscillates around z'_c. The absorption coefficient is given approximately by Equation 5.87.

The absorption coefficients of porous layers with various flow resistivities and a thickness of 30 mm are shown in Figure 5.41. Compared with the infinite layer shown in Figure 5.39, it is clear that the low-frequency absorption is small for all values of the flow resistivity.

To summarise, the absorption coefficient of porous materials is small at low frequencies because the material impedance becomes too large for $\Omega \ll 1$, and also because the layer is too thin compared with the wavelength for $|\bar{k}l| \ll 1$. At low frequencies, the impedance is stiffness dominated. This stiffness is determined by the layer thickness, and is almost independent of the material (see first term in Equation 5.89). At high frequencies, the attenuation is large if the depth l is at least a quarter of a wavelength, although

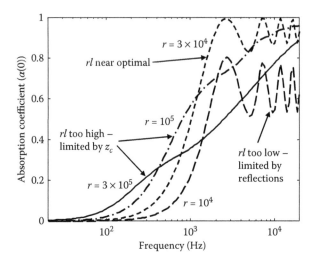

Figure 5.41 Predicted normal incidence absorption coefficient for layers of porous material of different flow resistivities (in Rayls m^{-1}) r and thickness 0.03 m.

this is the wavelength in the porous material, which can be considerably smaller than the wavelength in air.

For a finite layer, a compromise is required in terms of the flow resistivity, as shown in Figure 5.41. If r is too large, there will be too much reflection at the free surface, whereas if r is too small, there will be too little damping within the porous material, so the waves will not be effectively absorbed in the material, leading to large standing waves in the layer. A suitable compromise is obtained if the frequency at which $\lambda = 4l$ is similar to that at which $\Omega = 1$. This leads to a rule of thumb: $rl/\rho_0 c_0 \approx 2\text{--}3$. In practice, the scope for changing l is limited, so it is necessary to choose the material to give a suitable value for r.

5.9.6 Practical sound absorbers

There are various possible ways to improve the performance of a porous layer at low frequencies which are considered in the following paragraphs. Several of these utilise a resonant system to achieve improved absorption at low frequencies. However, these have a narrower bandwidth and reduced absorption at higher frequencies.

5.9.6.1 Porous layer in front of air cavity

At a rigid surface such as a wall or ceiling, the sound pressure has a maximum, whereas the particle velocity is zero. For a porous layer to be effective,

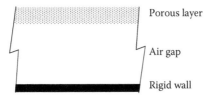

Figure 5.42 Porous layer separated from a rigid wall by an air gap.

a large particle velocity is required, which occurs (for normal incidence) at a quarter of a wavelength from the rigid surface. Where space permits it, mounting a porous layer at a distance from a rigid surface, as shown in Figure 5.42, enables increased low-frequency absorption; the effective depth of the porous layer is increased by such an arrangement and it is nearly as effective as filling the whole space with porous material at this frequency, although less effective at other frequencies.

5.9.6.2 Porous layer covered by thin surface layer

As seen in Figure 5.40, the impedance of a porous layer is dominated by its stiffness at low frequencies. By adding a mass in series with this stiffness, an impedance can be generated that is much closer to $\rho_0 c_0$, since the imaginary parts cancel out at some resonance frequency. To achieve this, layers of absorptive material are often covered by thin perforated surface layers, as shown in Figure 5.43, to improve the low-frequency behaviour (as well as to protect them).

The characteristic impedance of a thin porous surface layer which is free to move can be approximated at low frequencies by the impedance of a mass. For a limp porous surface layer, this is the mass per unit area of the layer, whereas for a rigid perforated cover, the equivalent mass is that due to the air within the holes.

By tuning this mass impedance to be equal and opposite to the stiffness impedance of the absorbent layer underneath, the combined impedance can be made real and close to $\rho_0 c_0$ at some frequency. However, the performance is degraded at higher frequencies.

Figure 5.43 Layer of absorptive material covered by thin perforated surface layer.

5.9.6.3 Helmholtz resonators

A Helmholtz resonator consists of an fluid volume, which acts as a stiffness, and a constricted neck, in which fluid behaves like a lumped mass. This is shown schematically in Figure 5.44. Its resonance frequency is given by

$$f_0 = \frac{c_0}{2\pi}\sqrt{\frac{S}{V(l + 16a/3\pi)}} \qquad (5.92)$$

where:
- S is the area of the opening (radius a)
- V is the volume of the resonator
- l is the length of the opening
- $16a/3\pi$ is the end correction (inner and outer) (see Section 2.6.2)

To be effective, the internal loss factor and the radiation loss factor have to be closely matched (Fahy and Gardonio 2006). This generally means that the openings should be small. Consequently, Helmholtz resonators are usually far more effective when used in arrays, which increases the radiation loss factor to match the typical internal loss factor.

5.9.6.4 Panel absorbers

Another way of achieving a resonant system is to mount a flexible panel over a thin layer of air, as shown in Figure 5.45. The air volume acts as a stiffness while the panel acts as a mass. Damping is introduced by the panel itself and by viscosity within the fluid. For a limp panel, the resonance frequency can be approximated by

$$f_0 = \frac{1}{2\pi}\sqrt{\frac{\rho_0 c_0^2}{\mu d}} \qquad (5.93)$$

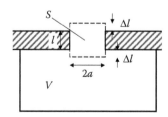

Figure 5.44 Helmholtz resonator.

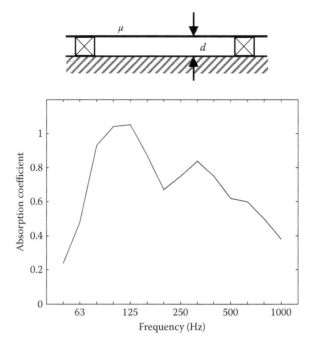

Figure 5.45 Panel absorber and example absorption coefficient. (Redrawn from Walker, R. and Randall, K.E., An investigation into the mechanism of sound-energy absorption in a low-frequency modular absorber. Report BBC RD 1980/12, BBC Research and Development, 1980. With permission.)

where d is the depth of the cavity and μ is the mass per unit area of the panel. In practice, the panel bending stiffness may also be important, as well as the stiffness of the air layer.

Panel absorbers are widely used in broadcasting and recording studios due to the possibility of introducing high absorption at low frequencies. Figure 5.45 shows an example of the absorption coefficient of a panel absorber used by the BBC (Walker and Randall 1980). In practice, it is difficult to predict the performance of a panel absorber as the stiffness is highly dependent on details of the installation. Decorative wall-mounted panels in auditoria can also introduce natural absorption by the same mechanism; in music auditoria, this absorption is unwanted and should be avoided by ensuring that, if required for aesthetic reasons, wall panels have a high mass and are firmly fixed to the supporting surface.

5.9.6.5 Micro-perforated panels

In recent years, the use of micro-perforated panels to increase sound absorption has gained popularity. A micro-perforated panel contains holes which are usually smaller than 1 mm in diameter. Due to high viscous losses in the holes, such a panel, mounted in front of a rigid wall, produces significant absorption without the need for any absorptive material. Figure 5.46 shows the absorption coefficient for a 0.5 mm thick steel micro-perforated plate with an array of 0.45 mm diameter holes (0.45% perforation) for different distances d from a rigid wall (Fuchs and Zha 2006). The main peak corresponds to the Helmholtz resonance, the frequency of which depends on the distance d. Many panel materials can be used, including, for example, transparent acrylic glass.

5.9.6.6 Duct attenuators

Sound-absorbing materials are widely used in attenuators installed in ducts which carry gas flow; for example, ventilation ducts and gas-turbine exhaust ducts. Special-purpose materials are necessary in ducts carrying hot gas.

Attenuation reaches a maximum in the frequency range where the acoustic wavelength is close to twice the width of the channel between

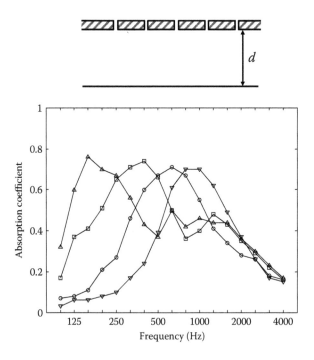

Figure 5.46 Absorption coefficient for an array of 0.45 mm holes in a 0.5 mm steel plate (0.45% perforation) for $d=40$ (△), 20 (□), 10 (○), 5 (▽) cm from a rigid wall. (Redrawn from Fuchs, H.V. and Zha, X., *Acta Acustica with Acustica* 92, 139–146, 2006. With permission.)

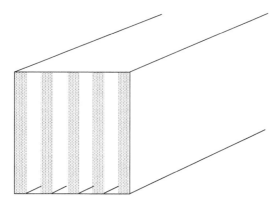

Figure 5.47 Typical arrangement of a splitter attenuator.

the splitters. It falls considerably towards lower frequencies and moderately towards higher frequencies. Thicker linings provide increased low-frequency absorption, but increase flow pressure losses. The attenuation is expressed in decibels per metre, so that the length of the attenuator has to be sufficient to obtain the required performance. The necessary length in ventilation ducts is often determined by the sound power of the fan in the 63 and 125 Hz octave bands.

It is often ineffective simply to line the wall surfaces of a duct with sound-absorbent material, except at sharp bends, because the ratio of the exposed absorbent surface to the cross-sectional area through which the sound passes is too small. Consequently, splitter attenuators are used, as illustrated in Figure 5.47. The flow resistance of the liner must be relatively high to provide sufficient attenuation. Optimum performance is generally obtained when the product of the flow resistivity and the liner thickness is around 3–4 times $\rho_0 c_0$.

5.9.7 Measurements of absorption

There are two main methods of measuring the absorption coefficient of a sample: using an *impedance tube* and using a reverberation room. Measurement methods are also available using sound intensity (see Section 8.8.4).

5.9.7.1 Impedance-tube method

The impedance tube, shown in Figure 5.48, consists of a tube terminated with a sample of absorbent material of known thickness (ISO 2001a). The tube diameter needs to be small enough so that only plane waves are sustained within it; for a circular tube this requires $ka < 1.84$, where k is the

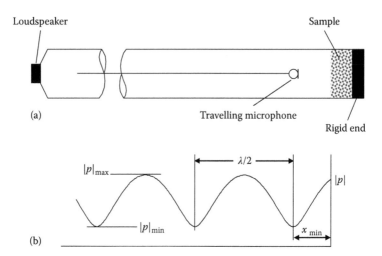

Figure 5.48 Impedance tube. (a) Schematic view; (b) pressure distribution.

acoustic wavenumber and a the tube radius. This forms the upper frequency limit. Its length should be at least three-quarters of the acoustic wavelength at the lowest frequency of interest.

The acoustic pressure is measured at different locations along the tube with a travelling microphone. It is also possible to use several fixed microphones instead of a travelling microphone. The incident and reflected waves can be separated by finding the maximum and minimum pressure amplitude and their locations. The ratio of maximum to minimum pressure amplitude, s, allows the magnitude of the pressure reflection ratio to be found:

$$|R_p| = \frac{s-1}{s+1} \qquad (5.94)$$

from which the absorption coefficient for normal incidence can be found, $\alpha(0) = 1 - |R_p|^2$. The distance of the first pressure minimum from the front face of the sample, x_{min}, also allows the phase of R_p to be found:

$$\angle R_p = 2kx_{min} - \pi \qquad (5.95)$$

From the complex reflection coefficient R_p, it is possible to determine the normal impedance for this thickness of sample from Equation 5.70:

$$z_n = \frac{(1+R_p)}{(1-R_p)} \rho_0 c_0 \qquad (5.96)$$

Hence, the absorption coefficient at other angles of incidence can be found from Equation 5.72. Methods are also available based on transfer functions between the acoustic pressure at different positions in the tube (ISO 2001b). The advantage of the impedance-tube method is that the impedance can be determined as well as the absorption coefficient, but it requires relatively small samples of the material, which is not always practical, or representative of practical installation.

5.9.7.2 Reverberation room

The main method of determining the random-incidence absorption coefficient uses a reverberation chamber (ISO 2003b) (see also Section 5.3.3). The method consists in measuring the reverberation time with and without one or more samples of the absorptive material installed on the floor or walls of the chamber. The total area of absorbent material should be at least 10 m².

If T_1 is the reverberation time without the absorbent treatment, this is related to the absorption area of the room (from Equation 5.10, including air absorption) by

$$S_1 a_1 = \frac{55.3V}{c_0 T_1} - 4Vm_1 \qquad (5.97)$$

where:
 V is the room volume
 S_1 is the surface area of the room
 m_1 is the air absorption

If T_2 is the modified reverberation time when the absorbent of area S_2 is introduced,

$$S_2 a_2 = \frac{55.3V}{c_0 T_2} - 4Vm_2 \qquad (5.98)$$

If the speed of sound and the air absorption are unchanged between the two tests, this allows the absorption coefficient of the sample a_2 to be approximated as

$$a_2 \approx \frac{0.16V}{S_2}\left(\frac{1}{T_2} - \frac{1}{T_1}\right) \qquad (5.99)$$

In practice, allowance can also be made for the differences in c_0 and m. This method determines the random-incidence absorption coefficient directly, but cannot be used to determine the impedance due to the lack of phase information. It should be noted that it is possible to obtain values of the absorption coefficient greater than unity from this method due to diffraction effects at the edges of the finite sample. Moreover, the assumption of a diffuse field may break down when the absorption coefficient is high. Nevertheless, the method is more practical for many applications than the impedance tube, since realistic sample sizes can be used.

5.10 VIBRATION CONTROL FOR NOISE REDUCTION

As noted in Section 5.2.2, a large proportion of noise radiators take the form of vibrating solid surfaces. This includes industrial machinery and plant, building components, vehicle engines and body structures. In most cases, reduction of the surface vibration amplitude will produce a proportionate change in the radiated sound – see Equation 5.17. It is, of course, preferable to reduce the excitation forces (or displacements) causing the vibration, but this is often impractical for technical or economic reasons. The following passive vibration-control strategies can be used to attenuate sound transmitted through structure-borne paths:

1. *Impedance mismatching*: this involves the design or modification of structures where they are attached to the directly excited structure to reduce the transfer of vibrational energy from one to the other.
2. *Vibration isolation*: this involves the insertion of resilient (flexible) elements between the structure which is directly excited and other structures to which it is attached. This can be seen as a special case of impedance mismatching.
3. *Damping*: this involves the application of highly dissipative materials to the vibrating structure to convert a large proportion of the vibrational energy into heat and thereby reduce the vibration level. Damping may also be increased by connecting ancillary structures to a structure so as to produce enhanced friction losses or by attaching damped resonators.
4. *Structural modification*: this usually involves the addition of mass or stiffness to change the level of the impedance or to shift resonance frequencies. The latter is particularly applicable in cases of single-frequency excitation.
5. *Dynamic neutralisers*: this involves the attachment of ancillary mechanical resonators to the structure. In a narrow frequency region around their resonance they apply strong dynamic reaction forces which suppress vibrational response in the structure to which they are attached. These are not considered further here but are treated by Brennan and Ferguson (2004).

Noise control

5.10.1 Impedance mismatch and vibration isolation

5.10.1.1 Low-frequency model of vibration isolation

The simplest model for vibration isolation represents the source structure as a mass, the receiver structure as a rigid foundation and the isolator as a massless spring and damper, as shown in Figure 5.49a for force excitation or Figure 5.49b for base excitation. In both cases, the transmissibility is given by (see Section 3.3.2)

$$T_R = \left| \frac{1 + 2i\zeta\Omega}{1 - \Omega^2 + 2i\zeta\Omega} \right| \quad (5.100)$$

where:
$\zeta = C/(2m\omega_n)$ is the damping ratio
$\omega_n = \sqrt{K/m}$ is the natural frequency
$\Omega = \omega/\omega_n$ is a non-dimensional frequency

For base excitation, T_R is the ratio of the displacement of the equipment to that of the base, whereas for force excitation it is the ratio of the transmitted force to that applied to the source.

The transmissibility predicted by Equation 5.100 is shown in Figure 3.23. As can be seen, at low frequencies $T_R \approx 1$ while at resonance ($\Omega = 1$), $T_R \approx 1/\zeta$, that is, the input can be amplified considerably. This implies that high damping is required to prevent large amplitudes at resonance. Isolation, that is, a reduction in force or vibration transmitted, only occurs for frequencies $\Omega > \sqrt{2}$. For low damping, the high-frequency asymptote is $T_R \approx 1/\Omega^2$, whereas for high damping it becomes $T_R \approx 2\zeta/\Omega$, which has a reduced frequency dependence. This suggests that low damping is required at high frequencies.

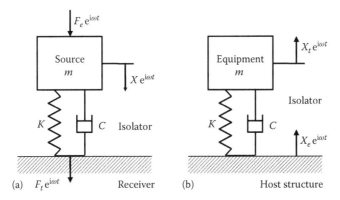

Figure 5.49 Simple models for vibration isolation.

This simple model of vibration isolation has a number of limitations for practical applications:

1. The foundation is never completely rigid.
2. At high frequencies, the source will have modal behaviour and can no longer be considered simply as a mass.
3. Isolators, especially those based on viscoelastic materials such as rubber, often behave as springs with hysteretic damping (see Section 5.10.2), in which case the dominance of the damping in the viscous model at high frequencies is overstated.
4. At high frequencies, internal resonances can occur in the isolators, leading to greater transmission.
5. There are usually multiple mounts and mounting points which may interact in complex ways.
6. Usually, more than one direction of motion is important, including rotations.

Some of these aspects are discussed further.

5.10.1.2 Mobility and impedance

To derive a general model for vibration isolation, the concepts of mechanical impedance and mobility are required (see Table 3.1). The point mobility* Y of a structure is the ratio of the complex amplitude of velocity response to a unit amplitude force at a given frequency:

$$Y(\omega) = \frac{v(\omega)}{F(\omega)} \tag{5.101}$$

This is the inverse of the mechanical impedance, Z:

$$Z(\omega) = \frac{1}{Y(\omega)} = \frac{F(\omega)}{v(\omega)} \tag{5.102}$$

Both quantities are complex and include the relative phase between force and velocity.

Some examples of impedance and mobility are given in Table 5.8. It may be noted that the real part of any point impedance (or mobility) is always positive, corresponding to energy dissipation (or radiation). The imaginary part can be positive or negative (corresponding to mass or stiffness; for example, a mass has a positive imaginary impedance and a negative imaginary mobility). The

---

* In Chapter 3, the symbol μ is used for mobility.

Table 5.8 Examples of impedance and mobility

	Impedance	Mobility
Mass m	$Z = i\omega m$	$Y = \dfrac{-i}{\omega m}$
Spring stiffness K	$Z = \dfrac{-iK}{\omega}$	$Y = \dfrac{i\omega}{K}$
Viscous damper C	$Z = C$	$Y = \dfrac{1}{C}$
Infinite beam (area A, second moment of area I)	$Z = 2(1+i)\omega^{1/2}(EI)^{1/4}(\rho A)^{3/4}$	$Y = \dfrac{1}{Z}$
Infinite plate (thickness h)	$Z = 8h^2 \sqrt{\dfrac{E\rho}{12(1-\nu^2)}} = 2.3 c'_L \rho h^2$	$Y = \dfrac{1}{Z}$

Note: Young's modulus = E, density = ρ, Poisson's ratio = ν, longitudinal wave speed in a plate is $c'_L = \sqrt{E/\rho(1-\nu^2)}$.

impedance of an infinite plate is real and independent of frequency (equivalent to a damper). The impedance of an infinite beam has a damper part and a mass part, both of which are frequency dependent.

As a rule of thumb (Cremer et al. 2005), the point impedance of a structure can be approximated from the mass of the structure that is located within a quarter of a wavelength of the excitation point ($m_{\lambda/4}$), that is, $|Z| \approx \omega m_{\lambda/4}$. It is the shortest wavelength present, usually that of the bending waves, that should be considered.

At high frequencies, the impedance or mobility of a finite structure tends to that of the equivalent infinite structure, especially where frequency-average values are considered. An example is shown in Figure 5.50 for a finite plate. At low frequencies, clear resonances are seen, but at high frequencies the resonances tend to overlap and the average tends to that for an infinite plate of the same thickness. This approach can be useful in estimating the mobility of complex structures at high frequency.

5.10.1.3 Input power

When an external excitation is applied to a structure at a point in a single direction, the power input to the structure is given by the product of the force and the velocity. For harmonic inputs with complex amplitudes F and v, the time-average input power is given by

$$W_{in} = \tfrac{1}{2} \mathrm{Re}(F^* v) \tag{5.103}$$

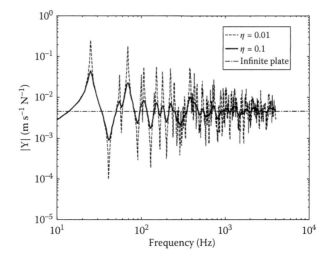

Figure 5.50 Magnitude of the point mobility of 1.5 mm steel plate 0.6×0.5 m, $\eta=0.01$ and 0.1.

where * denotes the complex conjugate. Since the force and velocity are related by the mobility and impedance, this can be written as

$$W_{in} = \text{Re}(Z)v_{rms}^2 \qquad (5.104)$$

or

$$W_{in} = \text{Re}(Y)F_{rms}^2 \qquad (5.105)$$

where the factor of ½ is accounted for by the use of the squared r.m.s. amplitudes.

5.10.1.4 Impedance mismatch

Consider two structures, a source and a receiver, coupled at a point, as shown in Figure 5.51. The degree to which their impedances $|Z_s|$ and $|Z_r|$ differ is called the impedance mismatch. The power injected to the receiver structure by the vibration of the source structure reaches a maximum when the impedances are similar ($|Z_s| \approx |Z_r|$) and reduces as the difference between them increases (Mondot and Petersson 1987). Although the impedances are complex, their relative phase is only important when their magnitudes are similar. Where there is an impedance mismatch, only a small part of the 'incident' power is transmitted to the receiver.

Noise control

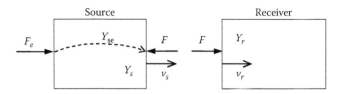

Figure 5.51 Source and receiver structures connected at a point.

To derive a simple model of a source structure coupled to a receiver, both structures will be modelled by their mobility at the interface (impedance could be used instead): Y_s for the source structure and Y_r for the receiver structure mobility (both functions of frequency). Suppose an external force F_e acts on the source at some arbitrary point. Y_{se} is introduced for the transfer mobility of the source (the response at the connection point due to a force at the excitation point and the structure is not connected to the receiver).

At the point of attachment, the velocities of the source and the receiver are identical, say v. The equations of motion of the source and receiver give

$$v = Y_r F \tag{5.106}$$

$$v = Y_{se} F_e - Y_s F \tag{5.107}$$

The source can be described by its *free velocity*, the velocity it would have at the interface if it were not connected to the receiver:

$$v_f = Y_{se} F_e \tag{5.108}$$

Hence, from Equations 5.106 and 5.107,

$$F = \left(\frac{1}{Y_r + Y_s}\right) v_f \text{ and } v = \left(\frac{Y_r}{Y_r + Y_s}\right) v_f \tag{5.109}$$

The source can also be defined by its *blocked force*, F_b, which is defined as the reaction force at the interface if it is connected to a rigid receiver. This gives

$$Y_s F_b = Y_{se} F_e \tag{5.110}$$

Consequently, $v_f = Y_s F_b$. Hence,

$$F = \left(\frac{Y_s}{Y_r + Y_s}\right) F_b \text{ and } v = \left(\frac{Y_r Y_s}{Y_r + Y_s}\right) F_b \tag{5.111}$$

5.10.1.5 Force and velocity excitation

In general, the input power is given by Equations 5.104 and 5.105. Combining these with Equation 5.109 gives

$$W_{in} = \frac{\text{Re}(Y_r)}{|Y_r + Y_s|^2} v_{f,\text{rms}}^2 \tag{5.112}$$

From this, it is clear that the power supplied to the receiver depends on the properties of both the source and the receiver structures; unlike airborne sources, there is no simple division possible into a 'source-strength' and a receiver transfer function. The same can be seen in Equations 5.109 and 5.111. However, it is possible to identify two extreme situations:

1. $|Y_s| \gg |Y_r|$. The force exerted on the receiver is unaffected by (small) changes in the receiver. The source can therefore be approximated by a constant force, $F \approx F_b$. This is known as *force excitation*. Equation 5.105 can be used for the input power together with the blocked force. Examples include light machinery on a heavy concrete floor.
2. $|Y_s| \ll |Y_r|$. The velocity imposed on the receiver is unaffected by (small) changes in the receiver. The source can therefore be approximated by a constant velocity, $v \approx v_f$. This is known as *velocity excitation*. Equation 5.104 can be used for the input power together with the free velocity. Examples include heavy machinery on a light wooden floor.

5.10.1.6 High-frequency model for vibration isolation

To ensure an impedance (or mobility) mismatch, resilient isolators are often introduced between the source and receiver structures. The model can now be extended, as shown in Figure 5.52, to cover this situation, including flexibility of the source and the receiver structures. A massless isolator is assumed (so that the forces F_1 and F_2 are equal) with a mobility defined by

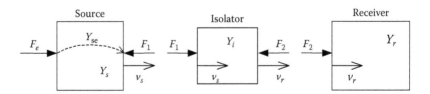

Figure 5.52 Source and receiver structures connected via an isolator.

$Y_i = (v_s - v_r)/F$. Using the same procedure as in Section 5.10.1.5, it is found that

$$F = \left(\frac{1}{Y_r + Y_i + Y_s} \right) v_f \quad \text{and} \quad v_r = \left(\frac{Y_r}{Y_r + Y_i + Y_s} \right) v_f \qquad (5.113)$$

These reduce to Equation 5.109 if the isolator is omitted, that is, the receiver is rigidly connected to the source, $Y_i = 0$.

The isolator effectiveness is defined as the ratio of the receiver velocity without the isolator to that with the isolator, which from Equation 5.113 is given by

$$E = \left| \frac{Y_s + Y_r + Y_i}{Y_s + Y_r} \right| = \left| 1 + \frac{Y_i}{Y_s + Y_r} \right| \qquad (5.114)$$

For good isolation, E should be as large as possible. This can also be written as an insertion loss in decibels: $IL = 20 \log_{10}(E)$. E depends on the mobilities of the source, isolator and receiver. For good isolation it is clear that the requirement is for:

- A stiff receiver (low Y_r)
- A stiff source (low Y_s)
- A flexible isolator (large Y_i)

However, note that the effect of a change in, say, Y_r should be determined in terms of the change in power input to the receiver, not the change in E. This is given by

$$W_{in} = \frac{\text{Re}(Y_r)}{|Y_r + Y_i + Y_s|^2} v_{f,\text{rms}}^2 \qquad (5.115)$$

5.10.1.7 Practical considerations

The high-frequency, low-amplitude dynamic stiffness of isolators, particularly those formed of viscoelastic materials, is usually much higher than the static or low-frequency, high-amplitude stiffnesses quoted by manufacturers. This is partly due to the frequency-dependent behaviour of elastomers (see Section 5.10.2.4).

Additionally, real isolators are not massless. This leads to internal resonances (standing waves) at high frequencies, as indicated in Figure 5.53. This shows the transfer stiffness (the force at the receiver side due to

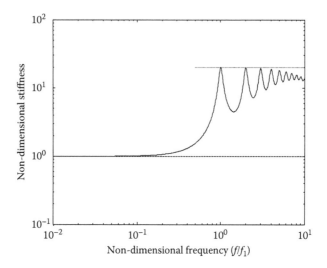

Figure 5.53 Increase in transfer stiffness of a column of elastic material due to internal resonances. The stiffness is normalised to the quasi-static value K (shown as dashed line). The frequency is normalised to the first internal resonance $f_1 = (E/\rho)^{1/2}/(2L)$. $\eta = 0.1$ in the present example.

a displacement at the source side) for a rod of elastic material (see e.g. Gardonio and Brennan 2004):

$$K_{12} = K \frac{kL}{\sin(kL)} \tag{5.116}$$

where:
$K = EA/L$, the quasi-static stiffness
E is Young's modulus
A is the cross-sectional area
L is the height
$k = \omega\sqrt{\rho/E}$ is the wavenumber in the material
ρ is the density

Damping is included by replacing E by $E(1+i\eta)$, where η is the damping loss factor (see Section 3.3.4 and Section 5.10.2.5). When the height of the isolator corresponds to one or more half-wavelengths in the material, the transfer stiffness has a maximum value of $2K/\eta$ and the transmitted power is also increased. The first such resonance occurs at $f_1 = \sqrt{E/\rho}/(2L)$.

In practical installations, sources have six degrees of freedom, leading to six rigid body resonances of the installation. To allow for multiple degrees of freedom at each isolator, a mobility *matrix* is required to describe the isolator properties, as well as for the source and receiver structures. A 12×12 mobility matrix is required for each isolator, and 6×6 matrices for the source and receiver structures. The direction with the largest amplitude of source motion is not necessarily dominant in transmitting power to the receiver; for example, if the isolator is stiffer in other directions or the corresponding receiver mobilities are higher. Rotational degrees of freedom can often become important at higher frequencies.

A machine is supported on multiple mounting points (usually three or four). At low frequencies, their interaction can be complex, but at high frequencies they can be considered as independent sources of power transmission to the receiver structure when they are further apart than about half a wavelength (in the receiver structure).

5.10.2 Damping

Damping is a measure of the energy loss (dissipation) in a structure. This can be quantified by the loss factor, see Section 3.3.4. While steel and aluminium have material loss factors of the order of 10^{-4}, building materials such as brick or masonry have loss factors between 3×10^{-3} and 2×10^{-2} and concrete between 5×10^{-3} and 5×10^{-2} (Vér and Beranek 2005). However, built-up metal structures often have much higher damping loss factors than the material itself, due to the influence of welded, bolted or riveted joints. These are typically of the order of 10^{-2}, and tend to decrease with increasing frequency in proportion to $f^{-1/2}$.

By adding damping treatments to a structure, it is possible to increase structural loss factors to the order of 10^{-1}, particularly for thin metal structures. Damping treatments such as constrained layer damping (see Section 5.10.2.4) are usually only effective when applied to thin structural components; for example, metal panels that are a few millimetres thick. Clearly, to be effective, damping treatments have to introduce much more damping than is already present in the structure, including the damping effects of installation.

5.10.2.1 Effects of added damping

Increasing the damping of a structure has a number of effects. First, it will affect the point mobility, reducing the height of resonance peaks and dips. However, this is only effective at low to middle frequencies where the response is dominated by individual resonances. Figure 5.50 shows an example of the point mobility of a steel plate with two different values of damping loss factor. Clearly, at high frequencies the mobility approaches that of the equivalent infinite plate more rapidly as the damping is increased and further increases in damping have no effect on the point mobility.

Second, structural waves are attenuated as they propagate through a structure. This can be quantified by the attenuation rate D (in decibels per metre). To reduce structure-borne noise transmission, D should be as high as possible. For example, for bending waves, $D = 13.6\eta/\lambda_B$, where λ_B is the structural wavelength; as λ_B decreases with increasing frequency, at high frequencies significant attenuation can be achieved, but at low frequencies the effect will be small.

The sound radiation is reduced by a combination of these two factors.

5.10.2.2 Mean-square velocity of a plate

An alternative way of considering the attenuation of vibration across a structure is to calculate the mean-square response. Recalling Equation 5.17 for the radiated sound power, it is clear that the spatially averaged mean-square velocity should be minimised. To estimate this, assuming steady state, the input power can be equated to the dissipated power. The dissipated power can be estimated from the product of the loss factor, the circular frequency and the vibration energy; at high frequencies (high modal overlap), the time-averaged kinetic and potential energies are equal. Thus, for a flat plate of area S, thickness h and density ρ_m, the dissipated power is given by

$$W_{\text{diss}} \approx \eta \omega \rho_m h S \langle \overline{v^2} \rangle \tag{5.117}$$

Hence,

$$\langle \overline{v^2} \rangle = \frac{W_{\text{in}}}{\eta \omega \rho_m h S} \tag{5.118}$$

The input power is given either by Equation 5.104 or Equation 5.105. For example, using the mobility of an infinite flat plate from Table 5.8, the mean-square velocity for velocity excitation (r.m.s. amplitude $v_{0,\text{rms}}$ at the drive point) is given by

$$\frac{\langle \overline{v^2} \rangle}{v_{0,\text{rms}}^2} = \frac{2.3 \, c_L' \, h}{\eta \, \omega \, S} \tag{5.119}$$

while for force excitation (r.m.s. amplitude F_{rms})

$$\frac{\langle \overline{v^2} \rangle}{F_{\text{rms}}^2} = \frac{1}{2.3 \eta \omega S c_L' \rho_m^2 h^3} \tag{5.120}$$

In both cases, the vibration, and hence the sound power, can be reduced by 10 dB by a tenfold increase in the damping loss factor.

5.10.2.3 Sound radiation from near field around forcing point

As the damping of plate-like structures is increased, the mean-square velocity is reduced. However, a vibrational near field remains around the forcing point, radiating sound which should be added to the estimates in Section 5.5 in the region below the critical frequency. From Cremer et al. (2005) and Fahy and Gardonio (2006), the radiated power can be calculated by modifying the radiation ratio to include an additional term:

$$\sigma_n = \frac{4f}{\pi f_c} \eta \quad \text{for } f < f_c \tag{5.121}$$

Although this increases with increasing damping, it is used in combination with the mean-square velocity of the whole plate (Equation 5.118) which, in turn, decreases with increasing damping. Hence, the corresponding component of the sound power due to the near field is independent of the damping and can be written as

$$W_{\text{rad},n} = \rho_0 c_0 S \sigma_n \langle \overline{v^2} \rangle = \frac{2\rho_0 c_0 W_{\text{in}}}{\pi^2 f_c \rho_m h} \quad \text{for } f < f_c \tag{5.122}$$

where the subscript n signifies near field. This component of sound radiation can limit the benefit of added damping when the loss factor is already high.

5.10.2.4 Methods of adding damping

There are many different means of increasing the damping of a structure. At low frequencies, viscous dampers such as shock absorbers or hydromounts are effective. However, at high frequencies other techniques are required. Added layers of material with a high loss factor (unconstrained layer or constrained layer damping treatments) are commonly used, as described in Section 5.10.2.6.

Other techniques include squeeze film damping: an air or fluid layer is trapped between two thin plates and energy is lost as the air/fluid is pumped to and fro within the thin cavity. Friction damping is important at bolted or riveted joints. Damping can also occur through impacts, as energy is converted into higher-frequency vibration, which is then dissipated more readily. Where possible, a change of structural material (e.g. composites) can be used to produce an increase in damping.

Finally, tuned mass-spring systems can be used to add damping to a structure. A tuned damper or absorber (damped mass-spring system) can be useful for increasing the damping, especially at lower frequencies where constrained layer treatments are less effective. This has been applied to a railway track in Thompson et al. (2007) and Thompson (2008). A dynamic neutraliser, a lightly damped added mass-spring system, does not strictly add damping but can be useful for dealing with tonal excitation by effectively pinning the host structure at the natural frequency of the neutraliser, which should be tuned to match the excitation frequency (see Section 3.6.5 and Brennan and Ferguson 2004).

5.10.2.5 Material properties of viscoelastic materials

Viscoelastic materials (plastics, polymers, rubbers, etc.) have nonlinear material behaviours.

For a harmonic input, we can define the dynamic properties of viscoelastic materials in terms of their complex Young's modulus $E(1+i\eta)$ or shear modulus $G(1+i\eta)$, where η is the damping loss factor. E, G and η are dependent on frequency, temperature, strain amplitude, preload and strain history. The frequency and temperature dependencies are equivalent (a higher frequency is equivalent to a lower temperature).

Figure 5.54 shows the typical behaviour of a viscoelastic material. This exhibits a 'glassy' behaviour at low temperature or high frequency with a high stiffness and low loss factor. At high temperature or low frequency, it exhibits 'rubbery' behaviour with a low stiffness and low loss factor. Between these two regimes is a transition region where the modulus is strongly temperature dependent, and to an extent frequency dependent,

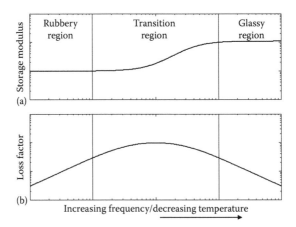

Figure 5.54 Typical behaviour of viscoelastic material: (a) storage modulus; (b) loss factor.

and the loss factor has a peak (Nashif et al. 1985). The height of the loss factor peak in the transition region is directly related to the steepness of the slope of the modulus, so effective damping materials tend to be limited in their temperature range of application.

5.10.2.6 Unconstrained and constrained damping layers

Surface damping treatments, in the form of layers of viscoelastic material, take two generic forms, shown in Figure 5.55. An 'unconstrained' damping layer is applied as a single layer, either sprayed, painted or bonded onto the surface of the structure. To be effective, such layers should be relatively stiff and not too thin. Damping losses are caused by extension of the viscoelastic layer.

A 'constrained' damping layer consists of a layer of viscoelastic material with a thin sheet of stiff material bonded onto its outer surface. This cover sheet constrains the viscoelastic material to deform in shear so that the energy dissipation is enhanced and can be considerable even for thin layers. For models of the behaviour of both of these systems (see Nashif et al. 1985; Mead 1998; Brennan and Ferguson 2004).

5.10.3 Structural modification

5.10.3.1 Added mass or stiffeners

At low frequencies, where individual resonances are far apart and play a large role in the response, it is possible to reduce resonance problems by structural modification. The purpose is to avoid the coincidence of a resonance with a strong excitation frequency. The stiffness can be increased to increase the natural frequencies, or the mass can be increased to reduce them. Added damping (see Section 5.10.2) can be used to reduce the response at resonance, or mass-spring systems can be used, tuned to deal with a problem frequency. At higher frequencies, where there is greater modal overlap, structural modification is aimed mainly at reducing the drive point mobility.

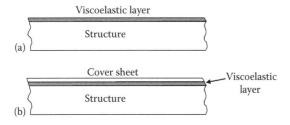

Figure 5.55 (a) Unconstrained damping layer; (b) constrained damping layer.

A structure can be stiffened by adding beams, ribs or flanges. As noted, in Section 5.10.1.2 these have to be located close to the excitation point (within a quarter of a structural wavelength) to have any significant effect on the mobility. It is thus important to ensure that mounting points are located over stiffeners, not between them. An increased plate or beam thickness is also effective in reducing the mobility, as the impedance is proportional to the square of the thickness (see Table 5.8).

If mass is added to a structure, the impedance of the combined structure is given by the sum of the impedances of the structure and the added mass. The impedance of a mass m is $i\omega m$, which increases in magnitude with increasing frequency, so added mass can be particularly effective at higher frequencies provided that it can be located close enough to the excitation point.

The effect of an added mass on a plate is shown in Figure 5.56. From this, it can be seen that a relatively small added mass can have a significant effect at high frequency. The real part of the mobility is reduced by more than its magnitude due to the fact that the mobility of the mass is imaginary and that of the plate is real valued. Thus, at high frequencies the combined mobility Y becomes both smaller and more imaginary:

$$|Y| = \left| \frac{1}{Z_p + i\omega m} \right| = \left(\frac{1}{|Z_p|^2 + (\omega m)^2} \right)^{1/2} \approx \frac{1}{\omega m} \quad (5.123)$$

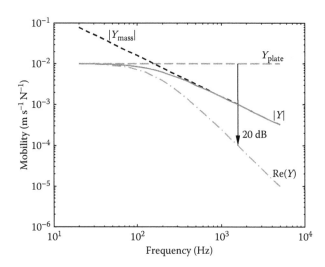

Figure 5.56 Example of added mass: 1 mm steel plate (7.8 kg m^{-2}) with 100 g added mass.

$$\text{Re}(Y) = \text{Re}\left(\frac{1}{Z_p + i\omega m}\right) = \left(\frac{\text{Re}(Z_p)}{|Z_p|^2 + (\omega m)^2}\right) \approx \frac{Z_p}{(\omega m)^2} \quad (5.124)$$

where Z_p is the plate impedance.

5.10.3.2 Effects of plate thickness and material

Throughout this chapter, a number of factors have been discussed involving plate thickness. In some situations it is advantageous to use thicker plates, whereas in other situations thinner plates are beneficial. These effects are summarised in Figure 5.57. Increased thickness is beneficial in reducing the vibration response due to a force excitation, but has a negative effect for a velocity excitation (see Section 5.10.1.5). The sound reduction index

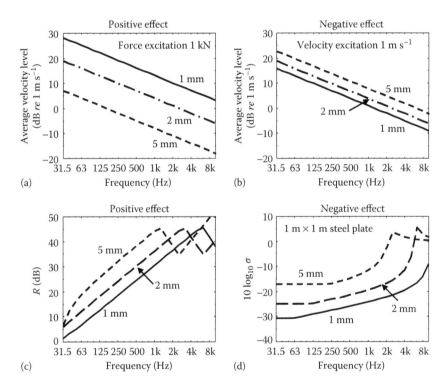

Figure 5.57 Summary of the effects of increasing plate thickness. (a) Vibration response decreased for force excitation, $\langle v^2 \rangle \propto h^{-3} F_{rms}^2$; (b) vibration response increased for velocity excitation, $\langle v^2 \rangle \propto h\, v_{0,rms}^2$; (c) sound reduction index increased in mass-law region (but note reduction at new critical frequency); (d) radiation efficiency increased in acoustic short-circuiting region and critical frequency is lowered.

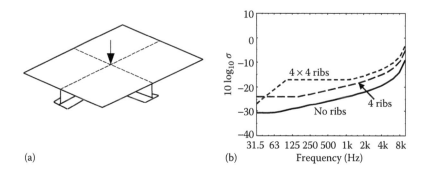

Figure 5.58 Summary of the effects of adding stiffeners to a plate. (a) Adding stiffeners where a force excitation acts (e.g. where a resilient mount is attached) is beneficial due to increase in impedance. (b) Adding stiffeners to a plate reduces the acoustic short-circuiting effect, hence increasing the radiation efficiency.

is increased (sound transmission is reduced) at most frequencies, although not all, by an increase in plate thickness (see Section 5.6), but the sound radiation ratio due to structural excitation is increased (see Section 5.5). These effects have to be considered together for any particular application to determine the overall effect.

Similarly, stiffening ribs can be beneficial in reducing the power input to a structure by a force excitation due to reduced mobility, although not for velocity excitation, but they have an adverse effect on the sound radiation ratio, as shown in Figure 5.58.

Changing the material of a plate will affect the critical frequency, the radiation ratio and also the mean-square velocity. As an example, Figure 5.59 shows measured results on three experimental engine valve covers (Thompson et al., 2002). These had the same geometry but were made from different materials. In the left-hand figure, the radiation ratio is smallest for the composite cover and highest for the aluminium cover. However, the average vibration for a given base input, shown in the right-hand figure, shows the opposite trend. From the combination of these results, it is the lighter, composite cover which has the highest overall radiated sound power.

5.11 FURTHER READING

For a more extensive treatment of engineering noise control, see the books by Bies and Hansen (2009) or Vér and Beranek (2005). The particular issues of sound radiation and transmission are dealt with in more detail in Fahy and Gardonio (2006) and Fahy (2000). The latter also includes

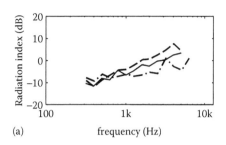 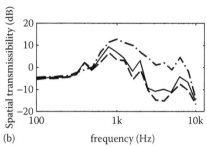

Figure 5.59 Acoustic performance of three geometrically similar valve covers made of different materials: (– – –): aluminium; (—): magnesium; (· — · — · — ·): composite. (a) Radiation index ($10 \log_{10} \sigma$); (b) spatial transmissibility. (Reprinted from Thompson, D.J., Grice, R.M. and Blaha, R., Assessment of the noise radiation from engine covers, *Proceedings of the European Conference on Vehicle Noise and Vibration*, London, 11–12 June 2002, 215–226 (C605/009/2002), 2002. With permission.)

a more extended discussion of absorptive materials; Cox and D'Antonio (2009) also discuss sound absorption in more detail. For coverage of vibration control, the chapter by Brennan and Ferguson (2004) and the book by Mead (1998) are recommended.

5.12 QUESTIONS

1. A manufacturer advertises an 'acoustic' door as providing 55 dB(A) noise reduction. Explain why this statement is scientifically unsatisfactory. Present an alternative, more satisfactory, form of statement of the acoustic performance of the door.
2. (a) The total sound power radiated by a petrol engine is known to be dominated by two mechanisms: mechanically induced noise increases at a rate of 6 dB(A) per doubling of speed, whereas combustion-induced noise increases at a rate of 15 dB(A) per doubling of speed. Sketch the speed dependence of the overall sound power due to the combination of these two sources.
   (b) At an engine speed of 1000 rev min^{-1}, the average sound pressure level at standard positions is measured as 70 dB(A). At the maximum engine speed of 6000 rev min^{-1}, it is measured as 98 dB(A). Estimate the speed at which the two source components are equal and the total sound pressure level at this speed.
   (c) Estimate the sound power level of the engine at its maximum speed, assuming that the microphones are located 1.2 m from the centre of the engine. Hence, estimate its acoustic efficiency if it produces a power of 100 kW at its top speed.

3. The noise in a particular situation is known to consist of several components from different paths and sources. The overall sound pressure level is measured as 80 dB, while the noise from the most important contribution is found to be 78 dB. Suppose that this source can be reduced by 10 dB. What is the likely reduction in overall noise level? Discuss the assumptions made and the likely margin of error in this estimate.

4. An enclosure is used to reduce the noise levels from a machine in a factory. It has dimensions of 3 m×2 m and a height of 2 m. There is an aperture of area 0.01 m². At the dominant frequency of the source, the absorption coefficient of the inner surface of the walls and top is 0.2 and their sound reduction index is 31 dB. Find the insertion loss of the enclosure at this frequency.

5. (a) A machine of mass 150 kg is excited by an out-of-balance force of amplitude 50 N at 200 Hz. Assuming that the machine mobility at this frequency is that of its mass, calculate its free-velocity amplitude.
   (b) The machine is mounted on a floor of mass per unit area 20 kg m⁻² and bending stiffness 8×10³ Nm. Estimate the mobility of the floor (you may assume it can be represented as an infinite plate). Hence, determine whether the machine acts as a force or velocity source. Determine the amplitude of the force acting on the floor.
   (c) Find the power input to the floor. Hence, assuming a damping loss factor of 0.03 and a surface area of 6 m², estimate the mean-square velocity of the floor. If it has a radiation ratio of 0.1 at 200 Hz, find the sound power radiated by the floor.

6. Measurements of the sound due to a source have been taken in a reverberation room of volume 200 m³. The average sound pressure levels in each octave band are listed below, along with the reverberation times. From these values, calculate the sound power levels in each band and hence the overall A-weighted sound power level.

Frequency band (Hz)	125	250	500	1000	2000	4000	8000
L_p (dB re 2×10⁻⁵ Pa)	78	82	81	79	78	75	72
A-weighting (dB)	−16.1	−8.6	−3.2	0.0	+1.2	+1.0	−1.1
Reverberation time (s)	2.2	2.4	2.3	2.2	2.0	1.8	1.5

7. A copper water pipe with an outer diameter of 22 mm and a wall thickness of 1 mm feeds a cistern. It is rigidly attached to a party wall and is causing excessive noise in the neighbouring property. It is required to reduce the noise at 500 Hz by 30 dB. Calculate the stiffness of a resilient vibration isolator required to achieve this. The properties of the

wall can be taken as: Young's modulus $E=2\times10^{10}$ N m^{-2}, Poisson's ratio $\nu=0.2$, density $\rho=1750$ kg m^{-3}, thickness $h=0.2$ m. (Hint: the pipe can be modelled as a beam with area $A=\pi(r^2-r_i^2)$ and second moment of area $I=\pi(r^4-r_i^4)/4$ where r is the outer radius and r_i is the inner radius.)

8. (a) Why does sticking acoustic tiles to the ceiling of a furnished room not significantly reduce noise created by upstairs neighbours?
   (b) Why does the sound from the music system inside a passing car appear to a pedestrian as though it only consists of bass notes and drum beats?
   (c) Why is typical double glazing effective in reducing high-frequency tyre noise and not low-frequency engine noise from traffic?
   (d) Explain why a person outside the closed door of a bathroom may understand the speech of a person inside the room when the shower is running, but it is unlikely that the person inside can understand the speech of the person outside.

9. For a surface with an impedance of $z_n/\rho c=1.8-2.5i$, calculate and plot the absorption coefficient for a range of angles of incidence. By evaluating $\partial\alpha(\varphi)/\partial\varphi=0$, find an expression for the angle of incidence at which the absorption is maximum. Compare this with the results in your graph.

10. A flat steel plate has dimensions 1 m $\times$ 0.8 m and thickness 1.5 mm. Estimate its radiation ratio at 1 kHz. If it is excited by a harmonic force of r.m.s. amplitude 10 N at 1 kHz, estimate its mean-square velocity assuming a damping loss factor of 0.03. Hence, estimate the radiated sound power from one side of the plate. If the plate is replaced by an aluminium one with thickness 3 mm, how do the radiation ratio, mean-square velocity and radiated sound power change?

REFERENCES

Allard, J.F. and Atalla, N. (2009) *Propagation of Sound in Porous Media: Modelling Sound Absorbing Materials*, 2nd edn, John Wiley & Sons, Chichester, UK.

Anderson, G.S. and Kurze, U.J. (2005) Outdoor sound propagation, in Vér, I.L. and Beranek, L.L. (eds), *Noise and Vibration Control Engineering: Principles and Applications*, 2nd edn, chapter 5, John Wiley & Sons, Hoboken, NJ.

Auweraer, H. van der, Wyckaert, K., Hendricx, W., and Linden, P. van der (1995) Noise and vibration transfer path analysis. *Proceedings of International Seminar on Applied Acoustics ISAAC 6*, Leuven, vol. IV, 1–22.

Beranek, L.L (1971) *Noise and Vibration Control*. McGraw-Hill, New York.

Bies, D.A. and Hansen, C.H. (2009) *Engineering Noise Control: Theory and Practice*, 4th edn, Spon Press, Abingdon.

Brennan, M.J. and Ferguson, N.S. (2004) Vibration control, in Fahy, F. and Walker, J. (eds), *Advanced Applications of Acoustics, Noise and Vibration*, chapter 12, Spon Press, London.

Cox, T.J. and D'Antonio, P. (2009) *Acoustic Absorbers and Diffusers*, 2nd edn, Taylor & Francis, London.
Cremer, L. (1942) Theorie der Schalldämmung dünner Wände bei schrägen Einfall, *Akustische Zeitung* 7, 81.
Cremer, L., Heckl, M. and Petersson, B.A.T. (2005) *Structure-Borne Sound: Structural Vibrations and Sound Radiation at Audio Frequencies*, 3rd edn, Springer, Berlin.
Delany, M.E. and Bazley, E.N. (1970) Acoustical properties of fibrous absorbent materials, *Applied Acoustics* 3, 105–116.
Elliott, S.J. and Nelson, P.A. (2004) Active noise control, in Fahy, F. and Walker, J. (eds), *Advanced Applications of Acoustics, Noise and Vibration*, chapter 8, Spon Press, London.
EU (2000) Directive 2000/14/EC of the European Parliament and of the Council of 8 May 2000 on the approximation of laws of the member states relating to the noise emission in the environment by equipment for use outdoors. *Official Journal of the European Communities* L162, 1–78.
EU (2006) Directive 2006/42/EC of the European Parliament and of the Council of 17 May 2006 on machinery, and amending directive 95/16/EC. *Official Journal of the European Communities* L157, 24–86.
EU (2014) Regulation (EU) no. 540/2014 of the European Parliament and of the Council of 16 April 2014 on the sound level of motor vehicles and of replacement silencing systems, and amending Directive 2007/46/EC and repealing Directive 70/157/EEC. *Official Journal of the European Union*, L158, 131–195.
Fahy, F. (1995) *Sound Intensity*, 2nd edn, Spon Press, London.
Fahy, F. (2000) *Foundations of Engineering Acoustics*, Academic Press, London.
Fahy, F.J. and Gardonio, P. (2006) *Sound and Structural Vibration*, 2nd edn, Academic Press, London.
Fuchs, H.V. and Zha, X. (2006) Micro-perforated structures as sound absorbers – A review and outlook, *Acta Acustica with Acustica* 92, 139–146.
Fuller, C.C., Elliott, S.J., and Nelson, P.A. (1996) *Active Control of Vibration*, Academic Press, London.
Gardonio, P. and Brennan, M.J. (2004) Mobility and impedance methods in structural dynamics, in Fahy, F. and Walker, J. (eds), *Advanced Applications of Acoustics, Noise and Vibration*, chapter 9, Spon Press, London.
Halliwell, R.E., Nightingale, T.R.T., Warnock, A.C.C., and Birta, J.A. (1998) *Gypsum Board Walls: Transmission Loss Data*, Internal Report IRC-IR-761, National Research Council, Canada.
Heckl, M. and Müller, G. (1994) *Taschenbuch der Technischen Akustik*, 2nd edn. Springer, Berlin.
Hopkins, C (2008) *Sound Insulation: Theory into Practice*. Elsevier, Oxford.
IEC (International Electrotechnical Commission) (2013) *IEC 61672-1:2013: Electroacoustics: Sound Level Meters. Part 1: Specifications*, International Electrotechnical Commission, Geneva.
ISO (International Organization for Standardization) (1993) *ISO 9614-1:2009: Acoustics: Determination of Sound Power Levels of Noise Sources Using Sound Intensity. Part 1: Measurement at Discrete Points*, International Organization for Standardization, Geneva.

ISO (International Organization for Standardization) (1995) *ISO/TR 11688-1:1995: Acoustics: Recommended Practice for the Design of Low-Noise Machinery and Equipment. Part 1: Planning*, International Organization for Standardization, Geneva.

ISO (International Organization for Standardization) (1996a) *ISO 9613-2:1996: Acoustics: Attenuation of Sound during Propagation Outdoors. Part 2: A General Method of Calculation*, International Organization for Standardization, Geneva.

ISO (International Organization for Standardization) (1996b) *ISO 9614-2:1996: Acoustics: Determination of Sound Power Levels of Noise Sources Using Sound Intensity. Part 2: Measurement by Scanning*, International Organization for Standardization, Geneva.

ISO (International Organization for Standardization) (2000) *ISO/TR 11688-2:2000: Acoustics: Recommended Practice for the Design of Low-Noise Machinery and Equipment. Part 2: Introduction to the Physics of Low-Noise Design*, International Organization for Standardization, Geneva.

ISO (International Organization for Standardization) (2001a) *ISO 10534-1:2001: Acoustics: Determination of Sound Absorption Coefficient and Impedance in Impedance Tubes. Part 1: Method Using Standing Wave Ratio*, International Organization for Standardization, Geneva.

ISO (International Organization for Standardization) (2001b) *ISO 10534-2:2001: Acoustics: Determination of Sound Absorption Coefficient and Impedance in Impedance Tubes. Part 2: Transfer Function Method*, International Organization for Standardization, Geneva.

ISO (International Organization for Standardization) (2002) *ISO 9614-3:2002: Acoustics: Determination of Sound Power Levels of Noise Sources Using Sound Intensity. Part 3: Precision Method for Measurement by Scanning*, International Organization for Standardization, Geneva.

ISO (International Organization for Standardization) (2003a) *ISO 15186-1:2003: Acoustics: Measurement of Sound Insulation in Buildings and of Building Elements. Part 1: Laboratory Measurements*, International Organization for Standardization, Geneva.

ISO (International Organization for Standardization) (2003b) *ISO 354:2003: Acoustics: Measurement of Sound Absorption in a Reverberation Room*, International Organization for Standardization, Geneva.

ISO (International Organization for Standardization) (2007) *ISO 362-1:2007: Acoustics: Measurement of Noise Emitted by Accelerating Road Vehicles: Engineering Method. Part 1: M and N Categories*, International Organization for Standardization, Geneva.

ISO (International Organization for Standardization) (2009) *ISO 3743-2:2009: Acoustics: Determination of Sound Power Levels of Noise Sources Using Sound Pressure: Engineering Methods for Small Moveable Sources in Reverberant Fields. Part 2: Methods for Special Reverberation Test Rooms*, International Organization for Standardization, Geneva.

ISO (International Organization for Standardization) (2010a) *ISO 10140-2:2010: Acoustics: Laboratory Measurement of Sound Insulation of Building Elements. Part 2: Measurement of Airborne Sound Insulation*, International Organization for Standardization, Geneva.

ISO (International Organization for Standardization) (2010b) *ISO 15186-2:2010: Acoustics: Measurement of Sound Insulation in Buildings and of Building Elements. Part 2: Field Measurements*, International Organization for Standardization, Geneva.

ISO (International Organization for Standardization) (2010c) *ISO 3741:2010: Acoustics: Determination of Sound Power Levels of Noise Sources Using Sound Pressure: Precision Methods for Reverberation Rooms*, International Organization for Standardization, Geneva.

ISO (International Organization for Standardization) (2010d) *ISO 3743-1:2010: Acoustics: Determination of Sound Power Levels of Noise Sources Using Sound Pressure: Engineering Methods for Small Moveable Sources in Reverberant Fields. Part 1: Comparison for Hard-Walled Test Rooms*, International Organization for Standardization, Geneva.

ISO (International Organization for Standardization) (2010e) *ISO 3744:2010: Acoustics: Determination of Sound Power Levels of Noise Sources Using Sound Pressure: Engineering Method Employing an Enveloping Measurement Surface in an Essentially Free Field over a Reflecting Plane*, International Organization for Standardization, Geneva.

ISO (International Organization for Standardization) (2010f) *ISO 3746:2010: Acoustics: Determination of Sound Power Levels of Noise Sources Using Sound Pressure: Survey Method Employing an Enveloping Measurement Surface over a Reflecting Plane*, International Organization for Standardization, Geneva.

ISO (International Organization for Standardization) (2010g) *ISO 3747:2010: Acoustics: Determination of Sound Power Levels of Noise Sources Using Sound Pressure: Survey Method Using a Reference Sound Source*, International Organization for Standardization, Geneva.

ISO (International Organization for Standardization) (2012) *ISO 3745:2012: Acoustics: Determination of Sound Power Levels of Noise Sources Using Sound Pressure: Precision Methods for Anechoic and Semi-Anechoic Rooms*, International Organization for Standardization, Geneva.

ISO (International Organization for Standardization) (2013) *ISO 3095:2013: Acoustics: Railway Applications: Measurement of Noise Emitted by Railbound Vehicles*, International Organization for Standardization, Geneva.

ISO (International Organization for Standardization) (2014) *ISO 16283-1:2014: Acoustics: Field Measurement of Sound Insulation in Buildings and of Building Elements. Part 1: Airborne Sound Insulation*, International Organization for Standardization, Geneva.

Janssens, M.H.A., Verheij, J.W., and Thompson, D.J. (1999) The use of an equivalent forces method for quantifying structural sound transmission in ships, *Journal of Sound and Vibration*, **226**(2), 305–328.

Kurze, U.J. and Anderson, G.S. (1971) Sound attenuation by barriers, *Applied Acoustics*, **4**, 35–53.

Laulagnet, B. (1998) Sound radiation by a simply supported unbaffled plate, *Journal of the Acoustical Society of America*, **103**, 2451–2462.

Leppington, F.G., Broadbent, E.G., and Heron, K.H. (1982) The acoustic radiation efficiency of rectangular panels, *Proceedings of the Royal Society of London A*, **382**, 245–271.

Lighthill, M.J. (1952) On sound generated aerodynamically, I. General theory, *Proceedings of the Royal Society of London*, **211A**, 564–587.
Maekawa, Z. (1968) Noise reduction by screens, *Applied Acoustics* **1**, 157–173.
Maidanik, G. (1962) Response of ribbed panels to reverberant acoustic fields, *Journal of the Acoustical Society of America*, **34**, 809–826 (see also Erratum: Maidanik, G. (1975), *Journal of the Acoustical Society of America*, **57**, 1552).
Mead, D.J. (1998) *Passive Vibration Control*. John Wiley & Sons, Chichester.
Mondot, J.M. and Petersson, B.A.T. (1987) Characterization of structure-borne sound sources: The source descriptor and the coupling function, *Journal of Sound and Vibration*, **114**(3), 507–518.
Morfey, C.L. (2001) *Dictionary of Acoustics*. Academic Press, London.
Nashif, A.D., Jones, D.I.G., and Henderson, J.P. (1985) *Vibration Damping*. John Wiley & Sons, New York.
Nelson, P.A. and Elliott, S.J. (1991) *Active Control of Sound*. Academic Press, London.
Petyt, M. and Jones, C.J.C. (2004) Numerical methods in acoustics, in Fahy, F. and Walker, J. (eds), *Advanced Applications of Acoustics, Noise and Vibration*, chapter 2, Spon Press, London.
Putra, A. and Thompson, D.J. (2010a) Sound radiation from rectangular baffled and unbaffled plates, *Applied Acoustics* **71**(12), 1113–1125.
Putra, A. and Thompson, D.J. (2010b) Sound radiation from perforated plates, *Journal of Sound and Vibration* **329**, 4227–4250.
Rayleigh, Lord (1896) *The Theory of Sound*, 2nd edn. Reprinted by Dover, New York, 1945.
Sewell, E.C. (1970) Transmission of reverberant sound through a single leaf partition surrounded by an infinite rigid baffle, *Journal of Sound and Vibration* **12**, 21–32.
Sharp, B. (1978) Prediction methods for the sound transmission of building elements, *Noise Control Engineering* **11**, 53–63.
Thompson, D.J. (2008) A continuous damped vibration absorber to reduce broadband wave propagation in beams, *Journal of Sound and Vibration* **311**, 824–842.
Thompson, D.J. (2009), *Railway Noise and Vibration: Mechanisms, Modelling and Means of Control*. Elsevier, Oxford.
Thompson, D.J., Grice, R.M., and Blaha, R. (2002) Assessment of the noise radiation from engine covers, *Proceedings of the European Conference on Vehicle Noise and Vibration*, London, 11–12 June 2002, 215–226 (C605/009/2002).
Thompson, D.J., Jones, C.J.C., Waters, T.P., and Farrington, D. (2007) A tuned damping device for reducing noise from railway track, *Applied Acoustics* **68**, 43–57.
Vér, I.L. and Beranek, L.L (2005) *Noise and Vibration Control Engineering: Principles and Applications*, 2nd edn. John Wiley & Sons, Hoboken, NJ.
Verheij, J.W. (1997a) Inverse and reciprocity methods for machinery noise source characterization and sound path quantification. Part 1: Sources, *International Journal of Acoustics and Vibration* **2**(1), 11–20.
Verheij, J.W. (1997b) Inverse and reciprocity methods for machinery noise source characterization and sound path quantification. Part 2: Transmission paths, *International Journal of Acoustics and Vibration* **2**(3), 103–112.

Villot, M., Guigou, C., and Gagliardini, L. (2001) Predicting the acoustical radiation of finite size multi-layered structures by applying spatial windowing on infinite structures, *Journal of Sound and Vibration* **245**, 433–455.

Walker, R. and Randall, K.E. (1980) *An Investigation into the Mechanism of Sound-Energy Absorption in a Low-Frequency Modular Absorber*. Report BBC RD 1980/12, BBC Research Department, Kingswood-Warren, UK.

Wallace, C.E. (1972) Radiation resistance of a rectangular panel, *Journal of the Acoustical Society of America* **51**, 946–952.

Xie, G., Thompson, D.J., and Jones, C.J.C. (2005) The radiation efficiency of baffled plates and strips, *Journal of Sound and Vibration* **280**, 181–209.

Chapter 6

Human response to sound

Ian Flindell

6.1 INTRODUCTION

6.1.1 Objective standards

Human subjective and behavioural response to sound is related to physical sound levels, but can also be considerably affected by many other variables according to individual sensitivities, attitudes and opinions. This can be a problem for standards and regulations which are intended to reduce annoyance, sleep disturbance or effects on health, but which have to be based on physical quantities. It is not practicable to use actual annoyance, sleep disturbance or effects on health when setting standards and regulations or defining contracts. It is desirable that the physical metrics used for noise limits, targets and benchmarks should reflect human response, even if they are not directly connected. Some physical metrics are better than others in this respect, and this chapter sets out to explain why this is so and to outline where the major sources of uncertainty can be found. Human response to vibration is dealt with separately in Chapter 7.

6.1.2 Uncertainties

Acoustic metrics can be used to predict or estimate average subjective or behavioural response, but only by assuming homogeneous populations. Real people are not homogeneous. Individual response varies considerably above and below the mean, and this means that, depending on the characteristics of the population, actual average response varies above and below any overall central tendency. Generally speaking, around one-third of the total statistical variance in human response can be attributed to differences in measured physical sound levels (depending on the metrics used); another third of the variance can be attributed to many so-called non-acoustic variables such as the situation and context in which the sound occurs, and psychological and physiological differences in sensitivity, susceptibility and attitudes towards the source of the sound; and the remaining third of the variance is effectively random and therefore unpredictable.

Many of these problems arise because sound is a naturally occurring and mostly useful physical phenomenon. Sound only becomes harmful when it has some negative or adverse effect and can then be defined as noise. Sound level meters cannot, unfortunately, tell the difference between sound and noise. Only people can do this.

Standard sound level metrics such as L_{Aeq} and L_{AFmax} (see Section 6.3.4) are administratively convenient but do not reflect differences in the character of the sound, or sound quality, which can be as or more important than sound level alone. People pay selective attention to different features of a sound in different situations. There is no universally applicable method for predicting which features will be attended to.

6.2 AUDITORY ANATOMY AND FUNCTION

6.2.1 Research methods

Human listeners perceive and respond to different kinds of sound and noise in different ways. Human response depends on the physical structures and physiological functions of the auditory system and on the different ways in which central neural and endocrine processing systems are connected to, and stimulated by, the peripheral auditory systems. The anatomy is well understood. However, understanding the detailed function of each component part is more difficult, because investigation requiring dissection unavoidably interferes with normal function.

Psychophysical methods are used to observe the time-, frequency- and sound-level-resolving powers of the ear by comparing subjective impressions between different types of sound. Subjective descriptions or reports of perceived magnitudes have greater uncertainties than *just-detectable thresholds* and *just-noticeable differences*. Just-noticeable thresholds and differences can be measured by asking a listener to press a button when they can detect an audible difference or identify which of two similar sounds is louder or differs in some other defined way. Performance can be strongly affected by motivation and generally improves with practice and then deteriorates with increasing fatigue. Experimental design requires compromise between conflicting sources of potential bias. Subjective reports are affected by additional uncertainties because of possible differences in interpretation of instructions and explanations of tasks. The results of psychophysical tests can be related to the underlying structures by comparing auditory capabilities before and after damage caused by disease or trauma, or by comparing auditory capabilities against theoretical models based on the anatomy. Nondestructive scanning techniques are increasingly being used to observe normal functions under different conditions and seem likely to make an increasing contribution in the future, but are outside the scope of this chapter.

Additional and often unknown variance is contributed by differences in physiological functions that are dependent on cell biochemistry, which can be very sensitive to various chemicals and hormones circulating in the blood. However, it is not usually practical to be able to measure or control hormone concentrations, except in specialist research.

It would be useful if classical introspection could provide insights, because this research method requires no equipment or special facilities. Unfortunately, people rarely have much conscious awareness of how they actually hear things, and instead just perceive the meanings of sounds to which they have actually paid attention. It can be very difficult to 'hear' sounds as they really are, particularly if they are unfamiliar. However, thinking about how particular anatomical structures might have evolved and to what biological purpose can sometimes contribute valuable and important insights. No single research method can provide complete understanding and experimental design always requires compromise to balance conflicting sources of bias.

The next sections describe the anatomy and how this is related to the observable functions. A simplified diagrammatic cross-section of the ear is shown in Figure 6.1.

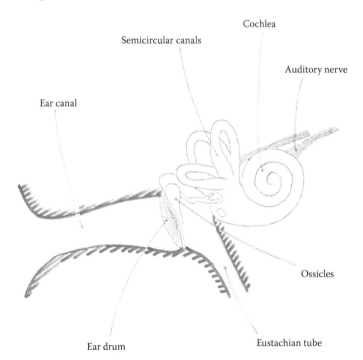

Figure 6.1 Simplified cross-section of the human ear.

6.2.2 External ear

Why do humans have two ears, one on each side of the head? Providing more than two auditory sensors could improve spatial perception and redundancy against damage or disease. Having only two ears is probably an accident of evolution. However, two ears are better than one, because *binaural hearing* improves sampling of complex sound fields, helps to discriminate between sounds coming from either side of the head and still provides some redundancy. Differences in relative amplitude and time of arrival between sounds arriving from either side help the central nervous system to localise sources on the left or the right much faster and with greater precision than would be possible with one ear alone. In fish ancestors living underwater, this function could have evolved from directional sensitivity to vibration inputs coming from different directions. Directional sensitivity confers a survival advantage from being able to quickly identify and then move away from, rather than towards, vibration that generates threats. Underwater predators additionally benefit from being able to quickly calculate the difference between threats and prey, so as to be able move either away from the threat or towards the prey, and thereby acquiring additional auditory neural processing.

Binaural recordings require two microphones in positions representative of the ears on a real head. Microphones can be put on a headband to record exactly the same sounds as heard by the person wearing the microphones, or they can be put on a dummy head, or simply spaced apart with either a solid disc or nothing (i.e. air) between them. On a real head or on a representative dummy head, the microphones should be capable of recording the full range of interaural differences in amplitude and time of arrival which would have been heard by a listener at the same position in space and time. Simplified binaural systems using an intervening solid disc or air gap between the two microphones reflect many, but not all, of the *interaural differences*. The main technical defect of binaural systems is that they cannot easily record the differential effects of head rotation and other movements on the interaural differences which provide important cues in real-life listening.

In real-life listening, if the head is fixed in position, it is difficult to tell whether sounds are coming from the front or the back. In real life, rotating the head helps listeners to resolve front-to-back and other directional ambiguities. Rotating the head when listening to binaural recordings using headphones has no effect on interaural differences. This encourages perceived stereo images to collapse into the centre of the head. When listening to binaural recordings, some listeners are better than others at disregarding or ignoring conflicting head rotation cues, leading to a wide range of subjective impressions. Binaural recording systems have been advocated for noise assessment and regulation purposes, on the premise that they are more representative of human hearing than standard measurements using single omnidirectional microphones, although the impossibility of being able to represent head rotation effects retrospectively somewhat undermines this philosophy.

In humans, the *external ear* or *pinna* is small compared with the size of the head. The pinna is not acoustically very significant except at higher audio frequencies, where the wavelengths are comparable. In human evolution, having an increasingly large brain sticking up above the ears has presumably been more beneficial for survival than having bigger ears sticking up above the head. The various folds and convolutions in the external ear provide directionally dependent acoustic filtering in both azimuth and elevation at higher audio frequencies where the wavelength is small compared with the physical dimensions but have no effect at lower frequencies. There is some evidence that elevation-dependent filtering by the external ear contributes to the perception of height, but only for sounds with significant high-frequency content. Readers can easily demonstrate this for themselves by holding their own ears flat to their head using their hands.

Binaural hearing helps to focus attention on sounds coming from directly in front while simultaneously helping to discriminate against background noise coming from all other directions. By cupping one hand behind each ear, readers can easily demonstrate that bigger ears can assist forward discrimination and can also 'improve' perceived stereo images when listening to two-channel loudspeaker stereo hi-fi systems. However, stick-on Mickey Mouse ears have not caught on as upgrades for stereo hi-fi systems, because it would be difficult to make them expensive enough.

It is interesting to speculate on what might have happened if humans had evolved bigger ears in proportion to their increasingly large brains. The very small contribution made by the current design of external ear to source height localisation at higher audio frequencies would have been extended lower down the audio-frequency scale, but the inconveniences of getting caught up on overhanging branches or being more visible to predators from greater distances might have offset these advantages. It is also possible that the shape and size of the external ear in humans could have been influenced by sexual selection without having very much to do with auditory function at all. The limited information available from the fossil record provides no definitive answers.

6.2.3 Middle ear

The middle- and inner-ear systems are protected within the lower part of the skull. Incident sound waves produce pressure disturbances at the entrance to the *ear canal* which are transmitted by wave motion to the sensitive *eardrum* which terminates the canal. The ear canal is acoustically resonant in the upper midrange of audio frequencies from around 3 to 4 kHz, depending on precise dimensions which vary between different individuals. Some authorities suggest that this resonance is the primary function of the ear canal, but a more plausible explanation is that the dimensions evolved simply to provide mechanical protection. Human infants with ear canals wider than their little finger would soon suffer damaged eardrums. In theory, fatter fingers could have provided an alternative evolutionary solution, but would have compromised operating mobile phones or picking one's nose.

The eardrum vibrates in response to rapid variations of sound pressure in the ear canal. The real-ear, head-related *transfer function* describes the relationship between the complex frequency spectrum of sound pressure at the position of the eardrum with no listener present and the spectrum at the eardrum. This transfer function varies for different angles of azimuth and elevation relative to the extended axis between the two ears, and between different individuals. This means that the frequency spectrum at the eardrum (which is what a person hears objectively) will be different from that recorded using an omnidirectional instrumentation-grade microphone as specified in most measurement standards. Most standards are intended to represent the preexisting sound field at the position in space where the listener's eardrum(s) would have been if the listener had been present and without any filtering due to the head-related transfer function. This difference is not a problem in practice, because human listeners each have a lifetime's experience of their own real-ear, head-related transfer functions, for which they make subconscious allowances when forming subjective impressions of the external acoustic environment.

There are special cases such as *bone conduction* or transmission via the *Eustachian tubes* at the back of the nasal cavity, or where hearing aids or headphones are used where a measurement in the ear canal may be more appropriate; but even in these special cases, it is often useful to be able to relate the measurements back to an equivalent free-field (listener not present) sound field for comparison purposes.

The air-filled *middle-ear cavity* behind the eardrum allows the eardrum to vibrate and to drive a chain of three small bones (the *ossicles*) to deliver vibration via a membrane (the *oval window*) to the fluid-filled cavities of the inner ear (the *cochlea*). The obvious function of the ossicles is to transmit vibrational motion from the eardrum across the air-filled middle-ear cavity, but they also provide leverage to convert motion from the relatively low *mechanical impedance** of the air outside to the relatively high mechanical impedance of the fluid-filled inner-ear cavities. The ossicles are thought to have evolved from jawbones in ancestor lizards which then became separated from the jaw to avoid interference when chewing food.

---

* For the purposes of this chapter, the mechanical impedance of fluids is considered as pressure over velocity rather than force over velocity in solids, as defined in Chapter 3.

Small muscles between the ossicles are able to tighten up the connections to provide limited protection against high-level transient sounds. This is known as the *acoustic reflex*, but it cannot protect against high-level continuous sounds. A fit person shouting can easily generate sound levels of 120 dBA or more at a distance of 1 m for up to a few seconds at a time. Close up to another person's ear canal, the sound levels are much higher and potentially damaging. The acoustic reflex provides some protection against this type of sound, but not against the type of continuous high-level sounds produced by machinery and heavy industry.

The middle-ear cavities are connected to the back of the throat and thence to the outside world via the Eustachian tubes, to allow drainage and equalisation of atmospheric pressure on both sides of the eardrums. The tubes can become blocked by inflammation and swelling of the lining, causing discomfort and reduced hearing sensitivity, which then recovers if the inflammation and swelling is temporary. Blockage is a common source of discomfort when the cabin pressure increases rapidly in a landing aircraft.

6.2.4 Inner ear

There are two flexible membranes in the dividing wall between the air-filled middle-ear cavity and the fluid-filled inner ear. The upper membrane, known as the *oval window*, accepts movement from the middle-ear ossicles. The lower membrane, known as the *round window*, vibrates in sympathy to release the internal pressure in the almost incompressible inner-ear fluid, which would otherwise prevent the oval window from moving. The inner ear comprises the parts concerned with hearing, the cochlea, and additional parts concerned with sensing motion, the semicircular canals. The cochlea is coiled through approximately two-and-a-half turns in a spiral. The *basilar membrane* divides the upper and lower chambers of the cochlea throughout its length, as shown in cross-section in Figure 6.2. It supports inner and outer rows of sensitive *hair cells* which are stimulated by the shearing action of the *tectorial membrane* across their tips whenever the basilar membrane as a whole is deflected up or down by vibrational waves in the cochlea fluid. The pattern of vibration along the basilar membrane represents the time-varying frequency content of incoming acoustic signals in the following way:

1. Incoming positive sound pressures deflect the eardrum, middle-ear ossicles and oval window inwards, creating a positive pressure in the upper chamber of the cochlea. The transmission of vibration from the ear canal to the cochlea involves a finite time delay, leading to frequency-dependent phase lag.
2. The positive pressure in the upper chamber causes downward deflection of the basilar membrane, with corresponding outward deflection of the round window directly beneath the oval window.

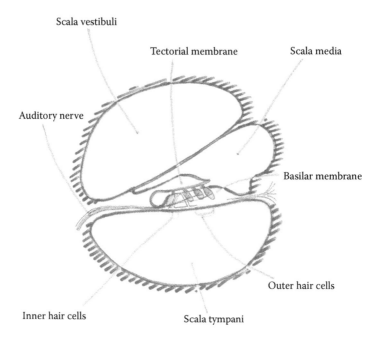

Figure 6.2 Cross-section of the basilar membrane.

3. The initial downward deflection of the basilar membrane progresses from the base end adjacent to the oval window towards the apex at the other end of the spiral as a travelling wave.
4. The travelling wave reaches a maximum amplitude at a position on the basilar membrane, which depends on the frequency content of the acoustic stimulus. The travelling wave is rapidly attenuated further along.
5. Low audio frequencies penetrate along to the apex of the cochlea spiral. The point of maximum amplitude moves further along from the base end (near to the oval window) as the frequency decreases. There is a physical connection between the upper and lower chambers at the apex known as the helicotrema which contributes to the low-frequency cut-off. Higher audio frequencies penetrate only a short distance from the base end along the basilar membrane. There is a frequency-dependent phase shift of the travelling wave along the basilar membrane.

The different vibrational patterns of the travelling wave at different frequencies act as a mechanical frequency analyser. The sensitive hair cells along the basilar membrane are organised such that auditory nerve impulses originating from different parts of the basilar membrane are associated

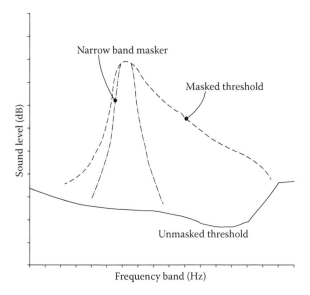

Figure 6.3 Upward spread of masking.

with different frequencies in the incident sound. The travelling wave passes through higher-frequency regions before reaching the point of maximum amplitude, and then decays rapidly. Lower-frequency sounds make it harder to hear higher-frequency sounds (this is known as the *upward spread of masking* – see Figure 6.3) but not the other way around. The travelling wave from a pure-tone test sound at a higher frequency than a masking noise reaches its maximum amplitude at a position along the basilar membrane, which is already stimulated by the lower-frequency masking noise, and this can make it harder to detect. The travelling wave from a pure-tone test sound at a lower frequency than a masking noise reaches its maximum amplitude at a position further along the basilar membrane beyond the area stimulated by the masking noise and can thereby be more easily detected or identified. One practical consequence of this design fault is that if the bass in a rock band is too loud, it can be difficult to hear the singer at higher audio frequencies.

The inner rows of hair cells appear to be mainly responsible for sending nerve impulses to the brain. Each nerve impulse is a single electrochemical discharge which travels from the cell body along the associated nerve fibres to the next connected neural interchange point. Every nerve impulse consumes biochemical resources which must be replenished before the cell can fire again. Consequently, there is a maximum firing rate for each nerve cell, and an associated and much lower resting firing rate, when the nerve cell fires spontaneously even when there is no input stimulation. The precise time at which each nerve cell is most likely to fire is uncertain, because it depends on the elapsed time since the last firing, the magnitude of input stimulation and the rate of biochemical recovery since the last impulse. Large-amplitude input signals generate stronger vibratory stimulation of hair cells and thus a higher firing rate for each associated nerve fibre. The dynamic range of individual hair cells is limited and the most sensitive connected nerve fibres become saturated at higher input sound levels. Other less sensitive combinations of hair cells and nerve fibres take over at higher sound levels to overcome the problems of the limited dynamic range of each hair cell and nerve fibre combination considered separately.

Wideband transient signals excite the basilar membrane throughout its length, causing complex patterns of nerve impulses to be sent along the auditory nerve, with the high-frequency-connected nerve fibres excited first. Continuous pure tones and narrowband frequency components stimulate continuous sequences of nerve impulses in the nerve fibres most closely associated with the part of the basilar membrane which is most strongly excited. The nerve-firing rate diminishes over time, as the biochemical resources of each hair cell and associated nerve fibres become depleted. The capabilities of the overall system with respect to relative phase information are limited by the general time uncertainty of separate nerve impulses, which exceeds the period of incident acoustic signals at middle and higher audio frequencies. There is some evidence that volleys of nerve impulses are phase-locked to the incident acoustic signal at lower audio frequencies, but it is not clear in what ways, if any, this information might be used.

For many practical purposes, the peripheral auditory system can be modelled as a bank of overlapping band-pass filters operating in real time, but it should be noted that there is no point in the real system at which the separate output of each modelled band-pass filter could be observed. The effective frequency selectivity of each modelled band-pass filter is of interest, and while this is generally assumed to be approximately equivalent to one-third of an octave over most of the audio-frequency range, the measured equivalent bandwidths vary depending on the method and conditions of measurement.

There is evidence that the outer rows of hair cells assist in fine frequency tuning. The outer rows are believed to vibrate in sympathy with incoming acoustic signals and thereby assist in tuning the system to narrowband and discrete frequency input signals. This enhances both frequency selectivity and sensitivity to weak sounds. It has been difficult to study this hypothesised active process *in vivo*, because access to the systems involved cannot be obtained without drastically interfering with or destroying them. The active process assists the ear to achieve the best compromise solution between time and frequency resolution that can be achieved. Because of *time–frequency uncertainty* (see Section 4.3.4), it is not possible to have both fine frequency resolution and fine temporal resolution in the same physical system at the same time. Fine frequency resolution requires that a sufficiently large number of cycles of a continuous waveform are observed to allow discrimination from slightly higher and slightly lower frequencies. For example, and in generic terms, it is necessary to observe 100 cycles of a cyclically repeating waveform within a fixed time period to be able to determine the frequency to within ±1%. Resolving the frequency of a 200 Hz tone to within 1% requires an observation time of around half a second, which could be too long to wait if an immediate reflex action is required.

In survival-critical situations, a rapid time response may be more important than fine frequency resolution. However, if a detected incoming acoustic signal is still present after the first few tens of milliseconds, an increasing capability to be able to resolve frequency content then becomes important for cognitive appraisal of the sound. Varying auditory capabilities in terms of frequency resolution increasing over short timescales from the initial transient appear to be consistent with the temporal and spectral variation of normal speech signals, suggesting synergies between the evolution of capabilities to produce speech sounds and the evolution of auditory systems intended for their resolution.

6.2.5 Auditory nerves

The final link in the chain is the system of auditory nerves which transmit nerve impulses from the hair cells in the basilar membrane to the brain. The auditory nerves pass through a number of interconnection points where nerves branch off to different parts of the brain and eventually reach the auditory cortex, where most of the complex processing involved in selective attention and cognitive appraisal is believed to take place. There is evidence of the presence of specialised nerve cells which are sensitive to specific acoustic features present in incoming acoustic signals, but it is not clear to what extent these systems are determined by genetic coding or develop naturally as a result of learning and experience. Multiple cross-connections between the left and right auditory nerves facilitate binaural and higher-order neural processing, which support source localisation and other complex perceptual functions. It seems unlikely that any simple model would be capable of representing the full complexity of the overall system.

6.2.6 Overview

The auditory system does not operate in the same way as a microphone. Incident sounds are represented by time-varying patterns of nerve impulses across parallel auditory nerve fibres, which are then interpreted by higher-order neural processing in the brain. People do not perceive incoming sounds as they really are. Instead, the brain constructs internal perceptual images of the external environment based on the overall pattern of incoming neural signals interpreted in the light of expectation and experience.

People are not always consciously aware of all sounds that are technically capable of being heard. Unfamiliar sounds might be completely unrecognisable. Conversely, it is also possible to perceive sounds that are not physically present, or to hear sounds not as they really are. Auditory consciousness remains a mystery.

6.3 AUDITORY CAPABILITIES AND ACOUSTIC METRICS

6.3.1 Auditory thresholds

The normal range of hearing is shown in Figure 6.4. The normal threshold-of-hearing curve at the bottom of the figure shows maximum sensitivity between 3 and 4 kHz, corresponding to the first quarter-wave resonance of the open-ended ear canal. In this frequency range, for healthy young adults with no hearing loss from any cause (defined as normal hearing), the minimum *r.m.s. sound pressure* (see Chapter 2) that can be just detected when listening in a diffuse field with both ears (defined as the *minimum audible field*) is around 10–20 μPa, which is equivalent to −6 to 0 dB sound pressure level (SPL or L_p) relative to 20 μPa when expressed in decibel terms. The corresponding *minimum audible pressure* measured in the ear canal is generally a bit higher than the minimum audible field and is similarly variable between different individuals. The differences between minimum audible field and minimum audible pressure may need to be taken into account

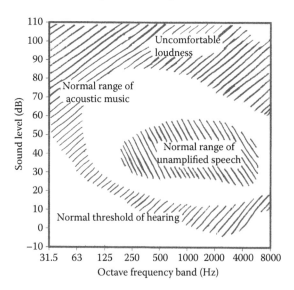

Figure 6.4 Normal range of hearing.

by audiometric equipment designers, but are not otherwise of any great concern.

'Just detected' means that the presence of the sound can be correctly identified at slightly better than chance level. Low-level sounds at the just-detected threshold would not necessarily be of any practical significance for everyday conditions, unless the listener had become particularly sensitised to those sounds for any reason. Individual thresholds vary above and below the mean at different frequencies, even within a sample of people where everyone has so-called normal hearing. It is not clear how much of this variability represents fundamental differences in hearing sensitivity or how much can be attributed to measurement uncertainty. Some people tend to be rather cautious when asked to press a button to show they can hear a particular sound, while others tend to press the button even when they are not quite sure if they can really hear the sound or not.

The normal hearing thresholds increase significantly below 300 Hz and above 8 kHz. There are upper and lower limits to the range of frequencies that can be heard by people with normal hearing. Children may be able to hear higher-frequency sounds up to around 20 kHz, and in some cases even higher, but a more practical upper limit for most adults would be around 15 kHz or less. It is not clear what survival advantages, if any, would accrue from being able to hear very high audio frequencies, which are mostly highly directional and potentially more useful for echolocation than general perception.

Most people are equally insensitive to very low audio-frequency sounds, particularly below 50 Hz. There is very little information content in speech sounds below about 300 Hz. Most natural low-frequency sounds are generated by distant sources which are less important for individuals than nearby sources. If there had been any significant survival advantages from being able to hear very low-frequency sounds generated by distant thunderstorms or by large predators stamping their feet, then this would have required larger ear structures to have evolved, which might in turn have had other disadvantages, such being more likely to be seen and eaten by those same predators.

The upper limit of normal hearing is often quoted as being somewhere around 120 dB L_p, although there is no clear definition of what the upper limit actually means. In the midfrequency range from around 1 to 5 kHz, a pure tone at 120 dB L_p can be extremely unpleasant and would certainly become strongly objectionable very rapidly if continued. Midfrequency pure tones could be increased to 140 dB and still be 'heard', that is, within the range of audibility, although would nevertheless be highly uncomfortable or even painful to listen to and would certainly lead to significant temporary threshold shift (TTS) (see Section 6.4.2). A continuous tone at 140 dB L_p and more could cause permanent damage in the most susceptible individuals, even after only relatively short exposure. However, a 50 Hz continuous tone at 140 dB would not be so uncomfortable and might even be inaudible if the frequency was below 20 Hz. It should be noted that the measurement of auditory thresholds at very low audio frequencies can be complicated by the need to avoid higher-frequency harmonic distortion products or nonlinear acoustic resonances in the testing loudspeakers or in nearby objects which might be relatively more audible than the original very low audio-frequency tones.

At higher sound levels, broadband sounds tend to be less uncomfortable than pure tones. This could be because the vibratory excitation pattern on the basilar membrane is not concentrated in one place, as it is for pure tones. High-level pure tones tend to increase in unpleasantness with increasing frequency, until the frequency eventually goes above the normal audible range and the sound becomes inconsequential because it cannot be heard. Similarly, the harmonic distortion products of higher-frequency tones reproduced through imperfect loudspeakers become inconsequential above the audible frequency range.

6.3.2 Just-noticeable differences

The smallest just-noticeable differences between otherwise similar pure tones or narrowband sounds occur in the middle range of sound levels from 40 to 80 dB L_p and frequencies from 300 Hz to 5 kHz. This usefully corresponds to the sound-level and frequency ranges which provide the most differentiation between different speech sounds. Under laboratory listening conditions, it is quite possible within this most sensitive range to be able to correctly identify at better than chance level the louder or higher-pitched of two otherwise similar sounds down to level and frequency differences of <1 dB or <1% respectively. Any significant time gap between the two sounds results in reduced discrimination performance, which might more properly be described as a failure of auditory memory. The just-noticeable differences for sound level and intensity become progressively wider outside the middle range of sound levels and frequencies.

There is a weak correlation between subjective loudness and objective sound level. People are usually much better at making relative or comparative judgements than at judging absolute magnitudes. The average human listener can judge which is the louder or higher-pitched of two sounds without necessarily being able to judge the absolute magnitude or frequency of either sound on its own. Relative loudness is similar to relative pitch, in that subjective magnitude tends to be proportional to the ratio of objective physical magnitudes rather than the linear-scale values. This is one of the main justifications for using a decibel scale for measuring sound level. A difference between two sounds of x decibels always represents the same ratio of physical magnitudes irrespective of absolute sound level. A just-detectable difference in subjective magnitude between two sounds of 1 dB could represent a very small linear-scale difference of 0.001 Nm^{-2} at low sound levels or a much larger linear-scale difference of 1 or 10 Nm^{-2} at higher SPLs (Morfey 2001).

Unfortunately, or perhaps unsurprisingly, the general public has almost no understanding of how decibel scales actually work when applied to sound levels. The public mostly understand the word 'decibel' simply as a synonym for 'loud'. If acoustical engineers and physicists use the word 'decibel' when describing or explaining noise management to the general public, not many people will actually understand what they are talking about. This can lead to outcomes not meeting expectations and can contribute to misunderstandings and even mistrust.

Just-noticeable differences are wider under real-life conditions than the typically 1 dB differences in sound level and 1% differences in frequency observable under controlled-listening laboratory test conditions. Under real-life conditions, it is unlikely that the average listener would be able to correctly identify at better than chance level the louder of two otherwise similar aircraft flyover or road-vehicle pass-by events which differed in maximum sound level by < 3 dB. If the second of the two events was delayed by more than a few minutes, an even bigger difference would be required to achieve reliable discrimination. The relative probability of judging the most recent of two sounds as being louder increases with time delay, simply because the auditory memory of the earlier sound fades. This could be one of the reasons why the significant reductions in aircraft-noise sound levels which have been achieved over the past 30–40 years have not led to expected increases in public satisfaction.

At planning inquiries where noise is an issue, the parties often disagree about the significance of small changes in sound levels. Residents are often more aware of other changes in their own right than any small resulting changes in sound levels. For example, residents are more likely to notice a doubling of road traffic than the 3 dB increase in long-time average sound levels caused by the increase in traffic. Alternatively, increasing the proportion of heavy vehicles with no change in overall traffic flow could increase long-time average sound levels by similar amounts, but in this case, it would be the noisier heavy vehicles which would be noticed rather than the difference in long-time average sound levels.

The smallest just-noticeable differences for frequency are concentrated in the frequency range where the ear has its highest sensitivity, from about 300 Hz up to about 5 kHz. It is presumably just as beneficial to be able to discriminate between similar frequencies in this range as it is to be able to hear them in the first place. The smallest just-noticeable differences for sound levels are concentrated in the middle region between quieter and louder sounds from about 40 to about 80 dB L_p. The human ear appears to be well adapted to be able to discriminate between the different sounds involved in human speech. Which evolved first, the auditory discrimination capabilities of the human ear, or the sound-level and frequency differences produced by the vocal tract when generating and differentiating different speech sounds? Either both systems evolved in tandem, or there was an intelligent designer at work, and for the purposes of this chapter, it does not much matter which.

According to evolutionary theory, it might be naively assumed that all features of the human auditory system have been optimised for biological survival. However, this is not necessarily true. One aspect of hearing that suggests that evolution does not always come up with the best design solution is the *localisation gap* in the middle of the audio-frequency range, where auditory localisation does not work as well as at higher and lower audio frequencies. Interaural phase differences become increasingly ambiguous at frequencies above about 750 Hz, where they can exceed 180°, because the wavelength is small compared with the angle-dependent differences in propagation distance. Interaural intensity differences are too small to be detectable at frequencies below about 2 kHz where the head is too small compared with the wavelength of the sound to have any significant screening effect. In the middle audio-frequency range from about 750 Hz up to about 2 kHz, neither interaural phase nor interaural intensity difference make much contribution to auditory localisation of pure tones. With the standard ear spacing, it is harder to localise the position in space of a nearby talker than it would be if the ears had been much further apart. Perhaps being able to localise the position in space of a nearby talker was less important for biological survival than the presumably negative effects of having wide-spaced ears on stalks on either side of the head.

6.3.3 Equal-loudness contours

The internationally standardised equal-loudness contours (BSI 2003b), plotted in Figure 6.5, show how the relative loudness of pure tones varies across the audio-frequency range and across the range of sound levels from 0 to 120 dB L_p. Listeners are asked to adjust the level of a pure tone at a given frequency until it matches the subjective loudness of a reference pure tone at a fixed reference frequency (normally 1 kHz). Different listeners give different results, so composite loudness contours are averaged across statistically representative groups of listeners. Alternative psychometric methods used in different listening laboratories also give different results, suggesting that the precise shape of the current contours is generically representative of, but not completely determined by, underlying auditory capabilities.

One of the methodological problems is that comparing the relative loudness of pure tones at different frequencies is more difficult than it seems. Different listeners probably adopt different strategies when attempting this task. Comparative loudness judgements are much more subjective than observing just-detectable thresholds at different frequencies. In addition,

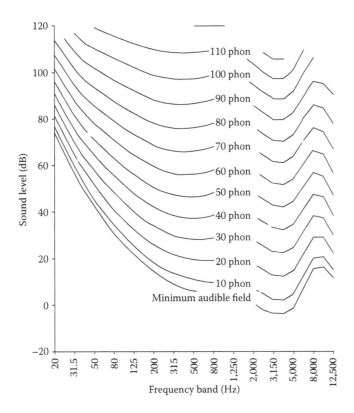

Figure 6.5 Standardised equal-loudness contours.

pure tones are rare in nature. Nobody has yet devised a method capable of deriving equivalent equal-loudness contours for more representative wide-band complex sounds.

Notwithstanding these uncertainties, equal-loudness contours can be assumed to represent fundamental properties of the peripheral auditory system as follows:

1. The higher hearing thresholds at the upper and lower extremes of the auditory frequency range (which are not in doubt because they can be measured objectively) appear to be reflected in reduced 'loudness' of very low and very high audio-frequency pure tones as compared with middle audio-frequency pure tones.
2. At the lower audio frequencies, the normal auditory threshold curve falls rapidly with increasing frequency and the corresponding equal-loudness contours above it flatten out at higher sound levels. This suggests that the relative or comparative loudness of low-frequency

sounds is proportionately greater at higher sound levels. However, it should be noted that the changing shape of the equal-loudness contours at higher sound levels could alternatively be an artefact of listeners preferring to avoid even higher sound levels when given experimental control.
3. Subject to the caveat noted in 2, the growth of subjective loudness with increasing sound levels appears to be much steeper at low and very high audio frequencies than in the middle frequency range. This is also associated with reduced sound-level and frequency discrimination outside the middle range of frequencies and sound levels.
4. Equal-loudness contours suggest that at 1 kHz, an average listener should be able to discriminate between 100 or more different sound levels between the auditory threshold at around 0 dB L_p and the so-called uncomfortable loudness level at around 120 dB L_p. At 50 Hz, the dynamic range between the auditory threshold and the uncomfortable loudness level is much less, and the just-noticeable difference for loudness is also much greater, suggesting that a far smaller number of different sound levels (and frequencies) should be separately distinguishable. This implies that bass players, for example, should be able to make more mistakes than violin players without the audience noticing.

The A-, B- and C-frequency weightings which were provided in traditional precision-grade sound level meters were intended to reflect the implied differences in the relative frequency response of the ear at different sound levels. The main function of a sound level meter is to measure sound levels as accurately and as repeatably as possible. These requirements are met by providing a precision condenser microphone, accurate and stable electronic systems, and appropriate and regular calibration. Problems arise when deciding which of the many possible acoustic metrics should be used for the measurement, each of which will give different results under different circumstances. A great deal of research effort has been expended over the past 50–60 years in various attempts to achieve higher correlations between sound level meter readings and subjective judgements to find the most appropriate acoustic metric. Hindsight has shown that this goal is unrealistic. It is not feasible for any single acoustic metric to represent all of the many variables which contribute to or determine subjective responses to noise, many of which are not even acoustic. On the other hand, the A-frequency weighting and the F- and S-time weightings are in general use for mainly practical reasons, and it is important that acoustical engineers and physicists should understand their advantages and limitations in everyday use.

For maximum objectivity, it is clearly desirable that the sound level meter should not impose any unnecessary time or frequency weighting or other nonlinearity on the measurement. Defined upper- and lower-frequency limits are desirable, so that inaudible infrasonic and ultrasonic frequency components can be excluded. Similarly, some form of averaging over short time periods is necessary, and it is desirable that the time weighting should not be unrepresentative of the equivalent time-averaging process in the human ear. In addition, any practical microphone imposes its own frequency response and high-frequency directivity onto the measurement. Precision condenser microphones can be set up in different ways to account for some, but not all, of these differences, but it is impossible to precisely match the frequency response and directivity of the human ear, particularly when binaural hearing is taken into account.

Most sound level meters available today are provided with a standardised *A-frequency weighting* setting in addition to an *unweighted* or *Lin* frequency response setting. Users are cautioned that unweighted or Lin frequency response settings in older equipment might not comply with current standards. Measurements using sound level meters with different unweighted or Lin frequency responses could give different results.

Most outdoor community noise standards specify the A-weighting. This reduces the sensitivity of the sound level meter at low and high frequencies compared with the upper-middle band of frequencies from around 1 kHz to around 4 kHz. The A-frequency weighting approximately follows the shape (inverted) of the 40 dB at 1 kHz equal-loudness contour. B- and C-frequency weightings may also be provided, particular on older instruments. The B- and C-weightings approximately follow the shape (inverted) of the 70 dB and 100 dB (respectively) at 1 kHz equal-loudness contours. The A-, B- and C-frequency-weighting curves are shown at Figure 6.6.

The original intention was that the A-, B- and C-frequency weightings should be used for the measurement of quiet, medium and loud sounds at around 40, 70 and 100 dB respectively. Three different frequency weightings were considered necessary, because the equal-loudness contours are steeply curved at the auditory threshold and become flatter at increasing sound levels. This well-intentioned scheme had two major defects. Approximating the assumed sound-level-dependent frequency response of the ear in this way does not account for many of the other acoustic (and non-acoustic) features that are likely to affect subjective impressions of the sound being measured. In addition, for sounds with prominent low- or high-frequency content, switching from one frequency weighting to another can have a considerable effect on the reading. To achieve consistency, selecting the appropriate frequency weighting for standardised measurements at different sound levels would have required arbitrary and precisely specified rules which have never been agreed.

Practical experience shows that the A-frequency weighting discriminates against predominately low-frequency wind noise generated at the microphone windshield when taking measurements outdoors and also

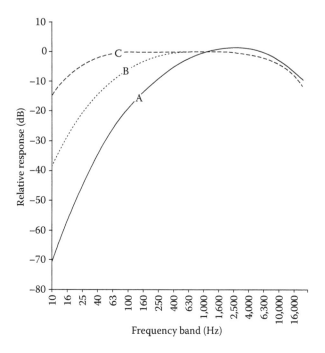

Figure 6.6 A-, B- and C-frequency weightings.

discriminates to some extent against low-frequency background noise from distant sources. These features alone are probably enough to explain its continued use. In addition, most field measurements carried out for research studies, and probably an even greater proportion of outdoor sound-level measurements carried out for many other purposes over the last 50–60 years, have used the A-weighting. Modest correlations have been obtained between A-weighted measurements of outdoor sound levels and subjective impressions of annoyance and disturbance across numerous noise-effects surveys. These correlations have often been wrongly accepted by many authorities as sufficient evidence that the A-weighting is the best weighting curve that could have been devised. In fact, for many field studies no other frequency weighting was tested, and for those studies where more than one frequency weighting was investigated, the two (or more) metrics are often so highly correlated with one another that no statistical distinctions could be drawn between them.

The B-frequency weighting has largely fallen out of use, mainly because it has not generally been found useful except for certain specialised applications. The C-frequency weighting is sometimes specified where there are low-frequency components present and there is concern that the

A-frequency weighting does not account for them properly. It should be noted that unweighted or 'flat' measurements are relatively more sensitive to extreme low- and high-frequency components (where present) than the human ear.

The A-frequency weighting has been compared against other frequency-weighting schemes in many laboratory-type listening studies. In theory, the derivation of the A-weighting from the 40 dB at 1 kHz equal-loudness contour means that it should become increasingly incorrect at higher sound levels, where it is most often used. It has been criticised for its indoor measurements, where typical outdoor-to-indoor building facade attenuation suppresses mid- and high-frequency components to a much greater extent than low-frequency components. A-weighted sound-level benchmarks and limit values for the interiors of houses might not, therefore, provide sufficient protection against low-frequency noise indoors for typical houses. The A-weighting is similarly inappropriate for measuring internal vehicle noise, because of the predominance of low frequencies inside the vehicle cabin, although it has been widely used for this purpose.

It should also be noted that the A-weighting has been criticised on the basis that on average, most people lose high-frequency hearing as they get older and that low frequencies might then become subjectively even more important than implied by the standard equal-loudness contours. While there is logic behind this suggestion, there is only very limited experimental evidence that any other frequency weighting provides a better match to older people's hearing.

The noise-rating (NR) frequency-weighting curves (BSI 1999) (see Figure 6.7) were devised to represent equal-loudness contours where measurements have been made using octave-band frequency-analysis filters. Many types of sound level meters can be fitted or supplied with octave and narrower bandwidth frequency filters to facilitate this kind of measurement. The generically similar noise-criteria (NC) and preferred-noise-criteria (PNC) curves were devised in the United States to meet the same requirement as the NR curves in Europe. NR curves have been widely used to define targets and limit values for sound levels inside buildings, taking into account both noise from building services and noise break-in from external noise sources. Typically, NR 10–20 is specified for audiometric test rooms, NR 20–30 for classrooms, NR 30–40 for quiet offices, and NR 60–70 for industry. NR 70 is suggested for industry as a basis for avoiding interference with speech communication, although in some sectors of manufacturing industry it might be considered rather optimistic in terms of what can be achieved in practice.

Of the frequency-weighting schemes which are widely known, the most complicated are the so-called loudness (ISO 1975) and noisiness (BSI 1979) calculation schemes. These schemes aggregate together the assumed separate contributions to overall subjective 'loudness' and 'noisiness' of each

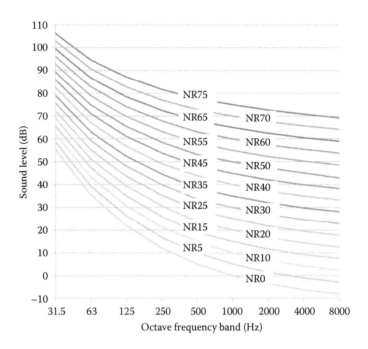

Figure 6.7 Noise-rating curves.

one-third octave band of the frequencies present. Zwicker's (European) and Stevens' (US) loudness-level calculation schemes have been the subject of considerable technical debate. All such schemes are based on more or less complex models of peripheral auditory processing. The results of experimental testing have not been universally conclusive. Unfortunately, no loudness calculation scheme yet devised seems capable of taking into account all possible acoustic features that might be present in any particular case.

Kryter's effective perceived noise level (EPNdB) calculation scheme has been the international standard for aircraft-noise certification purposes since the 1960s. It was originally modelled on Stevens' loudness-level calculation scheme, but was modified to deal with so-called perceived noisiness of aircraft flyover sounds rather than subjective loudness. Kryter's intention was that perceived noisiness should be understood as representing the objectionable or noisy features of aircraft flyover noise, in addition to the loudness, yet was not necessarily the same as disturbance or annoyance, which is felt by the individual person depending on the circumstances at the time, and could be argued to be outside the control of aircraft manufacturers. The EPNdB measurement scheme was developed to achieve a modest correlation with subjectively reported noisiness of aircraft types in service in the late 1950s and early 1960s, and takes particular account of the higher-frequency components common to the low-bypass-ratio turbojet engines used in that period. There is increasing evidence that this metric may not be optimum for representing the type of sound produced by more recent aircraft types powered by high-bypass-ratio engines or by advanced open-rotor-powered aircraft which might come into service in the future. Note that noise metrics such as EPNdB do not actually measure 'noise' anyway, because this is a matter of subjective perception.

Nevertheless, for assessment and regulation purposes, there is considerable merit in retaining standardised metrics such as EPNdB, even after they might have become less fit for purpose than at the time they were originally developed. Metrics become enshrined in legal contracts and specifications which could be difficult to change. In addition, it is not reasonable to expect any alternative metric to be able to perform any better over the longer term. Simply reducing the physical energy content in a sound will not necessarily lead to improvements in subjective sound quality, and if the wrong physical components or features of the sound are addressed, could even make matters worse. In some cases, annoyance can be reduced by addressing the psychological components of disturbance and annoyance without necessarily having to make any actual changes to the sound at all. There are also situations where actual physical reductions in sound level are too small to be directly noticeable and have no effect on disturbance and annoyance until people are told about them.

6.3.4 Time weightings and averaging

The selection of appropriate *time weightings* for objective measurements can be important. The time weighting determines the amount of time needed for a measurement. For completely steady, continuous sounds, the measurement averaging time has only to fit within the time for which the sound is present. However, most sounds that need to be assessed and possibly regulated are not steady. The ear requires a finite amount of time to register a sound. The various mechanical parts of the ear must first be set in motion. Then, the sensitive hair cells along the basilar membrane must be sufficiently stimulated to cause nerve impulses to be sent along the auditory nerve in sufficient quantity to alert the higher-level processing centres that something interesting is happening. Finally, the resulting neural activity must be sustained over a long enough fraction of a second that the central nervous system starts to process the new sensory input. If the sound then stops, any neural activity associated with the incoming sound then dies away quite rapidly.

The time-averaging response of the ear is best described by an approximately exponential averaging time of between 100 and 200 ms. Sound level meters can implement exponential or linear time averaging. Exponential averaging implies that the memory of recent events decays over time. Under linear averaging, however, everything happening within a fixed time interval makes an equal contribution to the average. A continuously updated exponential average or a sequence of very short-time linear averages is appropriate for showing a continuously varying sound-level history over time, whereas long-time linear averaging is appropriate for producing a single number to describe the average over longer time periods such as a 16 h day. Long-time-averaged sound levels such as $L_{Aeq,16h}$ are widely used for regulatory and contractual purposes for mainly administrative reasons, even though they do not reflect human hearing particularly well.

Transient sounds of shorter durations than around 100 ms are harder to detect and do not register the same loudness as continuous sounds of the same max or peak level. Figure 6.8 shows a generic curve of constant detectability for transient sounds with the sound level (when the sound is present) adjusted to maintain constant integrated acoustic energy (proportional to sound pressure squared × time) across the range of durations tested. Figure 6.8 shows that over the middle range of durations from about 15 ms up to about 150 ms, detectability appears to be a function of energy. As the duration increases beyond about 150 ms, detectability decreases. This is because the instantaneous sound level must decrease with increasing duration to maintain constant energy. Further increases in duration beyond 150 ms no longer offset the reduction in instantaneous sound level.

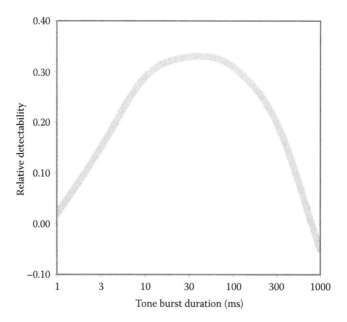

Figure 6.8 Relative detectability of constant energy short-duration sounds.

At very short durations below about 15 ms, detectability also decreases because time–frequency uncertainty causes the acoustic energy contained in the transient sound to be spread over an increasingly wide range of different frequencies. Consequently, there is less energy available to support detection in each separate part of the basilar membrane.

The time range of constant detectability varies with frequency, as might be expected, being shorter for high frequencies and longer for low frequencies. Similar curves can be plotted at sound levels above the detectability thresholds by using subjective loudness instead of detectability as the outcome variable. In this case, the precise shape of the curve varies, depending on the actual experimental method used. The comparative loudness of transient tone bursts of varying duration is essentially subjective and the judgement criteria used by different listeners is likely to vary. The subjective sound quality of short transient sounds varies with duration in addition to subjective loudness, and differences in sound quality can be easily confused with differences in loudness in this situation. This is not the case for detectability at the threshold, which is much more dependent on the fundamental capabilities of the auditory system, although differences in listener effort and motivation can still have significant effects.

One practical consequence of the finite integration time of the ear is that the subjectively perceived loudness of very short transient sounds might not be commensurate with the damage risk associated with high instantaneous peak sound levels. This is a good reason for setting the action levels for peak sound pressures in the *Control of Noise at Work Regulations* (2005) conservatively. These regulations specify that the peak SPLs should be measured using the C-weighting, presumably to discriminate against any very low- or very high-frequency content which might also be present, and expressed in decibels relative to 20 μPa.

The time weightings used in sound level meters (see Chapter 8) reflect what is known about the averaging time of the ear while avoiding unnecessary or impractical complications. Many sound level meters include both F(ast) and S(low) time weightings. The F-time weighting has a 125 ms exponential time constant and is intended to broadly represent the equivalent integration time of the ear, although without taking into account any differences at different frequencies. For rapidly fluctuating sounds such as normal speech, the instantaneous readings of a sound level meter set to the F-time weighting can change up and down too quickly to be followed by a human operator reading the sound level meter display. The S-time weighting, with an exponential time constant set at 1 s, averages out the instantaneous readings from a rapidly fluctuating sound, so that the slowed-down fluctuations can be followed by a human operator reading the sound level meter display visually. On the other hand, the S-time weighting is too slow for the sound level meter display to follow all of the rapid fluctuations in normal speech, and where these fluctuations are important, the F-time weighting should be used.

In practice, the maximum and minimum instantaneous sound levels of short duration or rapidly varying sounds will be underrepresented by the S-time weighting to a greater extent than by the F-time weighting. The peak (instantaneous) sound levels of short transient sounds with durations <125 ms will be underrepresented by both the F- and S-time weightings, but the maximum instantaneous sound levels of longer transient sounds with durations between 125 ms and 1 s will only be underrepresented by the S-time weighting. The different time weightings can be used for different purposes. For example, when measuring the maximum sound levels of aircraft flyover events, it is standard practice to use the S-time weighting. This is because the maximum S-weighted sound level is generally considered to provide the most reasonable indication of the sound level produced by the aircraft flyover event, where the effect of rapid random fluctuations caused by atmospheric turbulence along the acoustic propagation path from the aircraft source to the receiver has been largely averaged out. Rapid fluctuations in sound level associated with atmospheric turbulence cause the F-time weighted sound-level time history to fluctuate both above and below the equivalent S-time weighted sound-level time history and are considered (by the industry) to be outside the aircraft manufacturer's and operator's control. The amount of atmospheric turbulence varies depending on meteorological conditions at the time.

While long-time averages can be calculated from a series of readings taken from the time-varying display of a sound level meter, it is better to use digital averaging to obtain the most accurate results. Early analogue sound level meters have been superseded by digital sound level meters, which can now accomplish via software what used to require very expensive analogue circuitry to perform.

The widely used long-time-averaged sound level or L_{Aeq} metric is formally defined as the time-average or equivalent continuous sound level, and is widely specified in various standards and regulations. In the acronym for L_{Aeq}, the 'A' denotes the use of the A-frequency weighting. The formal definition recognises that a continuous sound with the same L_{Aeq} as a time-varying sound would have the same energy content as the time-varying sound, provided that the averaging time or measurement duration is the same in both cases. The L_{Aeq} is calculated by dividing the integrated A-weighted sound pressure squared by the measurement duration. If the same amount of sound energy or number of events is averaged over a longer measurement duration, the resulting average will be numerically lower, although the sound being measured will have been unchanged. Extending the measurement duration to include periods of relative silence before and after a single noise event reduces the average. Under these circumstances, doubling the measurement duration or halving the number of separate noise events included within it decreases the L_{Aeq} by 3 dB. The duration of the measurement must always be specified or implied by the context in which the metric is used. It is good practice to add the measurement duration as a suffix, for example $L_{Aeq,16\,h}$ or $L_{Aeq,5\,min}$. Where there is diurnal variation over the 24 h, it is good practice to state the start and end times of the measurement as well.

For environmental and community noise, it has become standard practice to report L_{Aeq} over 12 h or 16 h daytime periods and 8 h night-time periods, or averaged over the full 24 h. The European Environmental Noise Directive (European Parliament/Council of the European Union 2002) specifies L_{den} and L_{night} as harmonised acoustic metrics. L_{den} is the 24 h average L_{Aeq} with the evening period (19:00–23:00 h) weighted by +5 dB and the night period (23:00–07:00 h) weighted by +10 dB. The daytime period (07:00–19:00 h) is not weighted. Regulatory authorities in the United States use a generically similar metric, the L_{DN}, which is the 24 h average L_{Aeq} with the night period (2200–0700 h) weighted by +10 dB. In the United Kingdom, and for largely historical reasons, the Department of Transport has used the term L_{eq} instead of $L_{Aeq,16\,h}$ for many years to describe 16 h daytime aircraft-noise exposure.

It should be understood that, for the L_{den} and L_{DN} metrics, the precise weightings (+5 dB, +10 dB) and the precise time periods over which they are applied are essentially arbitrary committee decisions. They are not based on definitive research. It is difficult to conceive of any possible research design which would be capable of determining exactly which time periods and weightings should be used in any particular acoustic metrics in any case. The reason for this is that field measurements using similar metrics are normally so highly correlated together that it is impossible to tell them apart statistically.

The sound-exposure level, or L_{AE} (the traditional abbreviation, SEL, is more easily remembered as the single-event level), is a specific metric applied to separately identifiable noise events. Clearly, the maximum instantaneous sound level on its own cannot reflect the duration or time variation during the event. L_{AE} is a measure of the integrated acoustic energy in the event normalised to a standard averaging time of 1 s and is calculated in the same way as L_{Aeq}, except that the integrated A-weighted sound pressure squared is not divided by the measurement duration.

The best way to appreciate the difference between L_{AE} and L_{Aeq} is to consider two extreme examples: a completely steady sound, and a single transient event preceded and followed by extended periods of silence. For a completely steady sound, the L_{Aeq} does not change with different measurement durations, because the average sound level during the measurement duration is constant. The L_{AE}, however, increases logarithmically as the total acoustic energy received at the measurement point increases with measurement duration. For a single transient sound surrounded by extended periods of silence, the L_{Aeq} reduces logarithmically as the long-time average acoustic energy received at the measurement point reduces with increasing measurement duration. For the same situation, the L_{AE} is independent of measurement duration as the total acoustic energy received at the measurement point is independent of the total duration. There is no difficulty in translating between L_{AE} and L_{Aeq} measurements, provided that the relevant measurement durations are known. For the measurement of short-duration sounds, L_{AE} has an advantage, because the onset and offset times of the measurement are not critical, provided that the whole event is included. However, for people who are not acoustical engineers, neither metric is as easy to understand as the simple L_{AFmax} or A-frequency and F-time weighted maximum sound level during an event.

The L_{Aeq} and L_{AE} metrics have become increasingly popular with administrators and regulators in recent years and have largely superseded a number of older metrics. The two main reasons for this, lower cost and greater convenience, have nothing to do with subjective response. The inclusion of software capable of calculating L_{Aeq} and L_{AE} within digital sound level meters results in instruments that are much more effective than earlier instruments based on analogue electronics and that are readily available at much lower cost than when these metrics were first introduced. L_{Aeq} and L_{AE} measurements are simple and robust, and facilitate straightforward calculations of the likely effects of noise-management action, which will often be entirely consistent with actual measurement. This kind of rational simplicity does not apply to many of the more esoteric noise metrics and indicators which were invented in the past and have now fallen out of use. On the other hand, rational simplicity and engineering convenience do not automatically imply consistency with subjective impressions. For example, according to the physics, doubling the number of similar noise events within a defined time period doubles the acoustic energy received which will increase L_{Aeq} by 3 dB. The same physical effect of a doubling of energy could be achieved by increasing the sound levels of the separate events by 3 dB while keeping the number of events constant. Typical human listeners are less likely to perceive 3 dB changes in the sound levels of events than doublings or halvings of the numbers of events, particularly if the change happens gradually over time.

The interpretation of metrics such as L_{Aeq} and L_{AE} requires caution, particularly when the effect of adding a new noise source into an existing acoustic environment is considered. If the existing acoustic environment and a new noise source both generate the same L_{Aeq} at the measurement position when measured separately, adding the new noise source will only increase the overall L_{Aeq} at the measurement position by 3 dB compared with what it was for the existing situation. Depending on the character of the new noise source compared with the existing acoustic environment, the addition could be highly noticeable in terms of subjective impressions, whereas an equivalent 3 dB increase in the L_{Aeq} sound level of the existing noise source measured separately might be completely unnoticeable. Assessors and regulators often mistakenly assume that a 3 dB change in

L_{Aeq} always has the same effect in terms of subjective impressions, regardless of what has actually happened to cause the change.

Complex time-varying acoustic environments can also be described by using what are known as *statistical levels*, L_n. Any particular L_n is the sound level (note that this would normally be A-weighted – as in L_{An}) over the defined measurement duration, which is exceeded for $n\%$ of the measurement duration. The L_{A10} and L_{A90} are the A-weighted sound levels which are exceeded for 10% and 90% of the measurement durations respectively. These statistical levels were widely used in the days before reliable and stable integrating-averaging sound level meters or systems became available, because they could be calculated by plotting out the statistical distributions of instantaneous sound levels obtained by using a mechanical statistical distribution analyser attached to a pen-recorder system. Modern digital sound level meter systems can perform the same calculations in software, but the whole concept of statistical levels has become largely obsolete, except for defined specialist applications.

The main problem with statistical levels is that they do not behave rationally when arithmetically manipulated. For example, to calculate the combined L_{50} from the addition of two noise sources of known L_{50}, it is necessary to know the cumulative distributions of the two time histories separately. The L_{A90} is still widely used in the United Kingdom and in one or two other European countries to indicate the steady background sound level attributable to mostly distant noise sources, but cannot be arithmetically manipulated in any rational way. The L_{A90} provides no information about how far above or below that sound level (the L_{A90}) that instantaneous sound levels rise or fall during the 90% of the time that sound levels exceed the L_{A90} or during the 10% of the time that they fall below the L_{A90}. On the other hand, the L_{A90} does have some advantages over other methods for indicating background sound levels, because it does not require any subjective decisions to be made about which noise sources are part of the background and which are part of the foreground. Almost any other metric used for this purpose requires the user to make what are essentially arbitrary subjective decisions about which noise sources are included in the background and which are not. The subjective importance of the steady background sound level is that it determines an objective threshold, above which more prominent noise sources can be administratively deemed to be intrusive and below which they can be deemed to be not intrusive. Note that this is an objective definition which might not necessarily be very representative of actual subjective impressions, but it can at least be applied consistently.

In the United Kingdom, the L_{A10} metric is still used (despite being technically obsolete) to determine entitlement to a noise-insulation grant in mitigation of increased road-traffic noise. The L_{A10} metric can be considered as behaving semi-rationally when applied to busy road-traffic noise, which has similar statistical properties anywhere in the country, but it does not behave rationally when applied to any intermittent noise source such as railway noise or aircraft noise, which might only be present for less than 10% of the time. Comparisons between different intermittent noise sources using statistical-level metrics are meaningless.

Other metrics have been proposed for intermittent noise sources (such as aircraft flyover events) such as the time-above x dB L_{pA} and the number-above y dB L_{pA}. Within a defined measurement duration, the time-above metric indicates the proportion of time during which a defined sound level x dB L_{pA} is exceeded, and the number-above metric indicates the number of events which exceed a defined sound level y dB L_{pA}. Both alternative metrics reflect different acoustic features from the *de facto* standard L_{Aeq}-based metrics and could therefore be useful in some situations. However, neither behaves rationally when arithmetically manipulated. The time-above and number-above metrics provide no information about what happens during the time periods that the x and y dB L_{pA} thresholds are exceeded. For example, the N_{65} metric (the number of events within a defined measurement period with maximum A-frequency and F-time weighted sound level, L_{AFmax}, exceeding 65 – assuming the A-frequency and F-time weightings are used) provides no indication of the number of decibels by which the qualifying events exceed the 65 dB L_{pA} threshold. Suppose there are 100 events per day at 66 L_{AFmax} (i.e. just exceeding 65 dB), the N_{65} would be 100. If the maximum sound level for each event was increased by 15 dB, the N_{65} would still be 100. The change in event sound levels from 66 L_{AFmax} to 81 L_{AFmax} would be highly noticeable, but the N_{65} metric would not have changed at all. If the maximum sound level for each event was decreased by only 2 dB, the N_{65} would fall to zero. The change in event sound levels from 66 to 64 L_{AFmax} would hardly be noticeable, but the N_{65} metric would have changed from 100 to 0.

6.3.5 Time-varying frequency spectra

The most comprehensive representations of complex acoustic environments can be provided by showing successive time-varying frequency spectra, but these are of limited use for assessment and regulation because of their complexity. No meaningful method has yet been devised for setting sound-level targets and limit values in terms of time-varying frequency spectra. The main problem is that there are no simple acoustic metrics which can always reflect all acoustic features present in any particular case which are, or could be, important for subjective response. In addition, anything too complex will be unworkable for assessment and regulation purposes. All objective acoustic metrics which have been adopted to date are compromise solutions which can only ever take a proportion of potentially relevant variables into account.

While time-varying frequency spectra are too complicated for use as sound-level targets and limit values, they can nevertheless be useful for identifying specific acoustic features (normally, either discrete tonal or harmonic components in the frequency domain or various fluctuations or modulations in the time domain) which appear to be making a disproportionate contribution to overall disturbance and annoyance, and should perhaps be targeted for engineering noise control effort. There are two main types of frequency-analyser system in general use for such purposes, depending on whether the frequency-analysis bandwidths are distributed linearly or logarithmically across the audio-frequency range. The fast-Fourier-transform (FFT) frequency analyser is the most common type and generates linearly spaced frequency-analysis bands (see Section 4.4.5). The resulting spectra can be displayed using a linear or a logarithmically spaced frequency scale irrespective of the underlying operating principle. A typical 1024 point FFT analyser running at a 25 kHz audio sample rate will display 400 linearly spaced frequency-resolution bandwidths from 0 to 10 kHz at 25 Hz spacing. Note that the actual frequency resolution is usually wider than the line spacing on the display, because this also depends on the averaging time and the type of time-domain window filter applied. At low frequencies (say 100 Hz) the frequency resolution of around 25 Hz is much coarser (25%) than the frequency resolution of the human ear. At high frequencies (say 10 kHz), the frequency resolution is the same at around 25 Hz and is much finer (0.25%) than the frequency resolution of the human ear. FFT analysers are good at detecting linearly spaced harmonics but are not well adapted to reflect human hearing.

The constant percentage bandwidth frequency analyser comprises a series of parallel one-third octave or narrower fractional octave band-pass filters (see Section 5.2.2.2). Traditional analogue equipment required separate filter circuits for each filter band, but modern digital systems can operate multiple parallel filters in real time without difficulty, and are consequently much better adapted to human hearing than FFT analysers. One-third-octave band-pass filters have been an industry standard for many years, and while they mimic the approximately logarithmically scaled frequency resolution of the human ear, the actual frequency resolution of one-third-octave band filters is still very much coarser than that of the human ear. This is one of the problems of the EPNdB metric used for aircraft-noise certification discussed in Section 6.3.3. The EPNdB metric uses one-third-octave band-pass filters (as do Zwicker's and Stevens' methods for calculating loudness level), possibly because that was the state of the art, using analogue electronics, at the time these metrics were originally developed. However, the normal human ear has about 20–25 times narrower frequency resolution than one-third octave band-pass filters, each of which span four semitones of a 12-semitone musical octave scale. Traditional one-third-octave band-pass filters would be useless for transcribing music.

Much narrower bandwidth parallel filters can be modelled in software-based systems that can more closely reflect the frequency-resolution capabilities of the human ear. For frequency analysis of rapidly time-varying sounds, the extended averaging times required for narrower-bandwidth frequency resolution have to be balanced against the time resolution required. One possible advantage of constant percentage bandwidth filters is that each filter can be set up with different averaging times to best suit requirements across the entire frequency range of interest. Low-frequency filters require much longer averaging times than high-frequency filters to obtain the same degree of statistical accuracy across the frequency range. FFT filters operate on discrete blocks of sample data, providing less flexibility for selection of the optimum averaging time in each frequency range, which is normally constrained to multiples of sample data blocks. Other more sophisticated methods of frequency analysis have been developed to solve some of the problems outlined in this section. They are outside the scope of this chapter, but they are discussed to some extent in Chapter 4.

6.3.6 Spatial hearing

The mechanisms underlying binaural hearing are discussed in Section 6.2.2, but the implications for spatial hearing require further consideration in this section. It is obvious that no sound level meter measurement system using a single precision-grade omnidirectional condenser microphone can reflect the spatial resolution capabilities of human hearing. The technical issue is whether or not it is possible to devise an acoustic metric that would reflect human spatial hearing capabilities in addition to trying to reflect the time- and frequency-resolution capabilities of human hearing. Binaural systems using microphones installed at the ear positions of a dummy head have some spatial capability, but are still not entirely satisfactory for use in metrics for the following reasons:

1. Human spatial hearing relies on the spectral and interaural differences associated with different directions of the source and how these differences change with head movement or movement of the source relative to the head. No instrumentation systems have yet been devised which can reflect these differences in ways that might be useable in standards and regulations.
2. The real-ear, head-related transfer functions for individual listeners vary depending on individual physiognomy. Dummy head systems could be standardised so as to represent an average physiognomy, or there could be three or more standardised systems designed to represent ranges of physiognomies. Any system proposed would not be able to overcome the problem of individual differences in physiognomy between different listeners.

On the other hand, it seems likely that the continuing technical development of spatial audio systems will lead to increasingly realistic auralisations. *Auralisation* is the technical term used to describe the process of generating realistic simulations of actual or hypothetical sound sources using computer synthesis techniques, and can be extremely useful for simulating hypothetical situations and developments in advance of actual manufacture or construction. Depending on the situation, however, subjective realism is not necessarily the same as objective realism, because it depends on human perception, which can be coloured by selective attention and expectation. At present, it seems unlikely that auralisation techniques can be developed to provide simple metrics suitable for objective assessment and regulation.

6.4 RANGE OF NOISE EFFECTS ON PEOPLE

6.4.1 Direct health effects

Very high sound levels can potentially cause localised heating or physical disruption of the human body. However, and depending on acoustic conditions, even an SPL of 120 dB has an acoustic intensity of only around 1 Wm^{-2}, which, even if all this physical energy was absorbed by a human body, would not be enough to cause any measurable damage. At the highest levels of environmental and community noise normally encountered outside people's houses (not normally more than 75–80 dB $L_{Aeq,16h}$ with occasional events up to 110–120 dB L_{AFmax}), there are no direct effects on the human body and no permanent effects on the physical environment, because the amount of physical energy is far too small. Most, if not all, of the observed effects of environmental and community noise at these and lower sound levels are caused by the way that the human body and the human brain react to or respond to noise as a sensory stimulus, and not because of any direct effects of the sound energy on the human body.

6.4.2 Hearing damage risk

Transient sounds with peak overpressures reaching a fraction of one atmosphere within a few milliseconds can cause direct physical disruption, particularly to sensitive tissues such as the eardrum and the lungs, but these sounds require extremely powerful sources such as gunfire, explosives or rocket motors, which are not encountered in normal community situations. The lower and higher peak sound-pressure action values of 135 dB (C-weighted) and 137 dB (C-weighted) specified in the *Control of Noise at Work Regulations* (2005) (referred to in Section 6.3.4) are intended to protect workers against the risk of hearing damage caused by repeated exposure to high-level transient sounds, and do not necessarily imply significant risk of physical damage associated with single or infrequent exposures at such levels.

For continuous sounds, and depending on duration, even short-time exposures at lower sound levels can lead to temporary loss of hearing sensitivity, known as TTS, and which may also be accompanied by tinnitus. Tinnitus, or 'ringing in the ears', is an audible sensation occurring when there is no actual sound present. One of the possible causes of TTS and associated tinnitus is fatigue and possible damage to the sensitive hair cells in the inner ear. Fatigue caused by overstimulation could cause some nerve fibres to lose sensitivity to actual input sounds and to start firing spontaneously even after the acoustic stimulation has stopped, or it could cause a diminution of function in other nerve fibres which would normally have an inhibitory role. After TTS, hearing sensitivity begins to return to normal immediately after the acoustic stimulation stops, although it may take several hours for full recovery to take place, depending on the degree of threshold shift in the first place.

If the exposure is sufficiently prolonged, this can lead to permanent loss of hearing sensitivity, which may also be accompanied by permanent tinnitus. Permanent hearing loss caused by excessive noise exposure is associated with progressive atrophy of the sensitive hair cells in the inner ear. The three most important factors in hearing damage risk are the sound level, the duration of exposure, and individual differences in susceptibility, although the duration and frequency of recovery periods between periods of high noise exposure may also be important. It has not yet been possible to devise reliable methods for identifying individuals with higher susceptibility to noise-induced hearing loss (NIHL) in advance of exposure. The biological mechanisms underlying progressive loss of hair cells are not precisely understood, but the measurable consequences of losing both sensitivity and frequency selectivity are easy to understand. Since it is unlikely that prolonged noise exposure at the sound levels known to be associated with NIHL would have been a significant feature of the environment before the development of industrialised societies, it also seems unlikely that individual differences in susceptibility would have been directly associated with any evolutionary mechanism, although they could be associated with unrelated genetic factors connected with other functions and systems.

The *Control of Noise at Work Regulations* (2005) require both employers and employees to carry out a range of preventive actions to reduce the risk of NIHL in the workplace, depending on whether the specified lower and higher action values (80 and 85 dB $L_{Ep,d}$: equivalent to $L_{Aeq,8\ h}$) are exceeded. There is a general duty to reduce noise exposure as far as is practical; employers are required to offer advice and preventive options at the lower action value and above; and both employers and employees have specific duties to prevent the higher action value from being exceeded. There is no similar mandatory protection against NIHL in domestic and leisure activities. This means that in, for example, a live-music venue or discotheque, management would be required to protect employees such as waiters and bar staff against excessive noise, perhaps by being required to wear ear protection, while there is no obligation to protect customers against the same risk. Generally speaking, customers are assumed to be at much lower risk of NIHL than employees because of infrequent attendance, but there is no obligation for venues to assess this. Attendance and resulting noise exposure for customers is voluntary, and hence, any associated risk of NIHL is, presumably, a matter for the customers themselves, although the venue does have at least a moral duty to inform them of risk. Similarly, while manufacturers of personal stereo systems have some obligation to warn users against the risk of NIHL associated with excessive use, any actual risk is outside the manufacturer's control because the duration of exposure is entirely a matter for the user. Actual risk is assumed to be independent of whether the exposure is incurred through a voluntary leisure activity or as an unavoidable condition of employment, but is nevertheless treated differently under existing laws.

It is generally accepted that the risk of NIHL reduces at lower cumulative sound levels. The current World Health Organization (WHO) guidelines for community noise (WHO 2000) accepted the advice given in the then current version of ISO (1999) that long term exposure up to 70 $L_{Aeq,24\,h}$ will not result in significant hearing impairment and quote this value in the summary tables of guideline values. Note that 70 dB $L_{Aeq,24\,h}$ is approximately equivalent to 75 dB $L_{Aeq,8\,h}$ for an 8 h working day if there is no other significant exposure during the rest of the 24 h day.

6.4.3 Causal mechanisms for other adverse health effects

Outdoor measured sound levels at community and residential locations are normally in the range from 45 to 75 dB $L_{Aeq,16\,h}$. In this range, acoustic stimulation causes a direct response in the auditory nerve and higher neural processing centres and this can, in turn, lead to endocrine and autonomic system responses such as temporary increases in blood pressure and concentrations of stress hormones such as adrenaline and cortisol. Normal autonomic and endocrine responses are essentially temporary adaptations by the body to meet changes in the external environment and are not necessarily harmful in themselves. A number of researchers, mainly from Europe, have suggested that temporary or acute adaptations could lead to more prolonged or chronic negative effects such as raised blood pressure, cardiovascular disease or mental health problems. There are important differences between temporary or acute effects, which might possibly be harmful on a short-term basis if sufficiently severe but do not persist after the noise has ceased, and long-term permanent effects, which continue after the noise exposure has ceased, such as NIHL. It is not presently known to what extent noise-induced temporary or transient changes in physiological variables such as blood pressure or stress hormone concentrations contribute to permanent changes which are harmful. It seems likely that individual susceptibility to long-term permanent effects varies. If only a minority of the population are in fact susceptible, it might be very difficult to demonstrate any such hypothesised causal links by statistical means, or, on the other hand, to demonstrate that they do not exist.

Difficulties arise because of the large number of situational, dietary, lifestyle and other variables which are known to be associated with adverse health effects. Some evidence exists of increased prevalence of circulatory disorders in noisier areas (Niemann and Maschke 2004; Babisch 2006), but causal relationships remain unproven because of the large numbers of uncontrolled possible confounding factors. For example, it is usually difficult to eliminate the possibility that people with higher-risk factors for certain diseases might also tend to prefer different types of residential areas for reasons other than noise, but which happen to be noisy to a sufficient extent that statistical associations can be observed. Alternatively, it is equally possible that people with higher-risk factors for noise-related diseases might tend to avoid noisier areas, thereby reducing any statistical correlations that might otherwise have been observed. In addition, in most cities, houses in the noisier areas are usually also exposed to higher levels of air pollution, which might provide an alternative explanation for any adverse health effects found. In general, most city noise is a consequence of higher population density in areas where large numbers of people choose, or prefer, to live.

The technical uncertainties in this area create problems for policymakers and administrators. Noise control is not usually without cost, which could exceed the benefits if they have been overestimated by applying overprecautionary assumptions about the health effects of noise. On the other hand, it is also possible that significant numbers of citizens are seriously affected and not enough has been done to reduce the risks. Policymakers have to decide these issues on the basis of incomplete evidence and are rarely qualified to be able to make these assessments. Acoustical engineers and physicists may be asked to make recommendations without necessarily being aware of the wider consequences of those recommendations if followed up. Engineering noise control carried beyond what is strictly necessary to minimise disturbance and possible health risks could have economic consequences which could be more damaging than the noise effects to which it was addressed.

6.4.4 Speech interference

Speech communication is one of the most important characteristics of the human species. The evolutionary development of speech processing within the auditory system appears to be intimately linked to the development of the vocal tract and associated neural processing involved in producing speech. Two main methods exist for estimating speech intelligibility based on objective measurements of interfering background noise; the *speech intelligibility index* (SII) (ANSI 1997) and the related *speech transmission index* (STI) (BSI 2003a).

The classic articulation-index (AI) (ANSI 1969) calculation procedure, on which the current US SII calculation procedure is based, compared the long-time r.m.s. average one-third octave-band frequency spectrum of the wanted speech signal against the long-time r.m.s. average one-third octave-band frequency spectrum of the masking noise. The measured signal-to-masking noise ratio in each of the 14 one-third octave bands from 200 Hz to 5 kHz was assumed to contribute to overall intelligibility, according to weightings defined on the basis of laboratory listening tests. The standard weightings assumed that the 200 Hz frequency band contributed the least to overall intelligibility and that the 1.6 and 2 kHz frequency bands contributed the most. Note that the standard weightings were derived using native English-speaking talkers and listeners and might not be optimum for different languages or for talkers with unusually high- or low-pitched voices, or for electronically processed speech. The frequency bands and weightings were modified slightly for the current SII procedure.

The measurement of the speech signal-to-masking noise ratio has to take into account that the instantaneous sound level of any typical speech signal varies significantly over time. Obviously, the relative peaks in the speech signal contribute much more to overall intelligibility than the periods of relative silence in between the speech sounds. Similarly, and depending on their duration, any relative peaks in the masking noise contribute more to the masking effect than any quieter periods in between. The louder speech phonemes will be more easily distinguished during quieter periods in the background noise and vice versa. The maximum sound levels of consecutive peaks in the speech signal vary over a wide range, according to the phoneme being spoken and according to the talker. Standardised *peak programme meters* (PPM) have been used for monitoring speech sounds for specialist radio broadcasting and recording purposes, but the PPM standard has not been widely applied elsewhere. Most recording and broadcast monitoring is carried out using simple volume-unit short-time r.m.s. averaging meters, which operate similarly to conventional sound level meters used in the environmental and community noise-management fields. Long-time r.m.s. averages are used in the AI/SII measurement procedures to quantify both the speech-signal and masking-noise sound levels, assuming that instantaneous speech peaks rise 12 dB above the long-time r.m.s. average. In practice, this is a reasonable assumption for normal running speech, but might not apply to shouted or whispered speech, or for talkers who speak much faster or much slower than normal, and definitely does not apply where the dynamics of running speech have been compressed either by intentional audio processing or by unintended nonlinearities and distortions in an electro-acoustic speech communications system.

The unweighted contribution made by each one-third octave bandwidth of the speech signal to overall intelligibility (before multiplication by the standard weightings described) were assumed to vary from zero at speech signal-to-masking noise long-time r.m.s. ratio of −12 dB up to a maximum of unity at speech signal-to-masking noise long-time r.m.s. ratio of +18 dB. According to the assumption that the peaks rise 12 dB above the long-time r.m.s. average, this is equivalent to assuming a zero unweighted contribution at 0 dB speech peak-to-masking noise long-term r.m.s. ratio, increasing to a unity unweighted contribution at +30 dB speech peak-to-masking noise long-term r.m.s. ratio.

Guidance included with the original AI procedure assumes that an AI score of 0 represents zero intelligibility, that an AI score of 0.5 is acceptable for general purposes and that an AI score of 0.7 or above is desirable for electro-acoustic speech communications systems 'to be used under a variety of stress conditions and by a large number of different talkers and listeners with varying degrees of skill'. Clearly, the AI/SII calculation procedure can only deal with the masking effects of interfering steady background noise on undistorted normal running speech. It requires that the speech signal and the masking noise be measured separately, which would, of course, be rather difficult in a sports stadium or similar indoor situations where most of the masking effect is contributed by crowd noise or reverberation excited by the speech signal. The AI calculation procedure is of limited value in many practical situations because it does not address distortion of the wanted speech signals that might occur in electro-acoustic speech communications systems (such as telephone systems) other than reductions of speech signal-to-masking noise ratio caused by uncorrelated masking noise. It effectively assumes that the speech signal would be 100% intelligible if there were no masking noise. This is not always realistic. In addition, speech communications systems with no audible distortion and no background noise can still be ineffective if the talker does not speak clearly, breathes heavily into the microphone, or speaks in a language which is unknown to the listeners, when the actual intelligibility could then drop to zero.

The STI was invented by Houtgast and Steeneken at the Netherlands Organisation for Applied Scientific Research (TNO) (Houtgast et al. 1980) and is similar to the AI and SII but uses a special test signal to allow measurements to take place over a much wider range of circumstances. The STI test signal (and the simplified rapid speech transmission index (RASTI) test signal) is made up from cosine amplitude-modulated narrowband noise signals with similar dynamics and frequency content to normal running speech, but which allows speech signal-to-noise ratios to be inferred from measured modulation transfer functions in each frequency band. The STI measurement system is an improvement to the AI measurement and calculation procedure, because it does not require the speech signal and the masking noise to be measured separately, which is impossible where the masking noise is mainly reverberation. Modulation transfer functions can also be calculated from measured or calculated system impulse-response functions, which can be very useful in system design where the overall acoustic performance can be predicted in advance using image-source and ray-tracing models.

Note that the median AI or STI score of 0.5 can be obtained where the long-time r.m.s. average speech signal-to-masking noise ratio is only +3 dB across the one-third octave frequency bands from 200 Hz to 5 kHz. Experimental data reported in the original ANSI standard (ANSI 1969) shows that 95% sentence intelligibility (measured using real listeners) can be achieved where the AI or STI score is only 0.5. Sentence intelligibility is always higher than the intelligibility of separate words, which in turn is always higher than the intelligibility of random nonsense syllables. This is because of the redundancy and additional context information in meaningful words and sentences as compared with nonsense syllables. If a speech system is to be used where the intelligibility of unfamiliar words is a priority, a higher AI or STI score than 0.5 will be required.

Actual intelligibility varies above and below intelligibility estimated from objective measurements, reflecting individual differences in talking and listening skills, and also differences in the level of motivation among listeners. Selective attention and other higher-order perceptual processes affect individual test results. All objective speech intelligibility calculation and measurement procedures apply to ideal or standardised test conditions, which may not be representative of actual conditions.

Further uncertainties arise because real speech is far more complex than the one-third octave frequency bandwidths represented by the AI procedure or by the cosine amplitude-modulated one-third octave bands of random noise included in the STI procedure. Real speech comprises a sequence of separate phonemes, each of which is made up from a series of harmonically related voiced frequencies generated by the vocal cords, together with mostly higher-frequency broadband components generated by turbulent airflows through constrictions in the vocal tract. The overall speech frequency spectrum is then filtered through variable formant frequency bands generated by changing the articulation of the vocal tract. Transitions and silences between consecutive phonemes can be important for meaning. One-third octave-band spectrum methods such as AI and STI ignore the periodic and harmonic nature of typical vowel sounds, which contribute a large part of the intelligibility of normal running speech. The fundamental and harmonic frequency components in vowel sounds are controlled by the airflow through, and muscular tension in, the vocal cords, and can be heard as voice pitch, as in singing. Changes in voice pitch contribute information about the emotional state of the talker, in addition to the semantic content contributed by word and sentence intelligibility.

To summarise, objective procedures such as AI/SII and STI can be more useful for comparing the relative intelligibility capabilities between generically similar speech communications systems than for absolute assessments of overall performance where any significant degree of precision is required. The only way to determine actual speech intelligibility is to carry out behavioural measurements using real talkers and real listeners, and even then the results will vary depending on the actual talkers, listeners, word lists or sentence materials and test procedures used. Real speech using running sentences has considerable redundancy, which allows listeners to achieve relatively high intelligibility even where there are significant masking or distortion effects.

6.4.5 Activity disturbance and personal preference

It is well known that excessive noise can interfere with or disturb wanted activities such as rest and relaxation, private study or concentration on important tasks such as desk work. However, the amount of interference or disturbance can also be highly dependent on the sensitivities and sensibilities of the individuals concerned. Some people can ignore distracting noise to a much greater extent than others. Motivation can be strongly influenced by penalties and rewards. Because there are so many other factors involved, there are no reliable methods for predicting activity disturbance based on objective sound levels alone, except for specific listening tasks which depend on good speech intelligibility, in which case AI-type indicators may suffice.

In recent years, there has been a considerable amount of interest in the adverse effects of noise in the classroom. Excessive noise break-in from external sources such as aircraft, road traffic and industrial or commercial sources can interfere with speech and music in the classroom, and various authorities have been concerned about the extent to which this might affect academic progress. The recent European road-traffic and aircraft-noise exposure and children's cognition and health (RANCH) study (Stansfeld et al. 2005) found statistically significant associations between aircraft noise measured outside schools and reading comprehension and recognition memory, based on pooled analysis from survey data collected in three European countries. Higher external-noise sound levels (mainly road traffic) have also been associated with poorer performance on standardised assessment scores in primary-school children in London (Shield and Dockrell 2008). However, academic progress can also be affected by many other factors, and it is not yet clear to what extent these findings actually demonstrate cause and effect. It is difficult or impossible to control for all possible confounding variables such as differences in socioeconomic class and language spoken at home between different schools in different areas, or for exposure to other noise sources when not at school.

A number of different mechanisms have been proposed which could potentially explain the observed effects of external noise break-in on child development and attainment. External noise break-in can directly interfere with speech communications, and while this kind of disruption can be overcome by increased vocal effort or repeating sentences, it could nevertheless interfere with concentration and attention or reduce motivation. Complex tasks requiring high cognitive demands, as opposed to simple repetitive tasks, tend to be the most affected by environmental stressors such as noise, and it is possible that some children could be particularly sensitive to noise interference at key development stages. On the other hand, it is not known to what extent delayed development at key development stages can be recovered subsequently. Different children develop at different rates. The current WHO guidelines (WHO 2000) suggests that children and elderly people should be treated as 'vulnerable' groups, according to the assumption that both groups are more susceptible to adverse effects of noise, although the evidence to support this suggestion seems to be lacking.

In more general terms, people differ in their preferences for either complete silence or some form of masking noise or music which might prevent otherwise distracting noise from being heard. Depending on sound quality, a certain amount of noise can contribute to arousal and help to maintain alertness, whereas too much noise can be distracting. Feeling annoyed by the presence of what is assumed to be unnecessary or avoidable noise is unlikely to assist concentration, particularly on difficult tasks. The provision of artificial background noise in open-plan offices provides an interesting example. Depending on the type of work being done and the motivation of the people doing the work, introducing controlled background noise in open-plan offices at an appropriate level can increase performance. The secret is to provide just enough controlled background noise to mask distracting intrusions and maintain speech privacy without the controlled background noise becoming objectionable in itself. Most people have heard background music in public places such as railway stations and shopping malls. It is only ever provided because management perceive it to be beneficial in terms of sales or efficiency. The provision of military band-style music during the morning rush hour at railway stations can assist in encouraging people to disperse rapidly, while more restful music at less busy periods or in airport departure lounges can help to calm people who might otherwise find the experience stressful. Different types of music at the supermarket checkout can have significant effects on customer behaviour, but the effects are variable depending on individual personality and preference. If people were all the same, then they would all like to listen to the same things, which is clearly not the case.

The fashion industry is particularly sensitive to the type of music reproduced over public-address systems inside shops and shopping malls, and controlled experiments in these areas are rarely reported because of commercial sensitivity. Service companies provide online recorded music programmes tailored to specific markets. In addition to music-on-demand services for domestic consumers, they also provide specific programmes for different types of retail outlets based on market research. Retail management buys into these services because of the effect they have on sales. The particular types of music programme reproduced in different stores are designed to appeal to the type of customers who are most likely to purchase the type of products sold in that store. Management can immediately tell if the selected music is commercially beneficial or harmful because of the effect on sales. It is, of course, impossible to predict the likely effects of different types of music on sales turnover just by measuring the L_{Aeq} sound level. Personal preference is not an engineering quantity, and there is hardly any relationship between observed effects and measured sound levels in this area.

6.4.6 Sleep disturbance

Sleep disturbance is interesting, because everyone knows that noise can disturb sleep, but nobody really understands what sleep is actually for. The most important effect of loss of sleep from any cause seems to be increased sleepiness the next day. Anything that increases the risk of falling asleep while engaged in safety-critical tasks such as long-distance lorry driving is obviously dangerous. However, if the person has not actually fallen asleep, task performance also depends on application and motivation and can be relatively unaffected. Anything that interferes with normal sleep patterns can be very annoying, and may cause people to believe that their performance or mood the next day has been adversely affected, even if there is no objective evidence of any adverse effect. Disturbed sleep during one night can be compensated for by more time spent in deep sleep on following nights, and most people habituate to familiar sounds such as regular railway-train pass-bys to some considerable extent. Most people are more sensitive to noise while sleeping in an experimental laboratory situation than in their own homes, especially on the first night.

Most people sleep for around 7–8 h every night, although there is considerable variation above and below the average. Normal sleep passes through successive stages from light sleep to deep sleep and back to light sleep again in approximately 90 min cycles. People are least sensitive to external stimuli when in deep sleep, which can be identified from different brainwave patterns recorded via electroencephalogram (EEG) electrodes placed on the scalp. When a person is awake, EEG records tend to be random and irregular. EEG records become more regular with increasingly identifiable low-frequency wave patterns with increasing depth of either relaxation or sleep. There is normally a short awakening period between each sleep cycle which is not usually remembered the next day unless something interesting was happening at the time. Short periods of REM (rapid eye movement) sleep associated with dreaming are also most likely to occur in between the separate sleep cycles. The proportion of time spent in deep (slow-wave) sleep decreases with successive sleep cycles, while the proportion of time spent in REM (dreaming) sleep increases. Loss of sleep on previous nights is generally compensated for by an increased proportion of time spent in deep sleep on subsequent nights.

The peripheral auditory system remains active during sleep, but higher-level cognitive processing systems shut down unless an incoming signal is sufficiently intense or unexpected to require cognitive attention. Sensitivity varies depending on sleep stage and the number of sleep cycles which have already been completed at that time. Neural response to transient sound can lead to partial or complete arousal depending on the sound level and character of the sound. Transient arousals identified from changes in EEG records, other physiological measurements (heart rate, blood pressure, endocrine responses, etc.), changes in posture or other movements do not necessarily lead to behavioural awakening, although they may increase the probability that another sound following soon after causes behavioural awakening.

Habituation to familiar or expected sounds reduces the probability of behavioural awakening without necessarily reducing the prevalence of transient arousals. Scientific opinion is divided about the possible effects of transient arousals which do not progress to behavioural awakening. Transient arousals could be considered simply as indicators of normal biological function.

Babisch (2006) observed an increased risk of cardiovascular disease in people living in areas with high road-traffic noise sound levels, and speculated that the increased risk could have been associated with an increased prevalence of transient arousal while asleep. Repeated transient arousal leads to increased blood concentrations of cortisol and adrenaline hormones, the so-called stress hormones. The main function of these hormones is to facilitate action in response to an opportunity or threat. If there is no action because the person is asleep and has not actually been awakened, the stress hormones might still (in theory) have some effect on metabolic homeostasis (European Environment Agency 2010) and thereby contribute to an increased risk of cardiovascular disease over the longer term.

The current WHO guideline values (WHO 2000) for sleep disturbance at 30 dB L_{Aeq} and 45 dB L_{ASmax} are mostly based on laboratory data, which consistently shows transient arousals associated with noise events (and not necessarily leading to behavioural awakening) at quite modest sound levels down to around 40–50 dB L_{ASmax} measured indoors. The most recent

WHO Europe night noise guidelines document (WHO 2009) proposes even lower guideline values, based on recent field data showing small increases in the probability of transient awakening associated with aircraft-noise events (measured using EEG records) down to around 35 dB L_{ASmax} measured indoors. However, the most recent large-scale field study of aircraft noise and sleep (Jones and Rhodes 2013) carried out in residential areas around Heathrow, Gatwick, Manchester and Stansted airports in the United Kingdom found that disturbance measured by using wrist-worn actimeters was minimal where outdoor aircraft-event sound levels were below 80 dB L_{ASmax}. The probability of the average person being awakened at higher aircraft-event sound levels up to 95 dB L_{ASmax} measured outdoors was only about 1 in 75. A subsequent small-scale field and laboratory combined-methodology trial study (Robertson et al. 1999) using EEG recordings found similar results. Taking into account the typical sound-level difference between outdoors and indoors for average British houses of around 15 dB with the bedroom windows open for ventilation and around 25 dB with the bedroom windows closed, the current WHO guideline values appear to be somewhat conservative.

Which data should be relied on for assessment and regulation purposes? There is no clear-cut answer to this question. In the United Kingdom, Department of Transport policy on night-time aircraft noise continues to be informed by the Ollerhead et al. (1992) study. Airport noise objectors would probably prefer policy to be based on the WHO guidelines. One of the main difficulties here is that most kinds of transient arousals and other sleep disturbances associated with noise events also occur naturally throughout the night anyway, even if there is no noise at all. Even transient arousals or other sleep disturbances occurring at the same time as a noise event may have been about to happen anyway. A person who claims to have been woken up by an early-morning aircraft-noise event cannot be certain that it was actually the aircraft-noise event complained of that caused them to be woken. On the other hand, if there is a statistically significant increase in the overall amount of sleep disturbance in noisy areas compared with quiet areas, then this would be good evidence in support of an effect.

Policymakers tend to be influenced by public perceptions, individual complaints and community action unless there is scientific evidence to persuade them otherwise. The available evidence suggests that, while most people habituate to noise at night to a much greater extent than claimed by community action groups, there may also be a small minority who are persistently disturbed and may even be suffering from long-term health consequences as a result. Not surprisingly, not many people are very good at reporting what happened while they were asleep. It is very easy to misidentify the actual cause of any wakening event, particularly if the offending road vehicle or aircraft has moved on by the time the sleeper is sufficiently awake to try to work out what happened.

How can we best summarise what is known about noise and sleep disturbance? Sounds which cannot actually be heard because of masking by other sounds cannot disturb sleep, although it is possible that some other sensory input, such as car headlights shining into the bedroom, could be the actual cause of disturbance which might later be wrongly attributed to the noise made by the car. However, any sound which is audible to a person who is awake can potentially disturb sleep to an extent determined by the character of the sound, its relative prominence and onset rate compared with the background-noise sound level, and the unexpectedness or familiarity of the sound. Any audible sound can contribute to an observable transient arousal. Transient arousals may or may not lead to behavioural awakening, depending on the changing sensitivity of the person during different stages of the sleep cycle and the extent to which the sound can be recognised by subconscious processing as being of no interest without requiring cognitive appraisal, which can only be performed if the person is awake. Because of the number of variables involved it is almost impossible to predict whether any particular sound will actually wake a person up or not, except if the sound is very loud or particularly strident. This is the principle adopted for hotel fire alarms, which are required, according to a completely different concept from the WHO night noise guidelines, to have a minimum sound level of 75 dB L_{AFmax}, measured at the bedhead (BSI 2013), to have a reasonable probability of waking hotel guests.

6.4.7 Annoyance

By far the greatest amount of effort and resource expended on noise control and noise management is directed towards the reduction of noise annoyance. Somewhat curiously, noise annoyance is also the least well defined out of all possible adverse effects of noise. This is probably because there are no direct physical or physiological correlates of noise annoyance, so there are no possibilities of absolute calibration. As a subjective concept, noise annoyance is fundamentally intangible. Most people have a generic understanding of the concept, but there are large individual differences between different people in different situations. Actual annoyance (whatever that is) probably varies, and so also does different people's understanding of what the word really means. This is one of the main reasons why noise assessment and regulation is usually based on sound levels measured or calculated according to objective procedures rather than on subjective opinions, which can always be challenged on the basis of alleged or actual individual bias. Noise makers can usually estimate the engineering and financial costs and inconvenience of noise control and noise management action, whereas noise sufferers cannot quantify the reduction in annoyance likely to be achieved by noise control and noise management action in any comparable way.

The prevalence of noise annoyance can be measured by the numbers of noise complaints in any particular area. For example, most major airports record and tabulate noise complaints which are often about specific noise events, but can also be more general in nature or about other topics such as track keeping and air pollution. Not all complaints can be related back to the specific events complained of; for example, air traffic control radar records might show there were no aircraft flying in a particular area at the time of the complaint. Any sudden increase or decrease in the overall numbers of complaints could be associated with some actual change in operations, or alternatively with a change in awareness resulting from proposed developments or media attention. Complaint statistics can provide an indication of possible changes in the general attitudes of the population towards the organisation or facility receiving the complaints. However, the attitudes and opinions of the majority of the population who do not register complaints are often more interesting and could be as, or more, relevant for policy. People who do not complain may have nothing to complain about; or they could be annoyed but do not know how to complain or be concerned about possible victimisation, or they may simply feel it would be a waste of time. Sometimes people may be annoyed but are sufficiently reassured by the knowledge or belief that the noise maker has treated them fairly and is prepared to take complaints seriously that they are happy to accept the situation without actually complaining. This is the concept of perceived fairness, whereby people can be more inclined to accept a limited amount of disturbance and annoyance if they consider that preventive action has been commensurate and reasonable and that any residual disturbance is unavoidable.

Reported annoyance measured using standardised questionnaires administered to statistically representative samples of the population is assumed to be representative of underlying actual annoyance, whatever that might mean to the individuals concerned. Reported annoyance is a record of a person's objectively observable behaviour when responding to or completing a questionnaire. In cross-sectional studies, the effects of subjective bias can be controlled, at least to some extent, by averaging reported annoyance across statistically representative samples of survey respondents exposed to noise in different and objectively measurable ways, and by standardising the procedures according to the specification set out by an ISO committee established for that purpose.

The relevant ISO recommendation (ISO 2003) defines noise annoyance as 'a person's individual adverse reaction to noise'. The definition in the standard is amplified by two notes as follows: 'NOTE 1 The reaction may be referred to in various ways, including for example dissatisfaction, bother, annoyance, and disturbance due to noise; NOTE 2 Community annoyance is the prevalence rate of this individual reaction in a community, as measured by the responses to questions specified in clause 5 and expressed in appropriate statistical terms.'

The standard specifies two alternative direct questions which should be presented towards the beginning of any questionnaire to avoid bias from questions on other or more detailed topics which could then be presented later.

Verbal scale

Thinking about the last (... 12 months or so ...) when you are here at home, how much does noise from (... noise source ...) bother, disturb, or annoy you:

Not at all   Slightly   Moderately   Very   Extremely (annoyed)

Numerical scale

Next is a zero to ten opinion scale for how much (... source ...) noise bothers, disturbs or annoys you when you are here at home. If you are not annoyed at all choose zero, if you are extremely annoyed choose ten, if you are somewhere in between, choose a number between zero and ten.

Thinking about the last (... 12 months or so ...) what number from zero to ten best shows how much you are bothered, disturbed, or annoyed by (... source ...) noise:

Not at all annoyed 0 1 2 3 4 5 6 7 8 9 10 Extremely annoyed

The ISO committee was unable to agree on a single recommendation for either the verbal or the numeric scale and therefore offered both. The adjectives on the 5-point verbal scale were selected as being the most evenly spaced from 'not at all' to 'extremely', based on psycholinguistic testing. The main criticism of the 5-point verbal scale was that it does not allow finer discriminations between the scale points, so the 11-point numeric scale was offered as an alternative. However, statistical comparisons of the 11-point numeric scale against the 5-point numeric scale showed that the extra scale points in the numeric scale merely increase the variance without having much effect on statistical precision. The 11-point numeric scale is most useful for laboratory-type tests where listeners are asked to rate a series of different sounds, one after the other, in repeated measures experimental designs. For single one-off judgements of the sound outside a respondent's house, the 5-point verbal scale seems to be perfectly adequate and is generally easier to explain.

The ISO standard annoyance scale avoids the problem of not being able to define annoyance in any precise way by instead specifying the way in which it should be measured. Changing the wording of the questionnaire or changing the way in which the questionnaire is administered could change the results, but this uncertainty is avoided if the standard format for the questionnaire is always used in the same way. The ISO standard encourages further questionnaire items to investigate any specific issues in more depth, but only after the standard annoyance scale has been dealt with first to avoid bias.

The WHO defines health as 'a state of complete physical, mental and social well-being and not merely an absence of disease or infirmity'. Most people in the United Kingdom would not normally consider noise annoyance to be a health effect as such, but reported annoyance reflects (or is assumed to reflect) a degree of adverse reaction in terms of perceived 'physical, mental, and social well-being' as defined in the WHO constitution. This means that, according to the WHO, reported annoyance measured using the ISO standard annoyance scale can be considered as a legitimate health effect. The seemingly unlikely but nevertheless theoretical possibility that nonhabituated physiological response to noise while asleep or otherwise unaware could lead to potentially harmful long-term consequences was discussed in Section 6.4.6. A more plausible consideration is that high concentrations of stress hormones could arise simply from a person's continuing frustration and perceived lack of control to be able to do anything about their situation – and that these high concentrations could lead to more serious adverse health consequences over the longer term. For any individuals for whom this is the case, any measure which reduces annoyance would also reduce associated stress and any subsequent adverse health consequences. For the reduction of any direct effects, any action which reduces physical exposure could be effective. However, for the reduction of annoyance/stress-related effects, psychologically-based methods which do not necessarily involve any actual reduction in physical sound levels could nevertheless be equally or more effective.

What does noise annoyance really mean? In open-ended qualitative interviews, many respondents in areas where a majority report relatively high levels of aircraft-noise annoyance in standardised questionnaire surveys admit to having 'mostly got used to' the noise. The most straightforward explanation for this apparent inconsistency is that when reporting high annoyance in a standardised questionnaire, the respondent is expressing an opinion about the local noise environment, which, even if that respondent has personally adapted to it to at least some extent, does not necessarily mean that the reported annoyance should be ignored by policymakers. Most residents around airports (or near main roads or other noise sources) appreciate a range of both advantages and disadvantages of continuing to live where they do, and noise is usually only one of several disadvantages. It is perfectly reasonable for individual residents to describe the noise as very or extremely annoying (perhaps on a Sunday afternoon when they have friends round for an outdoor barbecue) but nevertheless acceptable within the overall balance of advantages and disadvantages of the place where they live. In the past, some research studies have attempted to conceal these inconsistencies by not disclosing the true purpose of the questionnaire until the last possible moment. In this way, they attempted to discover if respondents would mention aircraft noise as a disadvantage of living where they do without having been prompted by being told the true purpose of the research. Quite apart from the ethical issue of requiring informed consent before taking part in any research of this type, it is not obvious that this approach would have acquired data of any more use for informing policy. If people report that they are annoyed or even very or extremely annoyed by aircraft noise but not (separately) to the extent that it is unacceptable in terms of their everyday lives, then that should be of interest for policy.

Most surveys have shown a general tendency for average reported annoyance to be higher in higher-noise areas, and researchers have attempted to combine the various results to obtain aggregated and possibly more generally applicable noise dose–annoyance response relationships (Fidell 2003; European Commission 2000). In the past, researchers generally preferred to devise their own questionnaire wording and experimental designs, either to test particular hypotheses or simply because they did not particularly agree with other researchers' designs. The resulting diversity created particular difficulties for subsequent researchers attempting to combine different databases in the expectation of finding underlying trends not observable in individual studies because of sample size limitations. Even if exactly the same questionnaire wording has been used in different studies, this does not necessarily mean that the questionnaire responses are directly comparable, because there could have been other material differences in design and procedure. If different questionnaire wordings were used, the data becomes even less comparable. Meta-analysis of combined data sets is only feasible by making various normalising assumptions to enforce comparability of the data. Unfortunately, the only way to devise appropriate normalising assumptions is to assume underlying consistency between the different data sets. The normalising assumptions then become self-justifying. It is equally plausible that there is no *a priori* reason to expect consistency between different surveys, and that observed differences in noise dose–annoyance response relationships between different surveys represent normal and understandable diversity (Fidell et al. 2011).

In summary, it seems that the main causal factors responsible for reported noise annoyance can be divided into three main categories, in order of importance as follows:

1. The sounds themselves (this is *not* necessarily the same thing as a sound level attributed to the sound)
2. The psychological orientation of the listener towards the sound within the context in which it is heard
3. A large and essentially unpredictable or random component

Administrators and regulators may be able to devise and assess policy which can reduce the noisiness or potential to cause annoyance of objectionable or complained-of sounds, either by reducing sound levels measured according to established acoustic metrics, or by improving the sound character or sound quality. Whichever method is the most cost-effective should be used, but it should be noted that cost-effectiveness is difficult to assess in this area because of a lack of directly comparable metrics.

Alternatively, administrators and regulators may be able to devise and assess policy which addresses the psychological orientation of listeners towards objectionable or complained-of sounds. Methods which address psychological orientation can be as or more effective in reducing reported annoyance as methods addressed to physical or objectively measurable noise reduction. However, it seems unlikely that methods based solely on information exchange and good public relations would have any long-term effect unless they are also accompanied by physical measures that respondents can actually see and hear.

It is unlikely that administrators and regulators will ever be able to do anything about the unpredictable or random components of reported annoyance. There are many types of sound which some people enjoy and others hate, and this is often completely unpredictable. Exactly the same sound can be enjoyable one day and annoying the next, depending on whatever else is going on at the time. There would seem to be little or no point in trying to predict the unpredictable.

REFERENCES

ANSI (American National Standards Institute) (1969) S3.5 R 1986 *American National Standard Methods for the Calculation of the Articulation Index*, American National Standards Institute, New York.

ANSI (American National Standards Institute) (1997) S3.5 *American National Standard Methods for the Calculation of the Speech Intelligibility Index*, American National Standards Institute, New York.

Babisch, W. (2006) *Transportation Noise and Cardiovascular Risk*, Umweltbundesamt, Berlin.

BSI (British Standards Institution) (1979) *BS 5727: Method for Describing Aircraft Noise Heard on the Ground*, British Standards Institution, London.

BSI (British Standards Institution) (1999) *BS 8233: Sound Insulation and Noise Reduction for Buildings. Code of Practice: Annex B Noise Rating*, British Standards Institution, London.

BSI (British Standards Institution) (2003a) *BS EN 60268-16: Sound System Equipment Part 16: Objective Rating of Speech Intelligibility by Speech Transmission Index*, British Standards Institution, London.

BSI (British Standards Institution) (2003b) *BS ISO 226: Acoustics: Normal Equal Loudness Level Contours*, British Standards Institution, London.
BSI (British Standards Institution) (2013) *BS 5839-1: Fire Detection and Fire Alarm Systems for Buildings*, British Standards Institution, London.
Control of Noise at Work Regulations (2005) SI 2005/1643, Statutory Instruments, The Stationery Office, Norwich.
European Commission (2000) Position paper on dose-response relationships between transportation noise and annoyance. COM(2000) 468 final, Office for Official Publications of the European Communities, Luxembourg.
European Environment Agency (2010) Good practice guide on noise exposure and potential health effects, EEA Technical Report No. 11, European Environment Agency, Copenhagen.
European Parliament/Council of the European Union (2002) Directive 2002/49/EC of the European Parliament and of the Council of 25 June 2002 relating to the assessment and management of environmental noise, Office for Official Publications of the European Communities, Luxembourg.
Fidell, S. (2003) The Schultz curve 25 years later: A research perspective, *Journal of the Acoustical Society of America*, **114**(6) Part 1, 3007–3015.
Fidell, S. et al. (2011) A first-principles model for estimating the prevalence of annoyance with aircraft noise exposure, *Journal of the Acoustical Society of America*, **130**(2), 791–806.
Houtgast, T., Steeneken, H.J.M. and Plomp, R. (1980) Predicting speech intelligibility in rooms from the Modulation Transfer Function. I. General room acoustics, *Acta Acustica United with Acustica*, **46**(1), 60–72.
ISO (International Organization for Standardization) (1975) *ISO 532: Acoustics – Method for Calculating Loudness Level*, International Organization for Standardization, Geneva.
ISO (International Organization for Standardization) (2003) *ISO/TS 15666: Acoustics – Assessment of Noise Annoyance by Means of Social and Socio-Acoustic Surveys*, International Organization for Standardization, Geneva.
Jones, K. and Rhodes, D.P. (2013) Aircraft noise, sleep disturbance and health effects: A review, ERCD REPORT 1208, Civil Aviation Authority, The Stationery Office, Norwich.
Morfey, C.L. (2001) *Dictionary of Acoustics*, Academic Press, London.
Niemann, H. and Maschke, C. (2004) WHO large analysis and review of European housing and health status (LARES): Final report noise effects and morbidity, WHO Europe office, Berlin.
Ollerhead, J.B., Jones, C.J., Cadoux, R.E., Woodley, A., Atkinson, B.J., Horne, J.A., Pankhurst, F.L., et al. (1992) Report of a field study of aircraft noise and sleep disturbance, Department of Transport, London.
Robertson, K.A., Wright, N., Turner, C., Birch, C., Flindell, I., Bullmore, A., Berry, B., Jiggins, M., Davison, M., Dix, M. (1999) Aircraft noise and sleep 1999 UK trial methodology study, Report 6131 R01, ISVR Consultancy Services.
Shield, B.M. and Dockrell, J.E. (2008) The effects of environmental and classroom noise on the academic attainments of primary school children, *Journal of the Acoustical Society of America*, **123**(1), 133.

Stansfeld, S.A., Berglund, B., Clark, C., Lopez-Barrio, I., Fischer, P., Öhrström, E., Haines, M.M. et al. (2005) On behalf of the RANCH study team, Aircraft and road traffic noise and children's cognition and health: A cross-national study, *The Lancet*, **365**, 1942–1949.

WHO (World Health Organization) (2000) Guidelines for community noise, WHO, Geneva.

WHO (World Health Organization) (2009) Night noise guideline for Europe, WHO Regional Office for Europe, Geneva.

Chapter 7

Human responses to vibration

Michael J. Griffin

7.1 INTRODUCTION

The high sensitivity of the human body to oscillatory motion is such that human responses to vibration determine the acceptability of vibration in many environments. For example, vibration is felt and becomes annoying before it damages buildings, the vibration in transport causes discomfort or interferes with activities when it does not cause damage to the vehicle and the vibrations of tools and machines produce injuries and disease without breaking the tools or machinery.

It is convenient to consider human exposure to oscillatory motion in three categories.

Whole-body vibration occurs when the body is supported on a surface that is vibrating (e.g. sitting on a seat that vibrates, standing on a vibrating floor or lying on a vibrating surface). Whole-body vibration occurs in transport (e.g. road, off-road, rail, air and marine transport) and when near some machinery.

Motion sickness occurs when real or illusory movements of the body or the environment lead to ambiguous inferences as to the movement or orientation of the human body. The movements associated with motion sickness are always of very low frequency, usually less than 0.5 Hz.

Hand-transmitted vibration is caused by various processes in industry, agriculture, mining, construction and transport where vibrating tools or workpieces are grasped or pushed by the hands or fingers.

There are many different effects of oscillatory motion on the body and many variables influencing each effect. The variables may be categorised as extrinsic variables (those occurring outside the human body) and intrinsic variables (the variability that occurs between and within people), as in Table 7.1.

This chapter introduces the principal human responses to vibration, summarises current methods of measuring, evaluating and assessing exposures to vibration, and identifies methods of minimising unwanted effects of vibration (see also Griffin 1990).

Table 7.1 Variables influencing human responses to oscillatory motion

Extrinsic variables		Intrinsic variables	
Vibration variables	Other variables	Intrasubject variability	Intersubject variability
Vibration magnitude Vibration frequency Vibration direction Vibration input positions Vibration duration	Other stressors (noise, temperature, etc.) Seat dynamics	Posture Position Orientation (sitting, standing, recumbent)	Body size and weight Body dynamic response Age Gender Experience, expectation, attitude and personality Fitness

7.2 MEASUREMENT OF VIBRATION

The magnitudes of vibration to which the human body is exposed are usually measured with accelerometers and expressed in terms of the acceleration in metres per second per second (i.e. m s^{-2} or m/s^2).

The frequency of vibration is expressed in cycles per second using the SI unit, hertz (Hz). The frequency of vibration influences the extent to which vibration is transmitted to the surface of the body (e.g. through seating), the extent to which it is transmitted through the body (e.g. from seat to head) and the responses to vibration within the body. The three principal categories of human response to oscillatory motion are currently associated with three different ranges of frequency: whole-body vibration (from about 0.5 to 80 Hz), motion sickness (at frequencies less than 0.5 Hz), and hand-transmitted vibration (from about 8 to 1000 Hz).

The responses of the body differ according to the direction of the motion (i.e. axis of vibration). The three principal directions of whole-body vibration for seated and standing persons are fore-and-aft (x-axis), lateral (y-axis) and vertical (z-axis). The vibration is measured at the interface between the body and the surface supporting the body (e.g. on the seat beneath the ischial tuberosities for a seated person; beneath the feet for a standing person). Figure 7.1 shows the translational and rotational axes for an origin on a seat at the ischial tuberosities of a seated person. A similar set of axes is used for describing the directions of vibration at the backrest and feet of seated persons. Figure 7.2 shows a coordinate system used when measuring vibration in contact with a hand holding a tool.

Some human responses to vibration depend on the duration of exposure. In addition, the duration over which vibration is measured may affect the measured magnitude of the vibration when the motion is not statistically

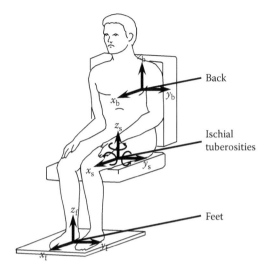

Figure 7.1 Axes of vibration used to measure exposures to whole-body vibration.

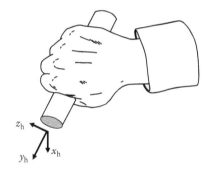

Figure 7.2 Axes of vibration used to measure exposures to hand-transmitted vibration.

stationary. The root-mean-square (r.m.s.) acceleration is often used, but it does not provide a good indication of vibration severity if the vibration is intermittent, contains shocks or otherwise varies in magnitude from time to time (see Sections 7.3.2.3 and 7.3.4.1).

7.3 WHOLE-BODY VIBRATION

Whole-body vibration can affect comfort, the performance of activities and health. Vibration of the whole body is produced by all forms of transport (including road, off-road, rail, sea and air transport) and by various types of industrial machinery.

7.3.1 Biodynamics

Biodynamics is the science of the physical responses of the body to force or motion. The human body is a complex mechanical system that does not, in general, respond to vibration in the same manner as a rigid mass: there are relative motions between the body parts that vary with the frequency and the direction of the applied vibration. Although there are resonances in the body, it is oversimplistic to summarise the dynamic response of the body by merely mentioning one or two resonance frequencies. The biodynamics of the body affect all human responses to vibration, but the discomfort, the interference with activities and the health effects of vibration cannot be predicted solely by considering the body as a mechanical system.

7.3.1.1 Transmissibility of the human body

The extent to which the vibration at the input to the body (e.g. the vertical vibration at a seat) is transmitted to a part of the body (e.g. vertical vibration at the head or the hand) is described by the transmissibility of the body. At low frequencies of oscillation (e.g. at frequencies less than about 1 Hz), the vertical oscillations of a seat and the body are very similar, and so the transmissibility is approximately unity. With increasing frequencies of oscillation, the motions of the body increase, so as to exceed those measured at the seat. The ratio of the motion of the body to the motion of the seat reaches a peak at one or more frequencies (i.e. resonance frequencies), with the frequencies of resonances varying according to the part of the body. At high frequencies, the motion of the body is less than the motion of the seat.

The resonance frequencies and the transmissibilities at resonance vary according to the direction (i.e. axis) of vibration, according to where the vibration is measured on the body and to the posture of the body. There can be large differences between people. For seated persons, there may be resonances to the head and the hand at frequencies in the range 4–12 Hz for vertical vibration, at <4 Hz for fore-and-aft vibration and <2 Hz for lateral vibration (see Figure 7.3 and Paddan and Griffin 1988a, b). A seat back can greatly increase the transmission of x-axis vibration to the head and upper body of a seated person (Paddan and Griffin 1988b), and bending of the legs can greatly affect the transmission of vertical vibration to the heads of standing persons (Paddan and Griffin 1993).

7.3.1.2 Mechanical impedance of the human body

Mechanical impedance reflects the relation between the driving force at the input to the body and the resultant movement of the body. If the human body were rigid, the ratio of the force to the acceleration at the vibration input to the body would be constant and indicate the mass of the body.

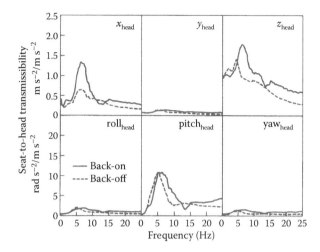

Figure 7.3 Transmissibility of vertical seat vibration to six axes of head vibration (median data from 12 males showing the effect of contact with a rigid vertical seat backrest). (Data from Paddan, G.S. and Griffin, M.J. *Journal of Biomechanics*, **21**(3), 191–197, 1988.)

Because the human body is not rigid, the ratio of force to acceleration (i.e. the *apparent mass*) only reflects the body mass at low frequencies (less than about 2 Hz with vertical vibration; less than about 1 Hz with horizontal vibration).

The apparent mass shows a principal resonance in the vertical vibration of seated persons at about 5 Hz, and sometimes a second resonance in the range 7–12 Hz (see Figure 7.4; Fairley and Griffin 1989). Although there can be large differences in the apparent masses of different people, the principal resonance in the vertical direction is always close to 5 Hz (Toward and Griffin 2011a). Unlike some of the resonances affecting the transmissibility of the body, these resonances are only influenced by the movement of large masses close to the input of vibration to the body. The large difference in response between a rigid mass and the human body means that the body cannot usually be represented by a rigid mass when measuring the vibration transmitted through seats (see Section 7.3.6). Many of the biodynamic responses of the body are nonlinear: the resonance frequencies decrease when the vibration magnitude increases (Figure 7.5; Nawayseh and Griffin 2003, 2005).

7.3.1.3 Biodynamic models

Many mathematical models of the responses of the body to vibration have been developed. A simple model with only one or two degrees of freedom

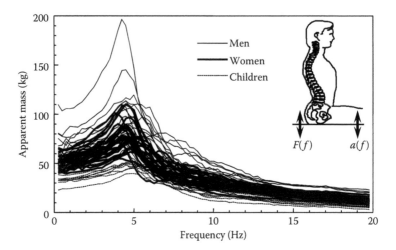

Figure 7.4 Apparent masses of 60 seated subjects (24 males, 24 females, 12 children) in the vertical axis measured during 64 s exposures to 1.0 m s^{-2} r.m.s. random vibration with simultaneous vibration of seat and feet. (Data from Fairley, T.E. and Griffin, M.J., *Journal of Biomechanics*, **22**, 81–94, 1989.)

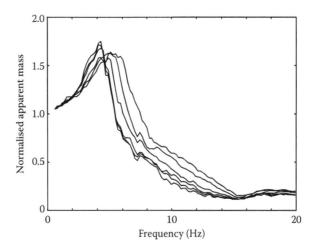

Figure 7.5 Median normalised apparent mass of 12 subjects measured at 0.25, 0.5, 1.0, 1.5, 2.0, 2.5 m s^{-2} r.m.s. Resonance frequencies decrease with increasing vibration magnitude. (From Mansfield, N.J. and Griffin, M.J., *Journal of Biomechanics*, **33**, 933–941, 2000.)

can accurately represent the vertical apparent mass of the body (e.g. Wei and Griffin 1998a) and mechanical dummies have been constructed to represent this dynamic response of the body for seat testing (e.g. Lewis and Griffin 2002). ISO 5982 (2001c) defines a model with three degrees of freedom for this purpose. In practice, the apparent mass of the body is influenced by contact with a backrest and the location of limbs (Toward and Griffin 2009, 2010), and also depends on the magnitude of vibration. The influence of these factors is not taken into account in the standard model.

The transmissibility of the body is affected by many variables and requires a more complex model reflecting the posture of the body and the translation and rotation of the body associated with the various modes of vibration. Multibody models are required to reflect the cross-axis responses of the seated and standing body (e.g. Matsumoto and Griffin 2001, 2003), and finite element models may be necessary to predict the forces within the body and the complex interaction between the body and seating (e.g. Kitazaki and Griffin 1997).

7.3.2 Vibration discomfort

For very low magnitudes of vibration, the percentage of persons likely to feel vibration can be estimated from the results of threshold studies (Parsons and Griffin 1988; Morioka and Griffin 2008). For higher magnitudes of vibration, subjective reactions can be estimated from various semantic scales of discomfort (e.g. BS 6841 and ISO 2631). In practice, the relative discomfort caused by different types of vibration or different magnitudes of vibration is often of greater interest (Griffin 2007). The appropriate limiting levels of vibration required to preserve comfort vary greatly between different environments (e.g. between buildings and transport) and between different types of transport (e.g. between cars and trucks) and within types of vehicle (e.g. between sports cars and limousines). The design limit depends on external factors (e.g. cost and speed) and the comfort in alternative environments (e.g. competitive vehicles).

7.3.2.1 Effects of vibration magnitude

The absolute threshold for the perception of vertical whole-body vibration in the frequency range 1–100 Hz is, very approximately, 0.01 m s^{-2} r.m.s. A magnitude of 0.1 m s^{-2} will be easily noticeable; magnitudes around 1 m s^{-2} r.m.s. are usually considered uncomfortable; magnitudes of 10 m s^{-2} r.m.s. are usually dangerous. The precise values depend on the vibration frequency and the exposure duration, and they are different for other axes of vibration (Griffin 1990; Morioka and Griffin 2006a, 2008).

Doubling the vibration magnitude (expressed in m s^{-2}) produces an approximate doubling of the sensation of discomfort. A halving of vibration magnitude can therefore produce a useful improvement in comfort. For some types of whole-body vibration, differences as small as 10% in the vibration magnitude can be detected, but for some people and in some situations, greater changes in magnitude are required for the change to be detected (Forta et al. 2009).

7.3.2.2 Effects of vibration frequency and direction

The extent to which a given vibration acceleration will cause a greater or lesser effect on the body at different frequencies is reflected in *frequency weightings*; frequencies capable of causing the greatest effect are given the greatest weight and other frequencies are attenuated in accordance with their decreased importance. Two frequency weightings (one for vertical vibration and one for horizontal vibration of seated or standing persons) were presented in ISO 2631 (ISO 1974, 1985). In 1987, BS 6841 defined weighting that extended the frequency range to 0.5 Hz, modified the sensitivity of the vertical weighting and added weightings for rotational vibration on the seat and translational vibration at the seat back and at the feet. This approach was later adopted in ISO 2631 (ISO 1997), although for political reasons the frequency weighting adopted for the evaluation of vertical vibration was slightly modified.

Frequency weightings have been derived from laboratory experiments in which volunteer subjects have been exposed to a variety of different motions varying in magnitude and frequency. The judgements of the subjects have been used to determine *equivalent-comfort contours*. The reciprocal of an equivalent-comfort contour forms the shape of the frequency weighting: the greatest weight is given to those frequencies that cause discomfort with the lowest vibration magnitudes. Figure 7.6 shows frequency weightings W_b to W_g defined in BS 6841 (BSI 1987) as they may be implemented by analogue or digital filters. ISO 2631-1 (ISO 1997) defines similar weightings, except that for some applications, the W_k weighting is used in preference to the W_b weighting (see also ISO 8041 [ISO 2005d]). Table 7.2 defines simple asymptotic (i.e. straight-line) approximations to these weightings and Table 7.3 shows how the weightings should be applied to the 12 axes of vibration illustrated in Figure 7.1. (The weightings W_g and W_f are not required to predict vibration discomfort: W_g is used to evaluate the severity of vibration with respect to interference with activities and is similar to the weighting for vertical vibration in ISO 2631 (ISO 1974, 1985) (see Section 7.3.3); W_f is used to predict motion sickness caused by vertical oscillation (see Section 7.4.2).

In BS 6841 (BSI 1987) and ISO 2631-1 (ISO 1997), some frequency weightings are used for more than one axis of vibration, with different multiplying factors allowing for overall differences in sensitivity between axes (Table 7.3). The frequency-weighted acceleration should be multiplied by the appropriate multiplying factor before being compared with components in other axes or included in any summation over axes or compared with

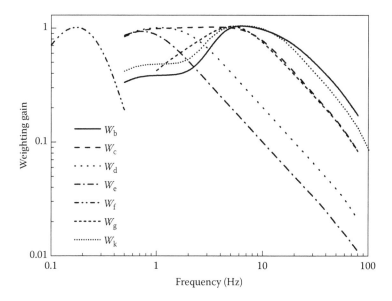

Figure 7.6 Acceleration frequency weightings for whole-body vibration and motion sickness. (Data from BSI, *BS 6841:1987: Measurement and Evaluation of Human Exposure to Whole-Body Mechanical Vibration and Repeated Shock,* British Standards Institution, London, 1987; ISO, *ISO 2631-1:1997: Mechanical Vibration and Shock: Evaluation of Human Exposure to Whole-Body Vibration. Part 1: General Requirements,* International Organization for Standardization, Geneva, 1997.)

any limit. The r.m.s. value of this acceleration (i.e. after frequency weighting and being multiplied by the multiplying factor) is called a *component ride value* (referred to as a *point vibration total value* in ISO 2631 [ISO 1997]). Figure 7.7 illustrates the component ride values for a small car. It may be seen that the greatest discomfort may be expected in this car from the vertical vibration on the seat, the vertical vibration at the feet and the fore-and-aft vibration on the backrest.

Vibration occurring in several axes is more uncomfortable than that occurring in a single axis. To obtain an *overall ride value* (called an *overall vibration total value* in ISO 2631 [ISO 1997]), the *root sums of squares* of the component ride values is calculated:

$$\text{overall ride value} = (\Sigma \, (\text{component ride values})^2)^{1/2} \qquad (7.1)$$

Overall ride values from different environments can be compared: a vehicle having the greatest overall ride value would be expected to be the most uncomfortable with respect to vibration. The overall ride values can also be compared with a scale suggesting possible degrees of discomfort associated with ranges of vibration magnitudes that might be experienced in vehicles (Figure 7.8).

Table 7.2 Asymptotic approximations to frequency weightings, $W(f)$, for comfort, health, activities and motion sickness in BS 6841 and frequency weighting W_k in ISO 2631

Weighting name	Weighting definition	
W_b	$0.5 < f < 2.0$	$W(f) = 0.4$
	$2.0 < f < 5.0$	$W(f) = f/5.0$
	$5.0 < f < 16.0$	$W(f) = 1.0$
	$16.0 < f < 80.0$	$W(f) = 16.0/f$
W_c	$0.5 < f < 8.0$	$W(f) = 1.0$
	$8.0 < f < 80.0$	$W(f) = 8.0/f$
W_d	$0.5 < f < 2.0$	$W(f) = 1.0$
	$2.0 < f < 80.0$	$W(f) = 2.0/f$
W_e	$0.5 < f < 1.0$	$W(f) = 1.0$
	$1.0 < f < 20.0$	$W(f) = 1.0/f$
W_f	$0.1 < f < 0.125$	$W(f) = f/0.125$
	$0.125 < f < 0.25$	$W(f) = 1.0$
	$0.25 < f < 0.5$	$W(f) = (0.25/f)^2$
W_g	$1.0 < f < 4.0$	$W(f) = (f/4)^{1/2}$
	$4.0 < f < 8.0$	$W(f) = 1.0$
	$8.0 < f < 80.0$	$W(f) = 8.0/f$
W_k	$0.5 < f < 2.0$	$W(f) = 0.5$
	$2.0 < f < 4.0$	$W(f) = f/4.0$
	$4.0 < f < 12.5$	$W(f) = 1.0$
	$12.5 < f < 80.0$	$W(f) = 12.5/f$

Source: BSI, BS 6841:1987: Measurement and Evaluation of Human Exposure to Whole-Body Mechanical Vibration and Repeated Shock. British Standards Institution, London, (1987); ISO, ISO 2631-1:1997: Mechanical Vibration and Shock: Evaluation of Human Exposure to Whole-Body Vibration. Part 1: General Requirements. International Organization for Standardization, Geneva, (1997).

Note: f = frequency, Hz; $W(f) = 0$ where not defined.

7.3.2.3 Effects of vibration duration

The discomfort caused by vibration tends to increase with the increasing duration of exposure to vibration. The rate of increase may depend on many factors, but a simple fourth-power time dependence is used to approximate how discomfort varies with duration of exposure (i.e. (acceleration)^4×duration = constant: see Section 7.3.4.1). Although originally based on variations in discomfort when the duration of vibration varies (Griffin and Whitham 1980), this simple procedure seems to provide useful evaluations of the severity of shocks (e.g. Howarth and Griffin 1991; Ahn and Griffin 2008) and also long durations of vibration. Consequently, it is now used to express the severity of motions from the shortest possible shock to a full day of vibration exposure.

Table 7.3 Application of frequency weightings for the evaluation of vibration with respect to discomfort in BS 6841

Input position	Axis	Frequency weighting	Axis multiplying factor
Seat	x	W_d	1.0
	y	W_d	1.0
	z	W_b	1.0
	r_x (roll)	W_e	0.63
	r_y (pitch)	W_e	0.40
	r_z (yaw)	W_e	0.20
Seat back	x	W_c	0.80
	y	W_d	0.50
	z	W_d	0.40
Feet	x	W_b	0.25
	y	W_b	0.25
	z	W_b	0.40

Source: BSI, BS 6841:1987: *Measurement and Evaluation of Human Exposure to Whole-Body Mechanical Vibration and Repeated Shock*. British Standards Institution, London, (1987).

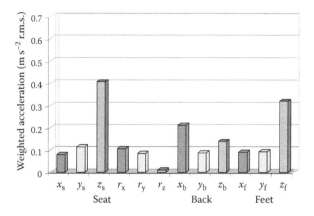

Figure 7.7 Component ride values for 12 axes of vibration for the driver of a small car. Axes defined in Figure 7.1. (Values shown after the application of appropriate frequency weighting and axis multiplying factors from BSI, BS 6841:1987: *Measurement and Evaluation of Human Exposure to Whole-Body Mechanical Vibration and Repeated Shock*, British Standards Institution, London, 1987.)

7.3.3 Interference with activities

Vibration can interfere with the acquisition of information (e.g. by the eyes), the output of information (e.g. by hand or foot movements) or the complex central processes that relate input to output (e.g. learning, memory, decision making).

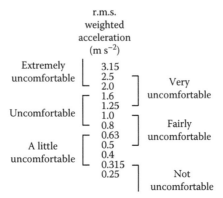

Figure 7.8 Scale of vibration discomfort. (From BSI, *BS 6841:1987: Measurement and Evaluation of Human Exposure to Whole-Body Mechanical Vibration and Repeated Shock*, British Standards Institution, London, 1987; ISO, *ISO 2631-1:1997: Mechanical Vibration and Shock: Evaluation of Human Exposure to Whole-Body Vibration. Part 1: General Requirements*, International Organization for Standardization, Geneva, 1997.)

The best understood effects of whole-body vibration are on input processes (mainly vision) and output processes (mainly continuous hand control). In both cases, there may be a disturbance occurring entirely outside the body (e.g. vibration of a viewed display or vibration of a handheld control), a disturbance at the input or output (e.g. movement of the eye or hand), and a disturbance within the body affecting the peripheral nervous system (i.e. the afferent or efferent system). Central processes may also be affected by vibration, but understanding is currently too limited to make confident generalised statements.

Vibration interferes with vision and manual control as a result of the movement of the affected part of the body (i.e. the eye or the hand). The interference may be reduced by reducing the transmission of vibration to the eye or the hand, or by making the task less susceptible to disturbance (e.g. increasing the size of a display or reducing the sensitivity of a control). Often, the effects of vibration on vision and manual control can be reduced by redesigning the task.

7.3.3.1 Effects of vibration on vision

Reading a book or newspaper in a vehicle may be difficult because the paper is moving, the eye is moving or both the paper and the eye are moving. There are many variables affecting visual performance in these conditions; it is not possible to adequately represent the effects of vibration on vision without considering the effects of these variables (Griffin and Lewis 1978).

When a stationary observer views a display moving with low velocity at frequencies of less than about 1 Hz, the eye can track the position of the display using smooth *pursuit eye movements*. At slightly greater frequencies of oscillation – the precise frequency depends on the predictability of the motion waveform – the eye may make *saccadic* eye movements to redirect the eye with small jumps. At frequencies greater than about 3 Hz, the eye will be directed to one extreme of the oscillation so as to view the image as it is temporarily stationary while reversing the direction of movement (i.e. at the *nodes* of the motion). The absolute threshold for the visual detection of the vibration of an object by a stationary observer occurs when the peak-to-peak oscillatory motion gives an angular displacement at the eye of approximately 1 min of arc (Griffin 1990). The acceleration required to achieve this threshold is very low at low frequencies but increases in proportion to the square of the frequency to become very high at high frequencies. If the vibration displacement is greater than the visual detection threshold, there will be a perceptible blur; the effects of vibration on visual performance (e.g. effects on reading speed and reading accuracy) may then be estimated from the maximum duration that the image spends over some small area of the retina (e.g. the period of time spent near the nodes of the motion with sinusoidal vibration). For sinusoidal vibration, this duration decreases (and so reading errors increase) in linear proportion to the frequency of the vibration and in proportion to the square root of the displacement of vibration (O'Hanlon and Griffin 1971). With dual-axis vibration (e.g. combined vertical and lateral vibration of a display), this duration is greatly reduced (unless there are identical in-phase motions in the different axes) and reading performance drops greatly (Meddick and Griffin 1976). With narrowband random vibration, there is a greater probability of low image velocity than with sinusoidal vibration of the same magnitude and predominant frequency, so the reading performance tends to be less affected by random vibrations than by sinusoidal vibrations (Moseley et al. 1982). Display vibration reduces the ability to see fine detail in displays while having little effect on the clarity of larger forms.

If an observer sits or stands on a vibrating surface, the effects of vibration on vision depend on the extent to which the vibration is transmitted to the eye. The head is likely to move in translation (i.e. in the x-, y- and z-axes) and in rotation (i.e. in roll, pitch and yaw). The dominant translational head motion is often vertical and the dominant rotational head motion is usually pitch. The pitch motion of the head is well compensated by the *vestibulo-ocular reflex* which serves to help stabilise the line of sight of the eyes at frequencies less than about 10 Hz (e.g. Benson and Barnes 1978). Hence, although there is often much pitch motion of the head at 5 Hz, there is less pitch motion of the eyes at this frequency. Pitch motion of the head, therefore, has a less than expected effect on vision, unless the display is attached to the head, as with head-mounted displays (see Wells and Griffin 1984). The effects on vision of translational motions of the head depend on viewing distance: the effects are greatest when close to a display. As the viewing distance increases, the retinal image motions produced by translational displacements of the head decrease until, when viewing an object at infinite distance, there is no retinal image motion produced by translational head displacement (Griffin 1976). For a vibrating observer, there may be little difficulty with low-frequency pitch head motions when viewing a fixed display and no difficulty with translational head motions when viewing a distant display. The greatest problems occur with pitch head motion when the display is attached to the head and with translational head motion when viewing near displays. Additionally, there may be resonances of the eye within the head, but these are highly variable between individuals and often occur at high frequencies (e.g. 30 Hz and greater), and it is often possible to attenuate the vibration entering the body at these frequencies.

When an observer and a display are supported so that they oscillate at low frequencies together and in phase, the retinal image motions (and decrements in visual performance) are less than when either the observer or the display oscillate alone (Moseley and Griffin 1986a). However, the advantage is lost as the frequency of vibration is increased, since there is increasing phase difference between the motion of the head and the motion of the display. At frequencies around 5 Hz, the phase lags between seat motion and head motion may be 90° or more (depending on seating conditions) and sufficient to eliminate any advantage of moving the seat and the display together.

When reading a newspaper on a train, the motion of the arms may result in the motion of the paper being different in both magnitude and phase from the motions of the seat and the motions of the head of the observer. The dominant axis of motion of the newspaper may be different from that of the person (Griffin and Hayward 1994). Increasing the size of detail in a display will often greatly reduce any adverse effects of vibration on vision (Lewis and Griffin 1979). Increasing the spacing between rows of letters and choosing appropriate character fonts may also be beneficial. The contrast of the display, or other reading material, also has an effect, but maximum performance may not occur with maximum contrast. The known influence of such factors is summarised in a design guide for visual displays to be used in vibration environments (Moseley and Griffin 1986b).

7.3.3.2 Manual control

Writing, and other complex control tasks involving hand or finger movement, can also be impeded by vibration (McLeod and Griffin 1989). The characteristics of the task and the vibration combine to determine the effects of vibration on performance: a given vibration may greatly affect one type of task but have little effect on another.

The direct mechanical jostling of the hand by vibration that causes unwanted movement of a control is called *breakthrough* or *feedthrough* or *vibration-correlated error*. The inadvertent movement of the pencil caused by jostling while writing in a vehicle is a form of vibration-correlated error. In a simple tracking task, where the operator is required to follow movements of a target, some of the error will also be correlated with the target movements. This is called *input-correlated error* and mainly reflects the inability of an operator to follow the target without delays inherent in visual, cognitive and motor activity. The part of the tracking error which is not correlated with either the vibration or the tracking task is called the *remnant*. This includes operator-generated noise and any source of nonlinearity: drawing a freehand 'straight line' does not result in a perfect straight line even in the absence of environmental vibration. The effects of vibration on vision can result in an increased remnant with some tracking tasks, and some studies show that vibration, usually at frequencies greater than about 20 Hz, interferes with neuromuscular processes (e.g. Goodwin et al. 1972; Ribot et al. 1986) and may be expected to result in an increased remnant. The causes of the three components of the tracking error are shown in the model presented as Figure 7.9.

The gain (i.e. sensitivity) of a control determines the control output corresponding to a given force, or displacement, applied to the control by the operator. The optimum gain in static conditions (high enough not to cause fatigue but low enough to prevent inadvertent movement) is likely to be too great during exposure to vibration where inadvertent movement is more likely (Lewis and Griffin 1977).

Vibration may affect the performance of tracking tasks by reducing the visual performance of the operator. Wilson (1974) and McLeod and Griffin (1990) showed that collimating a display by means of a lens so that it appears to be at infinity can reduce, or even eliminate, errors with some control tasks. It is possible that visual disruption has played a significant

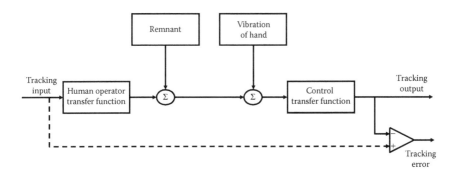

Figure 7.9 Linear model of a pursuit manual control system showing how tracking errors may be caused by the vibration (vibration-correlated error), the task (input-correlated error) or some other cause (remnant).

part in the performance decrements reported in other experimental studies of the effects of vibration on manual control.

The vibration transmissibility of the body is approximately linear (i.e. doubling the magnitude of vibration at the seat may be expected to approximately double the magnitude of vibration at the head or at the hand). Vibration-correlated error may therefore increase in approximately linear proportion to vibration magnitude. There is no simple relation between the frequency of vibration and its effects on control performance. The effects of frequency depend on the control order (which varies between tasks) and the biodynamic response of the body (which varies with posture and between operators). With zero-order tasks and the same magnitude of acceleration at each frequency, the effects of vertical seat vibration may be greatest in the range 3–8 Hz, since transmissibility to the shoulders is greatest in this range (see McLeod and Griffin 1989). In the horizontal axes (i.e. the x- and y-axes of the seated body), the greatest effects appear to occur at lower frequencies (e.g. 2 Hz or lower). Again, this corresponds to the frequencies at which there is the greatest transmission of vibration to the shoulders. The axis of the control task most affected by vibration may not be the same axis as that in which most vibration occurs at the seat. Often, fore-and-aft movements of a control (which usually correspond to vertical movements on a display) are most affected by vertical whole-body vibration.

Multiple-frequency vibration causes more disruption to performance than the presentation of any one of the constituent single frequencies alone. Similarly, the effects of multiple-axis vibration are greater than the effects of vibration in any of the single axes alone.

Few investigations of the effects of vibration have considered everyday tasks. Corbridge and Griffin (1991) showed that the effects of vertical whole-body vibration on spilling liquid from a handheld cup were greatest close to 4 Hz. They also found that the effects of vibration on writing speed and subjective estimates of writing difficulty were most affected by vertical vibration in the range 4–8 Hz. Although 4 Hz was a sensitive frequency for both the drinking and the writing task, the dependence on frequency of the effects of vibration was different for the two activities.

The first version of ISO 2631 offered a *fatigue-decreased proficiency* boundary as a means of predicting the effects of vibration on activities (ISO 1974, 1985). A complex time-dependent magnitude of vibration was said to be 'a limit beyond which exposure to vibration can be regarded as carrying a significant risk of impaired working efficiency in many kinds of tasks, particularly those in which time-dependent effects ("fatigue") are known to worsen performance as, for example, in vehicle driving' (ISO 1974, p. 4). There are no data justifying the complex fatigue-decreased proficiency boundary presented in this old standard; among researchers, it was discarded as unfounded committee speculation. Vibration may influence fatigue, but as yet there is little evidence on which to base a time-dependent influence of vibration on human performance.

7.3.3.3 Cognitive performance

Few investigators have addressed the possibility of vibration affecting cognitive performance. Simple cognitive tasks (e.g. simple reaction time) appear to be unaffected by vibration, other than by changes in arousal or motivation or by direct effects on input and output processes. This may also be true for some complex cognitive tasks. However, the scarcity and diversity of experimental studies allow the possibility of some real and significant cognitive effects of vibration (see Sherwood and Griffin 1990, 1992).

7.3.4 Health effects of whole-body vibration

Occupations, sports and leisure activities involving exposure to whole-body vibration have been reported to cause various disorders (see Dupuis and Zerlett 1986; Hulshof and van Zanten 1987; Griffin 1990; Bongers and Boshuizen 1990; Bovenzi and Betta 1994; National Institute for Occupational Safety and Health 1997; Bovenzi and Hulshof 1999; Bovenzi 2009). The studies do not all agree on either the type or the extent of disorders and the findings have rarely been related to measurements of the vibration exposures. However, it is widely believed that disorders of the back (especially increased reporting of low-back pain) are associated with exposure to whole-body vibration. There are alternative causes of increased back pain in persons exposed to vibration (e.g. poor sitting postures, heavy lifting) and low-back pain is very common in persons not exposed to whole-body vibration. It is, therefore, often difficult to conclude with confidence that a back disorder in a person exposed to whole-body vibration has been solely, or primarily, caused by vibration.

Other disorders that have been associated with occupational exposures to whole-body vibration include abdominal pain, digestive disorders, urinary frequency, prostatitis, haemorrhoids, balance and visual disorders, headaches and sleeplessness. Further research is required to identify the extent to which these signs and symptoms are causally related to the exposure to whole-body vibration.

7.3.4.1 Evaluation of whole-body vibration according to BS 6841 (1987) and ISO 2631 (1997)

The manner in which the health effects of whole-body vibration depend on the frequency, direction and duration of vibration is currently assumed to be similar to that for vibration discomfort (see Section 7.3.2). However, it is assumed that the total exposure rather than the average exposure is important, and so a dose measure is used.

ISO 2631 (ISO 1974, 1985) defined *exposure limits* in terms of the r.m.s. acceleration '... set at approximately half the level considered to be the threshold of pain (or limit of voluntary tolerance) for healthy human subjects ...'. Currently, BS 6841 (BSI 1987) defines an 'action level' based on vibration-dose values, and ISO 2631-1 (ISO 1997) offers two alternative 'health guidance caution zones': one based on r.m.s. acceleration and the other on vibration-dose values.

The *vibration-dose value* (VDV) of a motion can be considered to be the magnitude of a 1 s duration of vibration that is equally severe. The VDV uses a fourth-power time dependence to accumulate vibration severity over the exposure period from the shortest possible shock to a full day of vibration (see BS 6841 [BSI 1987]; ISO 2631-1 [ISO 1997]):

$$\text{Vibration dose value (VDV)} = \left[\int_{t=0}^{t=T} a_w^4(t) \, dt \right]^{1/4} \quad (7.2)$$

where:

$a_w(t)$ = frequency-weighted acceleration
T = exposure duration

The VDV is used to evaluate the severity of sinusoidal and random vibration, transients, shocks and repeated shock motions.

If the exposure duration (T, in seconds) and the frequency-weighted r.m.s. acceleration (a_{rms}, m s^{-2} r.m.s.) are known for conditions in which the vibration characteristics are statistically stationary, it can be useful to calculate the *estimated vibration dose value*, eVDV:

$$\text{Estimated vibration dose value} = 1.4 \, a_{rms} \, T^{1/4} \quad (7.3)$$

The eVDV is not applicable to transients, shocks or repeated shock motions in which the crest factor (the peak value divided by the r.m.s. value) is high.

7.3.4.2 Assessment of whole-body vibration according to BS 6841 (1987) and ISO 2631-1 (1997)

No precise limit can be offered to prevent disorders caused by whole-body vibration, but standards and directives define useful methods of estimating vibration severity and deciding on the need for preventive measures. With severe vibration exposures, prior consideration of the fitness of the exposed persons and the design of adequate safety precautions may be required. The need for regular warnings and checks on the health of routinely exposed persons may also be considered.

BS 6841 (BSI 1987, p. 18) offers the following guidance:

> Sufficiently high-vibration dose values will cause severe discomfort, pain and injury. Vibration dose values also indicate, in a general way, the severity of the vibration exposures which caused them. However there is currently no consensus of opinion on the precise relation between Vibration dose values and the risk of injury. It is known that vibration magnitudes and durations which produce Vibration dose values in the region of 15 m s$^{-1.75}$ will usually cause severe discomfort. It is reasonable to assume that increased exposure to vibration will be accompanied by increased risk of injury.

Figure 7.10 shows this action level for exposure durations from 1 s to 24 h.

ISO 2631-1 (ISO 1997) offers two different methods of evaluating vibration severity with respect to health effects, and for both methods there are two boundaries. When evaluating vibration using the VDV, it is suggested that below a boundary corresponding to a VDV of 8.5 m s$^{-1.75}$, 'health risks have not been objectively observed'; between 8.5 and 17 m s$^{-1.75}$, caution with respect to health risks is indicated; and above 17 m s$^{-1.75}$, 'health risks are likely'. The two boundaries define a 'VDV health guidance caution zone'. The alternative method of evaluation in ISO 2631-1 (ISO 1997) uses a time dependence, in which the acceptable vibration does not vary with a duration between 1 and 10 min and then decreases in inverse proportion to the square root of the duration from 10 min to 24 h. This method suggests an 'r.m.s. health guidance caution zone', but the method is not fully defined in the text; it allows very high accelerations at short durations, conflicts with the VDV method and cannot be extended to exposure durations of less than 1 min (Figure 7.10).

Unlike the use of r.m.s. acceleration, the use of the VDV does not allow extraordinarily high-vibration magnitudes at short durations (a few or many minutes). Any exposure to continuous vibration, intermittent vibration or repeated shocks may be compared with the action level by calculating the VDV. It would be unwise to exceed the guidance given by VDVs without consideration of the possible health effects of an exposure to vibration or shock.

ISO 2631-5 (ISO 2004) offers a method of estimating the daily equivalent static compression dose, S_{ed}, of the spine to vertical and horizontal vibration containing multiple shocks. It is suggested that for a lifetime exposure, if S_{ed} is <0.5 MPa, there is a low probability of adverse effects, whereas if S_{ed} is >0.8 MPa, there is a high probability of such effects. The vertical response of the spine is represented by a recurrent neural network (RNN), for which the standard says: 'The RNN for the z-axis

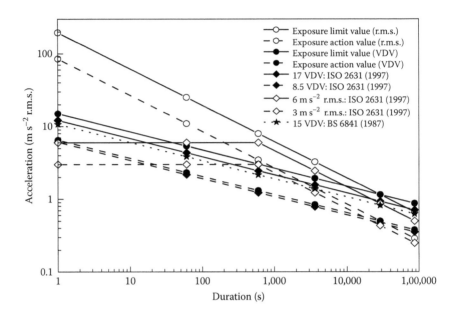

Figure 7.10 Comparison between the health guidance caution zones for whole-body vibration in ISO 2631-1 (ISO 1997) (3–6 m s^{-2} r.m.s.; 8.5–17 m s$^{-1.75}$), 15 m s$^{-1.75}$ action level implied in BS 6841 (BSI 1987), the 'exposure limit values' and 'exposure action values' for whole-body vibration in the EU Physical Agents (Vibration) Directive. (Data from BSI, *BS 6841:1987: Measurement and Evaluation of Human Exposure to Whole-Body Mechanical Vibration and Repeated Shock*, British Standards Institution, London, 1987; European Parliament/Council of the European Union, *Official Journal of the European Communities*, **L177**, 13–19, 2002; ISO, *ISO 2631-1:1997: Mechanical Vibration and Shock: Evaluation of Human Exposure to Whole-Body Vibration. Part 1: General Requirements*, International Organization for Standardization, Geneva, 1997.)

was trained using vibration and shocks in the range of -20 to 40 m s^{-2} and 0.5 to 40 Hz. As the model is nonlinear, this constitutes the range of applicability of this part of ISO 2631' (ISO 2004, p. 12). Many shock motions will contain components outside this range and may cause the output of the model to be suspect. This method of assessing shocks is not recommended in preference to the VDV, and it is likely to be superseded by other alternatives.

7.3.4.3 EU Machinery Safety Directive

The Machinery Safety Directive of the European Community (2006/42/EC) states: 'Machinery must be designed and constructed in such a way that risks resulting from vibrations produced by the machinery are reduced to the lowest level, taking account of technical progress and the availability of means of reducing vibration, in particular at source' (European Parliament/Council of the European Union 2006, p. 45). Instruction handbooks for machinery causing whole-body vibration should specify the frequency-weighted acceleration if it exceeds a stated value (currently, a frequency-weighted acceleration of 0.5 m s^{-2} r.m.s. when measured according to ISO 2631 [ISO 1997]). The vibration magnitude will depend on the test conditions and, although the test procedures must be specified, their selection leaves much control over the measured value to the manufacturer. Standardised test procedures are being prepared, but the magnitudes of whole-body vibration currently quoted by a manufacturer may not be representative of the vibration experienced when the machinery is operated in work.

7.3.4.4 EU Physical Agents (Vibration) Directive

The Parliament and Commission of the European Union have defined minimum health and safety requirements for the exposure of workers to the risks arising from vibration (European Parliament/Council of the European Union 2002). For whole-body vibration, the directive defines an 8 h-equivalent 'exposure action value' of 0.5 m s^{-2} r.m.s. (or a VDV of 9.1 m s$^{-1.75}$) and an 8 h-equivalent 'exposure limit value' of 1.15 m s^{-2} r.m.s. (or a VDV of 21 m s$^{-1.75}$). All member states of the European Union now have laws to enforce this directive.

The directive says that workers shall not be exposed to vibration in excess of the exposure limit value. If the exposure action value is exceeded, the employer shall establish and implement a programme of technical or organisational measures intended to reduce to a minimum exposure to mechanical vibration and the attendant risks. The directive says that workers exposed to vibration in excess of the exposure action value shall be entitled to appropriate health surveillance. Health surveillance is also required if there is any reason to suspect that workers may be injured by the vibration, even if the exposure action value is not exceeded.

The probability of injury arising from occupational exposures to whole-body vibration at the exposure action value and the exposure limit value cannot be estimated, because epidemiological studies have not yet produced dose–response relationships. However, it seems clear that the directive does not define safe exposures to whole-body vibration: the r.m.s. values are associated with extraordinarily high magnitudes of vibration (and shock) when the exposures are short, and these exposures may be assumed to be hazardous (see Figure 7.10; Griffin, 2004). The VDV procedure suggests more reasonable vibration magnitudes, especially for short-duration exposures and shocks.

7.3.5 Disturbance in buildings

The acceptable magnitudes for vibration in many buildings are close to the thresholds for the perception of vibration. However, human responses to the vibrations in buildings are assumed to depend on the use of the building, in addition to the magnitude, frequency, direction and duration of the vibration. ISO 2631-2 (ISO 2003a) provides some information on the measurement and evaluation of building vibration, but limited practical guidance. BS 6472-1 (BSI 2008a) offers guidance on the measurement, evaluation and assessment of vibration in buildings, and BS 6472-2 (BSI 2008b) defines a method currently used for assessing the vibration caused by blasting. Using the guidance contained in BS 6472-1 (BSI 2008a), it is possible to predict the acceptability of vibration in different types of building by reference to a simple table of VDVs (see Table 7.4 and BS 6472-1 [BSI 2008a]). The VDVs in Table 7.4 are applicable irrespective of whether the vibration occurs as a continuous vibration, intermittent vibration or repeated shocks.

7.3.6 Seating dynamics

Ideally, vibration should be reduced at source. This may involve minimising undulations of the terrain, reducing the speed of travel of vehicles, or improving the balance of rotating parts. For seated people, the seating dynamics can increase or decrease the vibration causing discomfort, or interference with activities, or contributing to injuries. Seats exhibit resonances at low frequencies that result in greater magnitudes of vertical vibration on the seat than on the floor supporting the seat. At high frequencies, there is usually attenuation of vibration. The resonance frequencies of common seats are usually in the region of 4 Hz (see Figure 7.11). The amplification at resonance is partially determined by the *damping* in the seat, but also influenced by the dynamic characteristics of the person sitting on the seat (Toward and Griffin 2011b). The variations in transmissibility

Table 7.4 Vibration dose value ranges expected to result in various degrees of adverse comment in residential buildings

Place	Low probability of adverse comment (m s$^{-1.75}$)	Adverse comment possible (m s$^{-1.75}$)	Adverse comment probable (m s$^{-1.75}$)
Residential buildings 16 h day	0.2–0.4	0.4–0.8	0.8–1.6
Residential buildings 8 h night	0.1–0.2	0.2–0.4	0.4–0.8

Source: BSI, *BS 6472-1:2008: Guide to the Evaluation of Human Exposure to Vibration in Buildings. Part 1: Vibration Sources Other Than Blasting*, British Standards Institution, London, 2008.

Note: For offices and workshops, multiplying factors of 2 and 4 respectively can be applied to the VDV ranges for a 16 h day.

between seats are sufficient to result in significant differences in the vibration experienced by people supported by different seats.

The transmissibility of a seat (i.e. the non-dimensional ratio of the vibration magnitude at the seat–subject interface to the vibration magnitude entering the seat at the floor, expressed as a function of vibration frequency) can be measured in various ways. The transmissibility is highly dependent on the apparent mass of the human body, so the transmissibility of a seat measured with a mass supported on the seat will be different from that with a human body sitting in the seat. The mechanical nonlinearity of the human body results in seat transmissibilities that vary with changes in the magnitude of the vibration and in the vibration spectra entering the seat. Measurements of seat transmissibility may be undertaken on laboratory simulators with volunteer subjects, but precautions are required to protect subjects from the risks of injury. Measurements may also be performed with drivers or passengers in vehicles. Anthropodynamic dummies are being developed to represent the average mechanical impedance of the human body, so that laboratory and field studies can be performed without exposing people to vibration; this reduces the costs and inherent risks and should provide a measurement of seat transmissibility unaffected by the intersubject variability in mechanical impedance (Lewis and Griffin 2002). Seat transmissibility may also be predicted using measurements of the impedance of a seat and the apparent mass of the human body (Wei and Griffin 1998b).

The suitability of a seat for a specific vibration environment depends on: (i) the vibration spectra present in the environment; (ii) the transmissibility of the seat; and (iii) the sensitivity of the human body to different frequencies of vibration. These three functions of frequency are contained within a simple numerical indication of the isolation efficiency of a seat called the *seat effective amplitude transmissibility* (SEAT) (Griffin 1978, 1990, 2007). A SEAT value compares the vibration severity on a seat with that on the floor beneath the seat:

$$\text{SEAT (\%)} = \frac{\text{ride comfort on seat}}{\text{ride comfort on floor}} \times 100 \tag{7.4}$$

A SEAT value greater than 100% indicates that, overall, the vibration on the seat is worse than that on the floor beneath the seat; SEAT values less than 100% indicate that the seat has provided some useful attenuation. Seats should be designed to have the lowest SEAT value compatible with other constraints. In conventional cars, SEAT values are often in the range of 60%–80%. In railway carriages, the SEAT value for vertical vibration is likely to be greater than 100%, because conventional seats cannot provide any attenuation of the low frequencies that are normally dominant in the vertical vibration of such vehicles. Consequently,

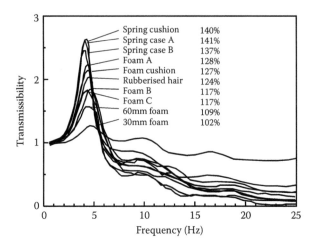

Figure 7.11 Comparison of the vertical transmissibility and SEAT values for ten alternative cushions of passenger railway seats. (Data from Corbridge, C. et al., *Proceedings of the Institution of Mechanical Engineers*, **203**, 57–64, 1989.)

Figure 7.11 shows high SEAT values in which the vibration discomfort experienced by the occupants of the seats will be degraded by the seat response. Selecting a seat so as to minimise the SEAT value in these circumstances may not result in lower vibration on the seat than on the floor, but should minimise exposure to vibration. The optimisation of the seating dynamics can be both the cheapest and the most effective method of improving vehicle ride.

The *ride comfort* that would result from sitting on the seat or on the floor can be predicted using the frequency weightings in the appropriate standard. The SEAT value may be calculated either from the frequency-weighted r.m.s. values (if the vibration does not contain transients) or from the VDVs of the frequency-weighted acceleration on the seat and the floor:

$$\text{SEAT (\%)} = \frac{\text{vibration dose value on seat}}{\text{vibration dose value on floor}} \times 100 \qquad (7.5)$$

The SEAT value is influenced by the vibration input, and not merely by the dynamics of the seat: different values are obtained with the same seat in different vehicles. The SEAT value indicates the suitability of a seat for attenuating a particular type of vibration.

Passenger seating (comprising some combination of foam, rubber or metal springing) usually has a resonance at about 4 Hz and, therefore, provides no attenuation at frequencies less than about 6 Hz. Attenuation can be provided at frequencies as low as about 2 Hz using a separate suspension mechanism beneath the seat pan. In such *suspension seats*, as provided for the drivers of some off-road vehicles, trucks and coaches, there are low resonance frequencies (sometimes < 2 Hz). This is beneficial if the dominant vibration is at higher frequencies, but of no value if the dominant motion is at very low frequencies. Standards for the testing of the suitability of suspension seats to specific classes of work vehicle have been prepared (see ISO 10326-1 [ISO 1992], ISO 7096 [ISO 2000], ISO 5007 [ISO 2003b]). One problem with suspension seats is that the suspension mechanism (comprising a spring and damper mechanism) has a limited travel, often 50–100 mm. If the relative motion of the suspension reaches this limit, there will be an impact which can be a greater cause of discomfort and hazard than would have been present with a conventional seat (Wu and Griffin 1997). If the suspension travel is too great, there will be discomfort or other problems arising from the relative motion between the seat and the vehicle (Jang and Griffin 2000). Suspension seats have been marketed showing the benefits of the suspension while assuming a linear single degree of freedom mechanical system. In practice, suspensions are nonlinear (having friction affecting the response with low magnitude motions, and hitting end-stops with high magnitudes) and their full response requires consideration of the dynamic response of the cushion and the impedance of the human body, in addition to the idealised response of the damper and spring forming the suspension mechanism.

7.4 MOTION SICKNESS

Motion sickness (e.g. vomiting, nausea, sweating, colour changes, dizziness headaches and drowsiness) is a normal response to motion that is experienced by fit and healthy people. Many different motions can cause sickness and reduce the comfort, impede the activities and degrade the well-being of those affected (Griffin 1990, 1991).

7.4.1 Causes of motion sickness

Translational and rotational oscillation, constant speed rotation about an off-vertical axis, Coriolis stimulation, movement of the visual scene and various other stimuli producing sensations associated with movement of the body can cause motion sickness. Although some motions can be predicted as being more nauseogenic than others, motion sickness is neither explained nor predicted solely by the physical characteristics of motion.

Motions of the body are detected by three sensory systems: the vestibular system, the visual system and the somatosensory system. The vestibular system is located in the inner ear and comprises the semicircular canals, which respond to the rotation of the head, and the otoliths, which respond to translational forces (either translational acceleration or rotation of the head relative to an acceleration field, such as the force of gravity). The eyes may detect relative motion between the head and the environment, caused either by head movements (in translation or rotation) or movements of the environment or a combination of the movements of the head and the environment. The somatosensory systems respond to force and displacement of parts of the body and give rise to sensations of body movement or force.

In normal environments, the movements of the body are detected by all three sensory systems and this leads to a sufficiently unambiguous indication of the movements of the body in space. In some other environments, the three sensory systems can simultaneously give signals corresponding to different motions (or motions that are not realistic) and lead to some form of uncertainty as to the true motion. This leads to the idea of a *sensory conflict theory* of motion sickness in which sickness occurs when the sensory systems disagree on the motions which are occurring. However, this implies some absolute significance to sensory information, whereas the meaning of the information is probably learned. This led to the *sensory rearrangement theory* of motion sickness which states that all situations which provoke motion sickness are characterised by a condition of sensory rearrangement in which the motion signals transmitted by the eyes, the vestibular system and the nonvestibular proprioceptors are at variance either with one another or with what is expected from previous experience (Reason 1970, 1978). Reason and Brand (1975) suggest that the conflict may be sufficiently considered in two categories: intermodality (between vision and the vestibular receptors) and intramodality (between the semicircular canals and the otoliths within the vestibular system). For both categories, it is possible to identify three types of situation in which conflict can occur (see Griffin 1990, chapter 7).

There is evidence that the average susceptibility to sickness among males is less than that among females, and susceptibility decreases with increased age among both males and females (e.g. Lawther and Griffin 1988a; Turner and Griffin 1999b) and differs between Asians and Europeans (Beard and Griffin 2014). However, there are larger individual differences within any group of either gender at any age: some people are easily made ill by motions that can be endured indefinitely by others. The reasons for these differences are not properly understood.

7.4.2 Sickness caused by oscillatory motion

Motion sickness is not caused by oscillation (however violent) at frequencies much greater than 1 Hz: the phenomenon arises from motions at the low frequencies associated with normal postural control of the body. Various experimental investigations have explored the extent to which vertical oscillation causes sickness at different frequencies. These studies have allowed the formulation of a frequency weighting, W_f (see Figure 7.6), and the definition of a *motion sickness-dose value*. The frequency weighting W_f reflects the greatest sensitivity to acceleration in the range 0.125–0.25 Hz, with a rapid reduction in sensitivity at greater frequencies. The motion sickness dose value predicts the probability of sickness from knowledge of the frequency and magnitude of vertical oscillation (see Lawther and Griffin 1987; ISO 2631-1 [ISO 1997]):

$$\text{Motion sickness-dose value (MSDV)} = a_{\text{rms}} T^{1/2} \quad (7.6)$$

where:

a_{rms} = root-mean-square value of the frequency-weighted acceleration (m s^{-2})
T = exposure period (s)

The percentage of unadapted adults who are expected to vomit is given by ⅓ MSDV. (These relationships have been derived from exposures in which up to 70% of persons vomited during exposures lasting between 20 min and 6 h.)

The motion sickness-dose value of the vertical oscillation seems to provide useful predictions of sickness where vertical oscillation has been shown to be the prime cause of sickness, such as on ships, hovercraft and hydrofoils (Lawther and Griffin 1988b) and in some aircraft (Turner et al. 2000). Vertical oscillation is not the principal cause of sickness in cars (Griffin and Newman 2004a) or in road coaches (Turner and Griffin 1999a) or in tilting trains (Donohew and Griffin 2007). The motion sickness in surface transport seems to be associated with horizontal acceleration arising from changes in speed and cornering (Griffin and Newman 2004b). The sickness caused by fore-and-aft and lateral oscillation has been explored (e.g. Donohew and Griffin 2004), but sickness in surface transport can also be influenced by roll and pitch motions (e.g. Donohew and Griffin 2007, 2009) and vision (Turner and Griffin 1999b; Griffin and Newman 2004a; Butler and Griffin 2009). Further research is required before sickness can be predicted with confidence in these environments.

7.5 HAND-TRANSMITTED VIBRATION

Hand-transmitted vibration can influence human comfort, the performance of activities and health. The various effects of hand-transmitted vibration are influenced by, but not solely determined by, the complex biodynamic responses of the fingers, hands and arms (Concettoni and Griffin 2009).

Whether vibration of the hand can be felt may be estimated from thresholds for the perception of vibration, and the discomfort or annoyance caused by hand-transmitted vibration may be estimated using frequency weightings determined from equivalent-comfort contours (Morioka and Griffin 2006b). The frequency dependence of the discomfort caused by vibration of the hand is dependent on the vibration magnitude, and depends on the location of contact with the hand, so precise estimates are not simple. Nevertheless, there is sufficient information to define methods of predicting the sensations, whether they are wanted (as with tactile feedback) or unwanted (as on car steering wheels) (Morioka and Griffin 2009).

Exposure of the fingers or the hands to vibration or repeated shocks can give rise to various signs and symptoms of disorder. The extent and interrelation between the various signs and symptoms are not fully understood, but five types of disorder may be identified (Table 7.6). More than one disorder can affect a person at the same time and it is possible that the presence of one disorder facilitates the appearance of another. The onset of each disorder is dependent on several variables, such as the vibration characteristics, the dynamic response of the fingers or hand, individual susceptibility to damage and other aspects of the environment. Confusingly, the terms *vibration syndrome* or *hand-arm vibration syndrome* (HAVS) are sometimes used to refer to an unspecified combination of one or more of the effects listed in Table 7.5.

7.5.1 Sources of hand-transmitted vibration

The vibration of tools varies greatly, depending on the tool design and the method of use, so it is not possible to categorise individual tool types as 'safe' or 'dangerous'. However, Table 7.6 lists tools and processes that are sometimes a cause for concern. Figure 7.12 shows vibration spectra measured on 24 tools that may sometimes be associated with injury.

Table 7.5 Five types of disorder associated with hand-transmitted vibration exposures

Type	Disorder
Type A	Circulatory disorders
Type B	Bone and joint disorders
Type C	Neurological disorders
Type D	Muscle disorders
Type E	Other general disorders (e.g. central nervous system)

Source: Griffin, M.J., *Handbook of Human Vibration*, Academic Press, London, 1990.

Note: Some combination of these disorders is sometimes referred to as the 'hand-arm vibration syndrome' (HAVS).

Table 7.6 Some tools and processes potentially associated with vibration injuries (standards identifying type tests indicated in parentheses)

Type of tool	Examples of tool type
Percussive metalworking tools	Riveting tools (ISO 28927-9) Caulking tools (ISO 28927-9) Chipping hammers (ISO 28927-9) Clinching and flanging tools Impact screwdrivers (ISO 28927-2) Wrenches, nutrunners and screwdrivers (ISO 28927-2) Scaling hammers (ISO 28927-9) Needle guns (ISO 28927-9) Nibbling machines and shears Swaging machines
Grinders and other rotary tools	Pedestal grinders Angle and vertical grinders (ISO 28927-1) Straight grinders (ISO 28927-4) Die grinders (ISO 28927-12) Handheld polishers and sanders (ISO 28927-3) Flex-driven grinders/polishers Rotary burring tools
Percussive hammers and drills used in mining, demolition, road construction, stone working, etc.	Drills and impact drills (ISO 28927-5) Road breakers (ISO 28927-10) Percussive hammers (ISO 28927-10) Rock drills (ISO 28927-10) Stone hammers (ISO 28927-11)
Forest and garden machinery	Chainsaws (ISO 22867) Antivibration chainsaws (ISO 22867) Brush cutters and grass trimmers (ISO 22867) Mowers (ISO 5395) Hedge trimmers (ISO 22867) Pole-mounted powered pruners (ISO 22867) Garden blowers/vacuums (ISO 22867) Barking machines Stump grinders
Other processes and tools	Rammers (ISO 28927-6) Reciprocating filing machines (ISO 28927-8) Reciprocating polishing machines (ISO 28927-8) Oscillating and reciprocating saws (ISO 28927-8) Nibblers and shears (ISO 28927-7) Nailing guns (ISO 28927-8) Stapling guns (ISO 8662-11) Scabblers (ISO 28927-9) Engraving pens Shoe-pounding-up machines Vibratory rollers Concrete vibro-thickeners Concrete levelling vibrotables Motorcycle handle bars Pedestrian-controlled machines

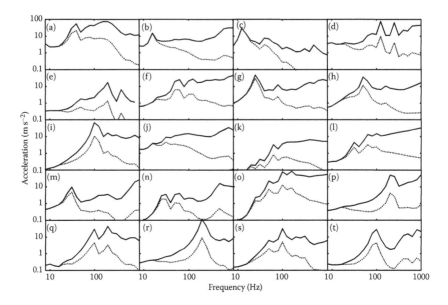

Figure 7.12 One-third octave band spectra on 20 (in ms^{-2}) powered hand tools: (———): unweighted spectra; (⋯⋯): weighted spectra (a: pneumatic rock drill; b: pneumatic road breaker; c: petrol-driven Wacker compressing road surface after mending; d: nonantivibration chainsaw; e: antivibration chainsaw; f: pneumatic metal-chipping hammer; g: pole scabbler; h: needle gun; i: random orbital sander; j: impact wrench; k: riveting gun; l: dolly used with riveting gun; m: nutrunner; n: metal drill; o: wire swager; p: etching pen; q: electric 9 in. angle grinder; r: pneumatic rotary file; s: pneumatic 5 in. straight grinder; t: pneumatic 7 in. vertical grinder). All spectra from the axis giving the highest weighted acceleration. (Data from Griffin, M.J., *Occupational and Environmental Medicine*, **54**, 73–89, 1997.)

7.5.2 Health effects of hand-transmitted vibration

7.5.2.1 Vascular disorders

In the United Kingdom, the possibility of 'neurosis due to vibration' was considered by the Departmental Committee on Compensation for Industrial Diseases in 1907. The first published cases of the condition now most commonly known as *vibration-induced white finger* (VWF) are generally acknowledged to be those reported in Italy by Loriga in 1911. VWF has subsequently been reported to occur in many other occupations in which there is exposure of the fingers to vibration (Taylor and Pelmear 1975; Griffin 1990).

VWF is characterised by intermittent whitening (i.e. blanching) of the fingers. The fingertips are usually the first to blanch, but with continued exposure to vibration, the affected area may extend to all of one or more fingers.

Attacks of blanching are precipitated by cold and, therefore, usually occur in cold conditions or when handling cold objects. The blanching lasts until the fingers are rewarmed and vasodilation allows the blood circulation to return. Many years of vibration exposure often occur before the first attack of blanching is noticed. Affected persons often have other signs and symptoms, such as numbness and tingling. Cyanosis and, rarely, gangrene have also been reported. It is not yet clear to what extent these other signs and symptoms are causes of, caused by, or unrelated to attacks of 'white finger'.

Vibration-induced white finger cannot be assumed to be present merely because there are attacks of blanching. It will be necessary to exclude other known causes of similar symptoms (by medical examination) and also necessary to exclude so-called primary Raynaud's disease (also called *constitutional white finger*). If there is no family history of the symptoms, if they did not occur before the first significant exposure to vibration, and if the symptoms and signs are confined to areas in contact with the vibration (e.g. the fingers, not the ears, etc.), they will often be assumed to indicate vibration-induced white finger. Diagnostic tests for vibration-induced white finger can assist the detection of the disease (Nielsen and Lassen 1977, Bovenzi 1993), and there are standardised tests for the measurement of finger systolic blood pressure following finger cooling (in ISO 14835-2 [ISO 2005b]) and the measurement of finger rewarming times following cooling (in ISO 14835-1 [ISO 2005a]). Normal responses to these tests are also available (Lindsell and Griffin 2002).

The severity of the effects of vibration are currently recorded by reference to the stage of the disorder, based on verbal statements made by the affected person. In the Stockholm Workshop staging system, the staging is influenced by both the frequency of attacks of blanching and the areas of the digits affected by blanching (Table 7.7).

Table 7.7 Stockholm Workshop scale for the classification of VWF

Stage	Grade	Description
0	–	No attacks
1	Mild	Occasional attacks affecting only the tips of one or more fingers
2	Moderate	Occasional attacks affecting distal and middle (rarely also proximal) phalanges of one or more fingers
3	Severe	Frequent attacks affecting all phalanges of most fingers
4	Very severe	As in Stage 3, with trophic skin changes in the fingertips

Source: Gemne, G. et al., *Scandinavian Journal of Work, Environment and Health*, 13, 275–278, 1987.

Note: If a person has Stage 2 in two fingers of the left hand and Stage 1 in a finger on the right hand, the condition may be reported as 2L(2)/1R(1). There is no defined means of reporting the condition of digits when this varies between digits on the same hand. The scoring system is more helpful when the extent of blanching is to be recorded.

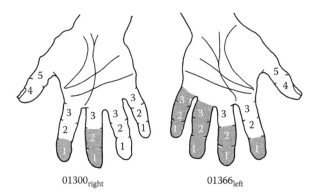

Figure 7.13 Method of scoring the areas of the digits affected by blanching. The blanching scores for the hands shown are 01300_{right}, 01366_{left}. (Adapted from Griffin, M.J., *Handbook of Human Vibration*. Academic Press, London, 1990.)

A scoring system is used to record the areas of the digits affected by blanching (Figure 7.13). The scores correspond to areas of blanching on the digits commencing with the thumb. On the fingers, a score of 1 is given for blanching on the distal phalanx, a score of 2 for blanching on the middle phalanx and a score of 3 for blanching on the proximal phalanx. On the thumbs, the scores are 4 for the distal phalanx and 5 for the proximal phalanx. The blanching score may be based on statements from the affected person or on the visual observations of a designated observer.

7.5.2.2 Neurological disorders

Numbness, tingling and elevated sensory thresholds for touch, vibration, temperature and pain are now considered to be separate effects of vibration, and not merely symptoms of vibration-induced white finger. There is a method of staging the extent of vibration-induced neurological effects of vibration (Table 7.8), but this is not currently related to the results of any specific objective test: the *sensorineural stage* is a subjective impression of a physician, based on the statements of the affected person or the results of any available clinical or scientific testing. The extent of the neurological dysfunction may be assessed by measures of sensory function, such as the thresholds for feeling vibration, heat or warmth on the fingers. The measurement of vibrotactile thresholds is partially standardised in ISO 13091-1 (ISO 2001a), although the precise implementation of the test varies. Although not fully standardised by the committee, both vibrotactile thresholds and thermal thresholds are widely used, and normal values are available for some tests (e.g. Seah and Griffin 2008). A scheme for improving the current methods of measuring, evaluating and reporting peripheral

Table 7.8 'Sensorineural stages' of the effects of hand-transmitted vibration

Stage	Symptoms
0_{SN}	Exposed to vibration but no symptoms
1_{SN}	Intermittent numbness with or without tingling
2_{SN}	Intermittent or persistent numbness, reduced sensory perception
3_{SN}	Intermittent or persistent numbness, reduced tactile discrimination or manipulative dexterity

Source: Brammer, A.J. et al., *Scandinavian Journal of Work, Environment and Health*, 13(4), 279–283, 1987.

neurological disorders associated with hand-transmitted vibration has been proposed (Griffin 2008).

7.5.2.3 Muscular effects

Workers exposed to hand-transmitted vibration sometimes report difficulty with their grip, including reduced dexterity, reduced grip strength and locked grip. Many of the reports are derived from symptoms reported by exposed persons rather than signs detected by physicians, and could be a reflection of neurological problems. Muscle activity may be of great importance to tool users, since a secure grip can be essential to the performance of the job and the safe control of the tool. The presence of vibration on a handle may encourage the adoption of a tighter grip than would otherwise occur, and a tight grip may increase the transmission of vibration to the hand. If the chronic effects of vibration result in reduced grip, this may help to protect operators from any further effects of vibration, but interfere with both work and leisure activities.

Grip strength is often tested using the Jamar grip meter and dexterity is measured using the Purdue pegboard, for which there are normal values for persons of various ages (Haward and Griffin 2002).

7.5.2.4 Articular disorders

Surveys of the users of handheld tools have found evidence of bone and joint problems, most often among workers operating percussive tools, such as those used in metalworking jobs and mining and quarrying. It is speculated that some characteristic of such tools, possibly the low-frequency shocks, is responsible. Some of the reported injuries relate to specific bones, and suggest the existence of cysts, vacuoles, decalcification or other osteolysis, degeneration or deformity of the carpal, metacarpal or phalangeal bones. Osteoarthrosis and olecranon spurs at the elbow and other problems at the wrist and shoulder are also documented (Griffin 1990). There is no universal acceptance that vibration is the cause of articular problems, and there is currently no dose–effect relation that predicts their occurrence. In the absence of specific information, it seems that current guidance for the prevention of vibration-induced white finger may provide reasonable protection for nonvascular disorders localised in the fingers, hands and arms.

7.5.2.5 Other effects

Hand-transmitted vibration may not only affect the fingers, hands and arms: some studies have found a high incidence of problems such as headaches and sleeplessness among users of vibratory tools, and have concluded that these symptoms are caused by hand-transmitted vibration (Griffin 1990). Although these are real problems to those affected, they are subjective effects that are not accepted as real by all researchers. It would appear that caution is appropriate, but it currently appears reasonable to assume that the adoption of the modern guidance to prevent vibration-induced white finger will also provide some protection from any other effects of hand-transmitted vibration within, or distant from, the hand.

7.5.3 Standards for evaluation of hand-transmitted vibration

International standards and government regulations provide guidance on the measurement, evaluation and assessment of the severity of exposures to hand-transmitted vibration.

7.5.3.1 Vibration measurement

General methods of measuring and evaluating hand-transmitted vibration on tools and processes are defined in ISO 5349-1 (ISO 2001b) and ISO 5349-2 (ISO 2002). Guidance for the measurement of vibration on specific tools is given elsewhere (e.g. in appropriate parts of ISO 8662 and ISO 28927). Some effort may be required to obtain measurements of tool vibration that are representative of appropriate operating conditions.

7.5.3.2 Vibration evaluation according to ISO 5349-1 (2001)

All current national and international standards use the same frequency weighting (called W_h) to evaluate hand-transmitted vibration over the approximate frequency range 8–1000 Hz (Figure 7.14; ISO 5349-1 [ISO 2001b]). This weighting is applied to measurements of vibration acceleration in each of the three translational axes of vibration at the point of entry of vibration to the hand (Figure 7.2). The standards indicate that the overall severity of hand-transmitted vibration should be calculated from the root sums of squares of the frequency-weighted acceleration in the three

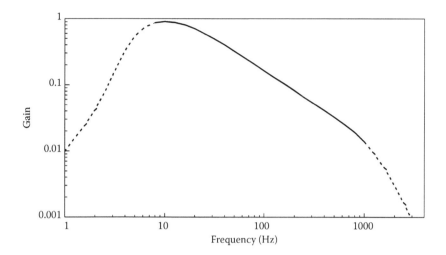

Figure 7.14 Frequency weighting, W_h, for evaluating hand-transmitted vibration. (Data from ISO, *ISO 5349-1:2001: Mechanical Vibration: Measurement and Evaluation of Human Exposure to Hand-Transmitted Vibration. Part 1: General Requirements*, International Organization for Standardization, Geneva, 2001.)

axes. The application of a single frequency weighting does not mean that the various physiological and pathological responses to hand-transmitted vibration have been shown to follow this frequency dependence (Griffin 2012). The weighting provides a simple and uniform method for evaluating vibration, but should not be expected to provide accurate predictions of the effects of hand-transmitted vibration.

If two tools expose the hand to vibration for the same period of time, the standards indicate that the tool with the lowest frequency-weighted acceleration will be least likely to cause injury or disease.

Occupational exposures to hand-transmitted vibration can have widely varying daily exposure durations – from a few seconds to many hours. Often, exposures are intermittent. To enable a daily exposure to be reported simply, the standards refer to an equivalent 8 h exposure:

$$a_{\text{hw}(eq,8h)} = A(8) = a_{\text{hw}} \left[\frac{t}{T_{(8)}} \right]^{1/2} \quad (7.7)$$

where:
 t = exposure duration to an r.m.s. frequency-weighted acceleration, a_{hw}
 $T_{(8)}$ = 8 h (in the same units as t)

7.5.3.3 Vibration assessment according to ISO 5349-1 (2001)

An informative annex to ISO 5349-1 (ISO 2001b) offers a relation between the lifetime exposure to hand-transmitted vibration, D_y, (in years) and the 8 h energy-equivalent daily exposure $A(8)$ for the conditions expected to cause 10% prevalence of finger blanching (Figure 7.15):

$$D_y = 31.8 \left[A(8)\right]^{-1.06} \tag{7.8}$$

The percentage of affected persons in any group of exposed persons will not always correspond to the values shown in Figure 7.15: the frequency weighting, time dependence and dose–effect information are based on less than complete information, and they have been simplified for practical convenience. Additionally, the number of persons affected by vibration will depend on the rate at which persons enter and leave the exposed group. The complexity of Equation 7.8 implies far greater precision than is possible: a more convenient estimate of the years of exposure (in the range 1–25 years) required for 10% incidence of finger blanching is

$$D_y = \frac{30.0}{A(8)} \tag{7.9}$$

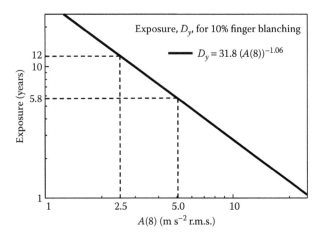

Figure 7.15 Relation between daily $A(8)$ and years of exposure expected to result in 10% incidence of finger blanching. A 10% probability of finger blanching is predicted after 12 years at the EU 'exposure action value' (i.e. when $A(8) = 2.5$ m s^{-2} r.m.s.) and after 5.8 years at the EU 'exposure limit value' (i.e. when $A(8) = 5.0$ m s^{-2} r.m.s.). (Data from ISO, *ISO 5349-1:2001: Mechanical Vibration: Measurement and Evaluation of Human Exposure to Hand-Transmitted Vibration. Part 1: General Requirements*, International Organization for Standardization, Geneva, 2001.)

This equation gives the same result as the equation in the standard (to within 14%) and there is no information suggesting it is less accurate.

The informative annex to ISO 5349-1 (ISO 2001b, p. 15) states: 'Studies suggest that symptoms of the hand-arm vibration syndrome are rare in persons exposed with an 8 h energy-equivalent vibration total value, $A(8)$, at a surface in contact with the hand, of less than 2 m/s^2 and unreported for $A(8)$ values less than 1 m/s^2'. The scientific evidence suggests very considerable doubts over the frequency weighting and time dependence in the standard, so it should not be expected that ISO 5349-1 will give precise predictions of risk (Griffin et al. 1997, 2003).

7.5.3.4 EU Machinery Safety Directive

The Machinery Safety Directive of the European Community (2006/42/EC) requires that instruction handbooks for handheld and hand-guided machinery specify the equivalent acceleration to which the hands or arms are subjected where this exceeds a stated value (currently a frequency-weighted acceleration of 2.5 m s^{-2} r.m.s.) (European Parliament/Council of the European Union 2006). Very many handheld vibrating tools can exceed this value. The standard test conditions for the measurement of vibration on many tools (e.g. chipping and riveting hammers, rotary hammers and rock drills, grinding machines, pavement breakers, chainsaws) have been defined (see ISO 8662 and ISO 28927).

7.5.3.5 EU Physical Agents Directive (2002)

For hand-transmitted vibration, the EU Physical Agents (Vibration) Directive defines an 8 h-equivalent exposure action value of 2.5 m s^{-2} r.m.s. and an 8 h-equivalent exposure limit value of 5.0 m s^{-2} r.m.s. (Council of the European Communities 1989). The manner in which the frequency-weighted acceleration increases with a reduced daily duration of exposure is shown in Figure 7.16. The directive says that workers shall not be exposed above the exposure limit value. If the exposure action values are exceeded, the employer shall establish and implement a programme of technical and organisational measures intended to reduce the exposure to mechanical vibration and the attendant risks to a minimum. The directive requires that workers exposed to mechanical vibration in excess of the exposure action values shall be entitled to appropriate health surveillance. However, health surveillance is not restricted to situations where the exposure action value is exceeded; it is required if there is any reason to suspect that workers may be injured by the vibration, even if the action value is not exceeded.

According to ISO 5349-1 (ISO 2001b), the onset of finger blanching would be expected in 10% of persons after 12 years at the EU exposure

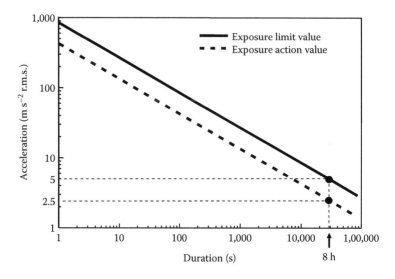

Figure 7.16 Magnitudes of hand-transmitted vibration required to reach the 'exposure limit value' (A(8) = 5.0 m s^{-2} r.m.s.) and 'exposure action value' (A(8) = 2.5 m s^{-2} r.m.s.) in the EU Physical Agents (Vibration) Directive with daily durations of exposure from 1 s to 24 h. (Data from European Parliament/Council of the European Union, *Official Journal of the European Communities*, **L177**, 13–19, 2002.)

action value and after 5.8 years at the exposure limit value. So, the exposure action value and the exposure limit value in the directive do not define safe exposures to hand-transmitted vibration (Griffin 2004).

7.5.4 Preventive measures

If a tool or machine causes hand-transmitted vibration, the minimisation of risk requires that a range of appropriate actions are taken by management, technicians, physicians and tool users at the workplace, and by tool manufacturers (Table 7.9). If there is reason to suspect that hand-transmitted vibration may cause injury, the severity of the vibration at tool–hand interfaces should be assessed, either by consulting information on the tool or by commissioning vibration measurements. It will then be possible to predict whether the tool or process is likely to cause injury and whether any other tool or process could give a lower vibration severity. The duration of exposure to vibration should also be quantified. The reduction of the exposure duration may include the provision of exposure breaks during the day and, if possible, prolonged periods away from vibration exposure. For any tool or process having a vibration magnitude sufficient to cause injury, there should be a system to quantify and control the maximum daily duration of

Table 7.9 Preventive measures to consider when persons are exposed to hand-transmitted vibration

Group	Action
Management	Seek technical advice
	Seek medical advice
	Warn exposed persons
	Train exposed persons
	Review exposure durations
	Policy on removal from work
Tool manufacturers	Measure tool vibration
	Design tools to minimise vibration
	Ergonomic design to optimise grip force, etc.
	Design to keep hands warm
	Provide guidance on tool maintenance
	Provide warning of dangerous vibration
Technical at workplace	Measure vibration exposure
	Provide appropriate tools
	Maintain tools
	Provide and maintain any vibration isolation device
	Inform management
Medical	Preemployment screening
	Routine health surveillance
	Record all signs and reported symptoms
	Warn workers of risks and symptoms
	Advise on consequences of exposure
	Inform management
Tool user	Use tool properly
	Avoid unnecessary vibration exposure
	Minimise grip and push forces
	Check condition of tool
	Inform supervisor of tool problems
	Keep warm
	Wear gloves when safe to do so
	Minimise smoking
	Seek medical advice if symptoms appear
	Inform employer of relevant disorders

Source: Adapted from Griffin, M.J., *Handbook of Human Vibration*, Academic Press, London, 1990.

exposure of any individual. The risks cannot be assessed accurately from an evaluation of the vibration: it contributes to an assessment of risk, but may not always be the best way of predicting risk. Risk may also be anticipated from the type of work and knowledge that the same or similar work has caused problems previously.

Gloves and cushioned handles may reduce the transmission of high frequencies of vibration, but current standards imply that these frequencies are not usually the primary cause of disorders. ISO 10819 (ISO 2013) defines the requirements of 'anti-vibration gloves', but the standard has limitations and cannot be considered a reliable indication of whether a glove is beneficial (Griffin 1998). Gloves may help to minimise pressure on the fingers, protect the hand from other forms of mechanical injury (e.g. cuts and scratches), and protect the fingers from temperature extremes. Warm hands are less likely to suffer an attack of finger blanching, and some consider that maintaining warm hands while exposed to vibration may also lessen the damage caused by the vibration.

Workers exposed to vibration suspected of being capable of causing injury should be warned of the possibility of vibration injuries and educated on the ways of reducing the severity of their vibration exposures. They should be advised of the symptoms to look out for and told to seek medical attention if the symptoms appear. There should be preemployment medical screening wherever a subsequent exposure to hand-transmitted vibration may reasonably be expected to cause vibration injuries. Medical supervision of each exposed person should continue throughout employment at suitable intervals, sometimes annually.

The current standards for measuring, evaluating and assessing the severity of hand-transmitted vibration provide guidance that should not be ignored. However, it would be wise to expect that future advances in the understanding of the relevant pathology, physiology and biodynamic responses are likely to result in changes to the methods of predicting the risks associated with hand-transmitted vibration.

REFERENCES

Ahn, S.-J. and Griffin, M.J. (2008) Effects of frequency, magnitude, damping, and direction on the discomfort of vertical whole-body mechanical shocks. *Journal of Sound and Vibration*, **311**, 485–497.

Beard, G.F. and Griffin, M.J. (2014) Motion sickness caused by roll-compensated lateral acceleration: Effects of centre-of-rotation and subject demographics, *Proceedings of the Institution of Mechanical Engineers, Part F: Journal of Rail and Rapid Transit*, **228**(1), 16–24.

Benson, A.J. and Barnes, G.R. (1978) Vision during angular oscillation: the dynamic interaction of visual and vestibular mechanisms. *Aviation, Space and Environmental Medicine*, **49**(1), Section II, 340–345.

Bongers, P.M. and Boshuizen, H.C. (1990) Back disorders and whole-body vibration at work. Thesis, University of Amsterdam, the Netherlands.

Bovenzi, M. (1993) Digital arterial responsiveness to cold in healthy men, vibration white finger and primary Raynaud's phenomenon. *Scandinavian Journal of Work, Environment and Health*, **19**(4), 271–276.

Bovenzi, M. (2009) Metrics of whole-body vibration and exposure-response relationships for low back pain in professional drivers: A prospective cohort study. *International Archives of Occupational and Environmental Health*, **82**, 893–917.

Bovenzi, M. and Betta, A. (1994) Low-back disorders in agricultural tractor drivers exposed to whole-body vibration and postural stress. *Applied Ergonomics*, 25(4), 231–241.
Bovenzi, M. and Hulshof, C.T.J. (1999) An updated review of epidemiologic studies on the relationship between exposure to whole-body vibration and low back pain (1986–1997). *International Archives of Occupational and Environmental Health*, 72(6), 351–365.
Brammer, A.J., Taylor, W., and Lundborg, G. (1987) Sensorineural stages of the hand-arm vibration syndrome. *Scandinavian Journal of Work, Environment and Health*, 13(4), 279–283.
BSI (British Standards Institution) (1987) *BS 6841:1987: Measurement and Evaluation of Human Exposure to Whole-Body Mechanical Vibration and Repeated Shock*. British Standards Institution, London.
BSI (British Standards Institution) (2008a) *BS 6472-1:2008: Guide to the Evaluation of Human Exposure to Vibration in Buildings. Part 1: Vibration Sources Other than Blasting*. British Standards Institution, London.
BSI (British Standards Institution) (2008b) *BS 6472-2:2008: Guide to the Evaluation of Human Exposure to Vibration in Buildings. Part 2: Blast-Induced Vibration*. British Standards Institution, London.
Butler, C.A. and Griffin, M.J. (2009) Motion sickness with combined fore-and-aft and pitch oscillation: Effect of phase and the visual scene. *Aviation Space and Environmental Medicine*, 80(11), 946–954.
Commission of the European Communities (1994) Amended proposal for a Council Directive on the minimum health and safety requirements regarding the exposure of workers to the risks arising from physical agents: Individual Directive in relation to Article 16 of Directive 89/391/EEC. *Official Journal of the European Communities*, C230, 3–29.
Concettoni, E. and Griffin, M.J. (2009) The apparent mass and mechanical impedance of the hand and the transmission of vibration to the fingers, hand and arm. *Journal of Sound and Vibration*, 325, 664–678.
Corbridge, C. and Griffin, M.J. (1991) Effects of vertical vibration on passenger activities: writing and drinking. *Ergonomics*, 34(10), 1313–1332.
Corbridge, C., Griffin, M.J., and Harborough, P. (1989) Seat dynamics and passenger comfort, *Proceedings of the Institution of Mechanical Engineers*, 203, 57–64.
Departmental Committee on Compensation for Industrial Diseases (1907) Report of the Departmental Committee on Compensation for Industrial Diseases. Cd. 3495, HMSO, London.
Donohew, B.E. and Griffin, M.J. (2004) Motion sickness: Effect of the frequency of lateral oscillation. *Aviation, Space and Environmental Medicine*, 75(8), 649–656.
Donohew, B.E. and Griffin, M.J. (2007) Low frequency motions and motion sickness on a tilting train. *Proceedings of the Institution of Mechanical Engineers, Part F: Journal of Rail and Rapid Transit*, 211, 125–133.
Donohew, B.E. and Griffin, M.J. (2009) Motion sickness with fully roll-compensated lateral oscillation: Effect of oscillation frequency. *Aviation, Space and Environmental Medicine*, 80(2), 94–101.
Dupuis, H. and Zerlett, G. (1986) *The Effects of Whole-Body Vibration*, Springer-Verlag, Berlin.

European Parliament/Council of the European Union (2002) On the minimum health and safety requirements regarding the exposure of workers to the risks arising from physical agents (vibration). Directive 2002/44/EC. *Official Journal of the European Communities*, L177, 13–19.

European Parliament/Council of the European Union (2006) Directive of the European Parliament and of the Council on machinery, and amending Directive 95/16/EC. *Official Journal of the European Communities*, L157, 24–86.

Fairley, T.E. and Griffin, M.J. (1989) The apparent mass of the seated human body: vertical vibration. *Journal of Biomechanics*, 22(2), 81–94.

Forta, N.G., Morioka, M., and Griffin, M.J. (2009) Difference thresholds for the perception of whole-body vertical vibration: dependence on the frequency and magnitude of vibration. *Ergonomics*, 52(10), 1305–1310.

Gemne, G., Pyykko, I., Taylor, W., and Pelmear, P. (1987) The Stockholm Workshop scale for the classification of cold-induced Raynaud's phenomenon in the hand-arm vibration syndrome (revision of the Taylor–Pelmear scale). *Scandinavian Journal of Work, Environment and Health*, 13(4), 275–278.

Goodwin, G.M., McCloskey, D.I., and Matthews, P.B.C. (1972) The contribution of muscle afferents to kinaesthesia shown by vibration induced illusions of movement and by the effects of paralysing joint afferents. *Brain*, 95, 705–748.

Griffin, M.J. (1976) Eye motion during whole-body vertical vibration. *Human Factors*, 18(6), 601–606.

Griffin, M.J. (1978) The evaluation of vehicle vibration and seats. *Applied Ergonomics*, 9(1), 15–21.

Griffin, M.J. (1990) *Handbook of Human Vibration*. Academic Press, London.

Griffin, M.J. (1991) Physical characteristics of stimuli provoking motion sickness. Paper 3 in: Motion sickness: Significance in aerospace operations and prophylaxis. AGARD Lecture Series LS-175.

Griffin, M.J. (1997) Measurement, evaluation and assessment of occupational exposures to hand-transmitted vibration. *Occupational and Environmental Medicine*, 54(2), 73–89.

Griffin, M.J. (1998) Evaluating the effectiveness of gloves in reducing the hazards of hand-transmitted vibration. *Occupational and Environmental Medicine*, 55(5), 340–348.

Griffin, M.J. (2004) Minimum health and safety requirements for workers exposed to hand-transmitted vibration and whole-body vibration in the European Union; a review. *Occupational and Environmental Medicine*, 61, 387–397.

Griffin, M.J. (2007) Discomfort from feeling vehicle vibration. *Vehicle System Dynamics*, 45(7–8), 679–698.

Griffin, M.J. (2008) Measurement, evaluation, and assessment of peripheral neurological disorders caused by hand-transmitted vibration. *International Archives of Occupational and Environmental Health*, 81(5), 559–573.

Griffin, M.J. (2012) Frequency-dependence of psychophysical and physiological responses to hand-transmitted vibration. *Industrial Health*, 50(5), 354–369.

Griffin, M.J., Bovenzi, M., and Nelson, C.M. (2003) Dose response patterns for vibration-induced white finger. *Occupational and Environmental Medicine*, 60, 16–26.

Griffin, M.J. and Hayward, R.A. (1994) Effects of horizontal whole-body vibration on reading. *Applied Ergonomics*, 25(3), 165–169.

Griffin, M.J. and Lewis, C.H. (1978) A review of the effects of vibration on visual acuity and continuous manual control, Part I: Visual acuity. *Journal of Sound and Vibration*, 56(3), 383–413.

Griffin, M.J. and Newman, M.M. (2004a) Visual field effects on motion sickness in cars. *Aviation, Space, and Environmental Medicine*, 75(9), 739–748.

Griffin, M.J. and Newman, M.M. (2004b) An experimental study of low frequency motion in cars. *Proceedings of the Institution of Mechanical Engineers, Part D: Journal of Automobile Engineering*, 218, 1231–1238.

Griffin, M.J. and Whitham, E.M. (1980) Discomfort produced by impulsive whole-body vibration. *The Journal of the Acoustical Society of America*, 68(5), 1277–1284.

Haward, B.M. and Griffin, M.J. (2002) Repeatability of grip strength and dexterity tests and the effects of age and gender. *International Archives of Occupational and Environmental Health*, 75(1–2), 111–119.

Howarth, H.V.C. and Griffin, M.J. (1991) Subjective reaction to vertical mechanical shocks of various waveforms. *Journal of Sound and Vibration*, 147(3), 395–408.

Hulshof, C. and Zanten, B.V. van (1987) Whole-body vibration and low-back pain. *International Archives of Occupational and Environmental Health*, 59, 205–220.

ISO (International Organization for Standardization) (1974) *ISO 2631:1974: Guide for the Evaluation of Human Exposure to Whole-Body Vibration.* International Organization for Standardization, Geneva.

ISO (International Organization for Standardization) (1985) *ISO 2631-1:1985: Evaluation of Human Exposure to Whole-Body Vibration. Part 1: General Requirements.* International Organization for Standardization, Geneva.

ISO (International Organization for Standardization) (1988) *ISO 8662-1:1988: Hand-Held Portable Tools: Measurement of Vibration at the Handle. Part 1: General.* International Organization for Standardization, Geneva.

ISO (International Organization for Standardization) (1992) *ISO 10326-1:1992: Mechanical Vibration: Laboratory Method for Evaluating Vehicle Seat Vibration. Part 1: Basic requirements.* International Organization for Standardization, Geneva.

ISO (International Organization for Standardization) (1997) *ISO 2631-1:1997: Mechanical Vibration and Shock: Evaluation of Human Exposure to Whole-Body Vibration. Part 1: General Requirements.* International Organization for Standardization, Geneva.

ISO (International Organization for Standardization) (2000) *ISO 7096:2000: Earth-Moving Machinery: Laboratory Evaluation of Operator Seat Vibration.* International Organization for Standardization, Geneva.

ISO (International Organization for Standardization) (2001a) *ISO 13091-1:2001: Mechanical Vibration: Vibrotactile Perception Thresholds for the Assessment of Nerve Dysfunction. Part 1: Methods of Measurement at the Fingertips.* International Organization for Standardization, Geneva.

ISO (International Organization for Standardization) (2001b) *ISO 5349-1:2001: Mechanical Vibration: Measurement and Evaluation of Human Exposure to Hand-Transmitted Vibration. Part 1: General Requirements.* International Organization for Standardization, Geneva.

ISO (International Organization for Standardization) (2001c) *ISO 5982:2001: Mechanical Vibration and Shock: Range of Idealized Values to Characterize Seated Body Biodynamic Response under Vertical Vibration.* International Organization for Standardization, Geneva.

ISO (International Organization for Standardization) (2002) *ISO 5349-2:2002: Mechanical Vibration: Measurement and Evaluation of Human Exposure to Hand-Transmitted Vibration. Part 2: Practical Guidance for Measurement at the Workplace.* International Organization for Standardization, Geneva.

ISO (International Organization for Standardization) (2003a) *ISO 2631-2:2003: Mechanical Vibration and Shock: Evaluation of Human Exposure to Whole-Body Vibration. Part 2: Vibration in Buildings (1 Hz to 80 Hz).* International Organization for Standardization, Geneva.

ISO (International Organization for Standardization) (2003b) *ISO 5007:2003: Agricultural Wheeled Tractors: Operator's Seat: Laboratory Measurement of Transmitted Vibration.* International Organization for Standardization, Geneva.

ISO (International Organization for Standardization) (2004) *ISO 2631-5:2004: Mechanical Vibration and Shock: Evaluation of Human Exposure to Whole-Body Vibration. Part 5: Method for Evaluation of Vibration Containing Multiple Shocks.* International Organization for Standardization, Geneva.

ISO (International Organization for Standardization) (2005a) *ISO 14835-1:2005: Mechanical Vibration and Shock: Cold Provocation Tests for the Assessment of Peripheral Vascular Function. Part 1: Measurement and Evaluation of Finger Skin Temperature.* International Organization for Standardization, Geneva.

ISO (International Organization for Standardization) (2005b) *ISO 14835-2:2005: Mechanical Vibration and Shock: Cold Provocation Tests for the Assessment of Peripheral Vascular Function. Part 2: Measurement and Evaluation of Finger Systolic Blood Pressure.* International Organization for Standardization, Geneva.

ISO (International Organization for Standardization) (2005c) *ISO 20643:2005: Mechanical Vibration: Hand-Held and Hand-Guided Machinery: Principles for Evaluation of Vibration Emission.* International Organization for Standardization, Geneva.

ISO (International Organization for Standardization) (2005d) *ISO 8041:2005: Human Response to Vibration: Measuring Instrumentation.* International Organization for Standardization, Geneva.

ISO (International Organization for Standardization) (2013) *ISO 10819:1996: Mechanical Vibration and Shock: Hand-Arm Vibration: Method for the Measurement and Evaluation of the Vibration Transmissibility of Gloves at the Palm of the Hand.* International Organization for Standardization, Geneva.

Jang, H.-K. and Griffin, M.J. (2000) Effect of phase, frequency, magnitude and posture on discomfort associated with differential vertical vibration at the seat and feet. *Journal of Sound and Vibration,* **229**(2), 273–286.

Kitazaki, S. and Griffin, M.J. (1997) A modal analysis of whole-body vertical vibration, using a finite element model of the human body. *Journal of Sound and Vibration,* **200**(1), 83–103.

Lawther, A. and Griffin, M.J. (1987) Prediction of the incidence of motion sickness from the magnitude, frequency, and duration of vertical oscillation. *The Journal of the Acoustical Society of America,* **82**(3), 957–966.

Lawther, A. and Griffin, M.J. (1988a) A survey of the occurrence of motion sickness amongst passengers at sea. *Aviation, Space and Environmental Medicine*, 59(5), 399–406.

Lawther, A. and Griffin, M.J. (1988b) Motion sickness and motion characteristics of vessels at sea. *Ergonomics*, 31(10), 1373–1394.

Lewis, C.H. and Griffin, M.J. (1977) The interaction of control gain and vibration with continuous manual control performance. *Journal of Sound and Vibration*, 55(4), 553–562.

Lewis, C.H. and Griffin, M.J. (1979) The effect of character size on the legibility of numeric displays during vertical whole-body vibration. *Journal of Sound and Vibration*, 67(4), 562–565.

Lewis, C.H. and Griffin, M.J. (2002) Evaluating vibration isolation of soft seat cushions using an active anthropodynamic dummy. *Journal of Sound and Vibration*, 253(1), 295–311.

Lindsell, C.J. and Griffin, M.J. (2002) Normative data for vascular and neurological tests of the hand-arm vibration syndrome. *International Archives of Occupational and Environmental Health*, 75(1–2), 43–54.

Loriga, G. (1911) Il lavoro con i martelli pneumatici. The use of pneumatic hammers. *Bollettino del Ispettorato del Lavoro*, 2, 35–60.

Mansfield, N.J. and Griffin, M.J. (2000) Non-linearities in apparent mass and transmissibility during exposure to whole-body vertical vibration. *Journal of Biomechanics*, 33, 933–941.

Matsumoto, Y. and Griffin, M.J. (2001) Modelling the dynamic mechanisms associated with the principal resonance of the seated human body. *Clinical Biomechanics*, 16, S31–S44.

Matsumoto, Y. and Griffin, M.J. (2003) Mathematical models for the apparent masses of standing subjects exposed to vertical whole-body vibration. *Journal of Sound and Vibration*, 260, 431–451.

McLeod, R.W. and Griffin, M.J. (1989) A review of the effects of translational whole-body vibration on continuous manual control performance. *Journal of Sound and Vibration*, 133(1), 55–115.

McLeod, R.W. and Griffin, M.J. (1990) Effects of whole-body vibration waveform and display collimation on the performance of a complex manual control task. *Aviation, Space and Environmental Medicine*, 61(3), 211–219.

Meddick, R.D.L. and Griffin, M.J. (1976) The effect of two-axis vibration on the legibility of reading material. *Ergonomics*, 19(1), 21–33.

Morioka, M. and Griffin, M.J. (2006a) Magnitude-dependence of equivalent comfort contours for fore-and-aft, lateral and vertical whole-body vibration. *Journal of Sound and Vibration*, 298(3), 755–772.

Morioka, M. and Griffin, M.J. (2006b) Magnitude-dependence of equivalent comfort contours for fore-and-aft, lateral and vertical hand-transmitted vibration. *Journal of Sound and Vibration*, 295, 633–648.

Morioka, M. and Griffin, M.J. (2008) Absolute thresholds for the perception of fore-and-aft, lateral, and vertical vibration at the hand, the seat, and the foot. *Journal of Sound and Vibration*, 314, 357–370.

Morioka, M. and Griffin, M.J. (2009) Equivalent comfort contours for vertical vibration of steering wheels: effect of vibration magnitude, grip force, and hand position. *Applied Ergonomics*, 40, 817–825.

Moseley, M.J. and Griffin, M.J. (1986a) Effects of display vibration and whole-body vibration on visual performance. *Ergonomics*, **29**(8), 977–983.

Moseley, M.J. and Griffin, M.J. (1986b) A design guide for visual displays and manual tasks in vibration environments. Part I: Visual displays, Technical Report No. 133, Institute of Sound and Vibration Research, University of Southampton, Southampton.

Moseley, M.J., Lewis, C.H., and Griffin, M.J. (1982) Sinusoidal and random whole-body vibration: comparative effects on visual performance. *Aviation, Space and Environmental Medicine*, **53**(10), 1000–1005.

National Institute of Occupational Safety and Health (NIOSH), Bernard, B.P. (ed.) (1997) Musculoskeletal disorders and workplace factors: A critical review of epidemiologic evidence for work-related disorders of the neck, upper extremities, and low back. U.S. Department of Health and Human Services, National Institute of Occupational Safety and Health, DHHS (NIOSH) Publication No. 97-141.

Nawayseh, N. and Griffin, M.J. (2003) Non-linear dual-axis biodynamic response to vertical whole-body vibration. *Journal of Sound and Vibration*, **268**, 503–523.

Nawayseh, N. and Griffin, M.J. (2005) Non-linear dual-axis biodynamic response to fore-and-aft whole-body vibration. *Journal of Sound and Vibration*, **282**(3–5), 831–862.

Nielsen, S.L. and Lassen, N.A. (1977) Measurement of digital blood pressure after local cooling. *Journal of Applied Physiology, Respiratory Environment and Exercise Physiology*, **43**(5), 907–910.

O'Hanlon, J.G. and Griffin, M.J. (1971) Some effects of the vibration of reading material upon visual performance, Technical Report No. 49. Institute of Sound and Vibration Research, Southampton.

Paddan, G.S. and Griffin, M.J. (1988a) The transmission of translational seat vibration to the head – I. Vertical seat vibration. *Journal of Biomechanics*, **21**(3), 191–197.

Paddan, G.S. and Griffin, M.J. (1988b) The transmission of translational seat vibration to the head – II. Horizontal seat vibration. *Journal of Biomechanics*, **21**(3), 199–206.

Paddan, G.S. and Griffin, M.J. (1993) The transmission of translational floor vibration to the heads of standing subjects. *Journal of Sound and Vibration*, **160**(3), 503–521.

Parsons, K.C. and Griffin, M.J. (1988) Whole-body vibration perception thresholds. *Journal of Sound and Vibration*, **121**(2), 237–258.

Reason, J.T. (1970) Motion sickness: A special case of sensory rearrangement. *Advancement of Science*, **26**, 386–393.

Reason, J.T. (1978) Motion sickness adaptation: a neural mismatch model. *Journal of the Royal Society of Medicine*, **71**, 819–829.

Reason, J.T. and Brand, J.J. (1975) *Motion Sickness*. Academic Press, London.

Ribot, E., Roll, J.P., and Gauthier, G.M. (1986) Comparative effects of whole-body vibration on sensorimotor performance achieved with a mini-stick and a macro-stick in force and position control modes. *Aviation, Space and Environmental Medicine*, **57**(8), 792–799.

Seah, S.A. and Griffin, M.J. (2008) Normal values for thermotactile and vibrotactile thresholds in males and females. *International Archives of Occupational and Environmental Health*, **81**(5), 535–543.

Sherwood, N. and Griffin, M.J. (1990) Effects of whole-body vibration on short-term memory. *Aviation, Space and Environmental Medicine*, **61**(12), 1092–1097.
Sherwood, N. and Griffin, M.J. (1992) Evidence of impaired learning during whole-body vibration. *Journal of Sound and Vibration*, **152**(2), 219–225.
Taylor, W. and Pelmear, P.L. (eds) (1975) *Vibration White Finger in Industry*. Academic Press, London.
Toward, M.G.R. and Griffin, M.J. (2009) Apparent mass of the human body in the vertical direction: Effect of seat backrest. *Journal of Sound and Vibration*, **327**, 657–669.
Toward, M.G.R. and Griffin, M.J. (2010) Apparent mass of the human body in the vertical direction: Effect of footrest and a steering wheel. *Journal of Sound and Vibration*, **329**, 1586–1596.
Toward, M.G.R. and Griffin, M.J. (2011a) Apparent mass of the human body in the vertical direction: Inter-subject variability. *Journal of Sound and Vibration*, **330**(4), 827–841.
Toward, M.G.R. and Griffin, M.J. (2011b) The transmission of vertical vibration through seats: Influence of the characteristics of the human body. *Journal of Sound and Vibration*, **330**(26), 6526–6543.
Turner, M. and Griffin, M.J. (1999a) Motion sickness in public road transport: The effect of driver, route and vehicle. *Ergonomics*, **42**(12), 1646–1664.
Turner, M. and Griffin, M.J. (1999b) Motion sickness in public road transport: The relative importance of motion, vision and individual differences. *British Journal of Psychology*, **90**, 519–530.
Turner, M., Griffin, M.J., and Holland, I. (2000) Airsickness and aircraft motion during short-haul flights. *Aviation, Space, and Environmental Medicine*, **71**(12), 1181–1189.
Wei, L. and Griffin, M.J. (1998a) Mathematical models for the apparent mass of the seated human body exposed to vertical vibration. *Journal of Sound and Vibration*, **212**(5), 855–874.
Wei, L. and Griffin, M.J. (1998b) The prediction of seat transmissibility from measures of seat impedance. *Journal of Sound and Vibration*, **214**(1), 121–137.
Wells, M.J. and Griffin, M.J. (1984) Benefits of helmet-mounted display image stabilisation under whole-body vibration. *Aviation, Space and Environmental Medicine*, **55**(1), 13–18.
Wilson, R.V. (1974) Display collimation under whole-body vibration. *Human Factors*, **16**(2), 186–195.
Wu, X. and Griffin, M.J. (1997) A semi-active control policy to reduce the occurrence and severity of end-stop impacts in a suspension seat with an electrorheological fluid damper. *Journal of Sound and Vibration*, **203**(5), 781–793.

Chapter 8

Measurement of audio-frequency sound in air

Frank Fahy

8.1 INTRODUCTION

Sound in fluids involves time-dependent variations of pressure. In atmospheric air, the acoustic pressure variations are superimposed on the local 'static' atmospheric pressure. Even at 'high' noise levels, for example at 200 m from a large passenger aircraft taking off, the maximum sound pressures are only about one-hundred-thousandth of the atmospheric pressure. Pressure is only one of the quantities that varies in time in a fluid medium carrying a sound wave, the others being density and temperature, together with particle position, velocity and acceleration. Sound pressure is the most commonly measured of these quantities because it is the easiest to measure and, most importantly, it is the physical agent that activates the human auditory system.

Sound recordings have been made since 1857 when Edouard-Leon Scott de Martinville patented his 'phonautograph'. The first electrical (carbon) microphone was designed by David Hughes in 1876 for use in the telephone. The condenser microphone was invented in 1916 by Christopher Wente and was greatly improved by him in 1922. The design, materials and precision of manufacture and assembly, together with amplification technology, have been steadily improved over the last 100 years, but Wente's design principle still forms the basis for the majority of microphones used for scientific measurement. This chapter will describe the transduction principle and performance characteristics of typical measurement microphones, their uses in the sound level meter (SLM), and recent developments in microphone and microphone array technology that enable sound fields to be investigated in great detail, a major application of which is the *in situ* location, identification and quantification of noise sources.

Human ears, together with the brain, form a very sophisticated sensing instrument having an enormous dynamic range of normal function, but which is vulnerable to damage by exposure to high sound levels, especially in cases of extended periods of exposure. Experiments have shown that the smallest root-mean-square (r.m.s.) acoustic pressure that is, on average, just audible to young, healthy ears is about 20 micropascals (20 μPa) at 1 kHz; this value has been adopted by international standards as the reference level for scientific measurements in air. The dynamic range of noise levels which is experienced in everyday life can be very high. As an extreme example of non-military noises, the popping of a highly inflated balloon can generate instantaneous sound pressures in very close proximity as high as 200,000,000 μPa. This range of a factor of 10 million is conveniently handled by defining the sound pressure level (SPL) of continuous sound on a logarithmic scale (see Chapter 2).

The principal audiological justification for defining a logarithmic metric of sound pressure is that, as explained in Chapter 6, the response of the auditory system to sound pressure is not linear, but closely logarithmic. Instantaneous values of transient pressure signals, such as those of gunshots, for which r.m.s. quantification is not appropriate, may also be expressed in decibels. The peak sound pressure of a balloon popping at 200,000,000 μPa can be expressed as an SPL of 140 dB *re* 20 μPa; a busy office where the r.m.s. sound pressure is 35,000 μPa can be expressed as 65 dB *re* 20 μPa. Because the decibel scale is based on a ratio, the SPL should strictly always be quoted together with the reference value. However, it is common practice to quote the SPL in air in decibels without including the reference value. Note: the reference pressure for underwater SPL is normally 1 Pa r.m.s.

8.2 CONDENSER MICROPHONE

Most measurement microphones are based on the principle of the condenser microphone. The construction of a typical half-inch (12.7 mm) condenser microphone is shown in Figure 8.1. The microphone capsule has a very thin diaphragm (about 5 μm) on its outer surface. This fragile element is normally protected by a grid cap, which can be designed to influence the directional response of the microphone. The exposed diaphragm should never be touched! The diaphragm forms a capacitor with a backplate which, in most older models, is charged by an externally supplied direct current (dc) voltage. The backplate is conventionally penetrated by a set of holes which are designed to damp the mechanical resonance of the diaphragm by means of their resistance to airflow. The diaphragm is displaced by the pressure of the sound field in which it is placed and it therefore moves relative to the backplate and changes the capacitance of the circuit. This change in capacitance is reflected by a change in voltage across the capacitor, since the charge is constant. The resulting voltage signal can then be measured and analysed. A dc polarisation voltage of 200 V is used to provide sufficient dynamic range of the microphone to accommodate the very large dynamic range of the human ear. A major advance in measurement microphone

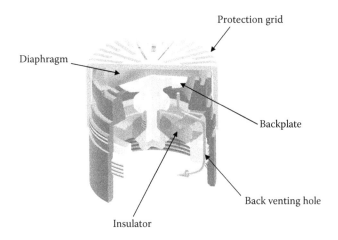

Figure 8.1 Cross section of a condenser measurement microphone. (Courtesy of G.R.A.S. Sound & Vibration A/S, Holte, Denmark.)

technology which has enabled its use with low-voltage electronic instrumentation was the introduction of a permanent polarising electret layer on the backplate, so that the instrumentation does not have to supply an external voltage. These microphones are designated as prepolarised (or electret) types.

The thin tube vent shown in Figure 8.1 equalises the pressures on either side of the diaphragm to minimise static pressure differences across the diaphragm caused by changes in atmospheric pressure. The microphone can be vented to the back, as in this case, or to the side. The diameter of the vent affects the low-frequency performance of the microphone since the resistance of the vent and the capacitance of the cavity act as a low-frequency cut-off filter. Therefore, a condenser microphone does not have a dc response; the vent can be used to control the lower cut-off frequency. Typically, a condenser microphone designed for measurement has a limit of flat frequency response as low as 3 Hz. Such microphones have directional responses that are uniform over a wide frequency range and they are referred to as omnidirectional microphones. For historical reasons, the diameters of the capsules that are mounted on coaxial cylindrical bodies are specified in inches, also denoted by the symbol ″.

The presence of a microphone and of any ancillary supporting body, such as an attached SLM, which will simply be referred to as 'the system', scatters and diffracts an incident sound field, so that the pressure acting on the diaphragm is different from that which would exist on the sensing area in the absence of the system. The degree of distortion of the acoustic field acting on the sensing area of the microphone is a complex function of the geometry and the dimensions of the system, together with its orientation relative to the wavefronts of the incident waves. The vibrational response of the diaphragm also distorts the field, but this is generally a minor effect. The only general remarks that can usefully be made are that the field distortion becomes substantial as the wavelength of the sound field becomes commensurate with the largest dimension of the system, and that the reflection of sound waves incident on a microphone diaphragm from a direction normal to the plane of the diaphragm can substantially increase the measured sound pressure. However, this enhancement decreases as the angle of incidence changes towards a direction parallel to the plane of the diaphragm. This is a high-frequency effect which largely determines the free-field directional sensitivity (directivity) of the system response. Consequently, manufacturers distinguish those systems that are designed to operate in free field, and which are approximately corrected for scattering and diffraction effects by means of the form of a protection grid, from those that are required to sense correctly the average pressure on the diaphragm: hence, 'free-field' and 'pressure' microphones. Examples of typical frequency responses of quarter-inch (6.3 mm) free-field and pressure microphones are shown in Figure 8.2. An additional feature of condenser microphones is that diaphragms exhibit high-frequency mechanical resonance, which can be seen as the broad peak around 50 kHz in Figure 8.2. This is controlled by backplate damping, as mentioned in Section 8.2. Free-field and pressure microphone diaphragms are damped by different amounts around their resonance frequencies.

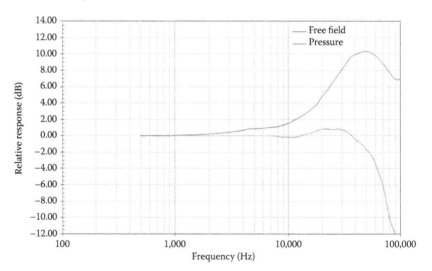

Figure 8.2 Examples of the frequency responses of a quarter-inch instrument condenser microphone. Upper curve: free field. Lower curve: pressure. (Courtesy of G.R.A.S. Sound & Vibration A/S, Holte, Denmark.)

A typical half-inch diameter *free-field* microphone has an upper limit of flat frequency response of about 20 kHz. A typical half-inch *pressure* microphone has a limit for flat frequency response that may extend to 40 kHz in special circumstances. Therefore, for most practical purposes where the main concern is sound that can be heard by a human, a half-inch free-field microphone is ideal.

The dynamic range of such a simple transducer is also very high. Fortunately, this means that microphones can be designed to measure accurately over the range of amplitudes corresponding to human hearing. The lower limit of the dynamic range is limited by thermal noise in the materials. In a half-inch microphone this is typically equivalent to an SPL of about 15 dB. In most practical circumstances there is very little concern about such low noise levels when applied to human response. The upper dynamic limit occurs at an SPL of between 140 and 150 dB, when membrane distortion limits the linearity of the response. Since these levels exceed those to which the human ear can safely be exposed, measurements above these levels are limited to research on physical phenomena rather than work which relates to the human perception of sound.

8.3 PREAMPLIFIERS AND TRANSDUCER ELECTRONIC DATA SYSTEMS (TEDS)

The condenser microphone has very high electrical impedance and generates a very small signal. The typical sensitivity of a half-inch microphone is 50 mV Pa^{-1}. This means that at the bottom of the dynamic range (SPL of 15 dB) the output is only 5 μV r.m.s. and even at 120 dB the output is only 1 V r.m.s. It is clear that the dynamic range is reflected in the voltage output since the transducer is linear, and that high-quality, low-noise electronics are therefore needed to convert the signal and process it into the form with which we are familiar: a display in decibels. The very high impedance of the microphone capsule means that the output is highly susceptible to electrical noise and therefore the preamplifier is always mounted immediately adjacent to the microphone. 'Preamplifier' is indeed a misnomer, since the part attached to the microphone is simply an impedance converter with a gain very close to unity. A very small signal from a low impedance source can be transmitted along lengthy cabling without the same susceptibility to electrical noise. However, low-noise cabling is still required and this can be expensive.

Alternative preamplifier technologies have come into use. These are mainly based on constant-current line-drive (integrated electronic piezoelectric [IEPE]) preamplifiers. Constant-current systems have a major advantage in terms of the cabling required. A simple coaxial cable is all that is required between the preamplifier and subsequent instrumentation.

These are not only considerably less expensive but they can also be much longer than conventional microphone cables. In fact, cables of up to 1 km can be used without loss of signal quality. The early IEPE preamplifiers had limited dynamic range and higher noise floors than conventional preamplifiers, but the technology is improving and they now challenge the conventional approach for all except the most demanding applications.

The other major advance that arrived with IEPE amplification, but is now available on traditional preamplifiers, is transducer electronic data system (TEDS). This enables the serial numbers of the transducer, all the calibration data, the sensitivity and the frequency response to be read by the instrumentation when the preamplifier is connected. As shown in Section 8.2, the microphone cartridge is a simple passive device and therefore TEDS cannot be implemented in the microphone capsule, only in the preamplifier. There is a resulting disadvantage in terms of flexibility and cost since each microphone has to have a dedicated preamplifier for the system to be implemented. The amplification and conditioning of the signal usually occur in the measuring amplifier, signal analyser or SLM. Most modern instrumentation has TEDS enabled within the instrument and transducer details are automatically available as soon as the transducer is attached.

8.4 CALIBRATION

A microphone capsule can be calibrated in a number of different ways, both in the field and in the laboratory. The initial calibration performed by the manufacturer is usually implemented by an electronic actuator. This involves removing the protection grid and attaching a plate to the front of the microphone in place of the grid. The actuator can then be excited electronically. A precise deflection of the microphone diaphragm can be achieved and this method has the advantage that it can be applied across the complete frequency range of interest. It can therefore be used to generate the frequency response plots (see e.g. Figure 8.2), which are provided by the manufacturer, and can be stored in TEDS.

A simpler, but less accurate, calibration method is to use a pistonphone. The microphone is placed in a cavity and a motor drives a piston to generate very precise changes in the volume. As the cavity is sealed there is a resulting precise pressure variation. The pistonphone is usually provided with a barometer because the pressure variation depends to a small extent on the mean (static) pressure of the enclosed air, which is at atmospheric pressure. The level generated is usually equivalent to 120 dB. This high level makes the calibration largely immune from external acoustic interference. The major disadvantage of the pistonphone is that the mechanics of the motor and piston arrangement limit the frequency of actuation. Typically, the pistonphone operates at 250 Hz. This means that if the device which is being used to monitor the level is a simple instrument with a frequency weighting incorporated in the instrument, such as the 'A' weighting (Figure 6.6), a level correction must be made equivalent to the value of the weighting at 250 Hz.

For field calibration it is common to use an electronic calibrator to perform a calibration check. The calibration should be checked before and after a series of measurements in the field. If no significant changes are seen, this confirms that the microphone and instrument have been stable and that no damage occurred during the measuring process. The electronic calibrator uses an electromagnetic actuator with a feedback circuit which is monitored by a reference microphone in the calibrator itself. The microphone is inserted into a cavity in the same way as with a pistonphone, and the calibrator generates an SPL inside the volume. Because of the electronic actuation, the frequency of the tone is usually 1 kHz. This is the frequency at which all of the frequency weightings have an attenuation of 0 dB. Therefore, it is irrelevant which weighting is in use when the calibration is made. The electronic calibrator often has two output levels at 94 and 114 dB. The lower of these two levels is too low to be used reliably in a noisy environment and care must be taken to perform the calibration in a quiet location; if not, the higher level must be used.

8.5 SPECIALISED MICROPHONE APPLICATIONS

8.5.1 MEMS microphones

The acronym MEMS stands for *micro-electromechanical systems*. Complete electret-type microphones can now be constructed in miniature units (Figure 8.3). They are very widely used in mobile phones and other items of information technology (IT) equipment. At present, they cannot be used as instrumentation microphones in SLMs that have to meet instrumentation standards because they have a considerably smaller dynamic range than conventional microphones. However, it is likely that technological solutions will be soon be found to overcome this deficiency.

8.5.2 Probe microphone

There are a number of applications where a conventional condenser microphone cannot be used. The physical size of a microphone capsule can be too large for it to be able to measure the pressure at a precise location in scale models, or other small devices. Normal microphone capsules are also too delicate to use in hot environments such as internal combustion engine and gas turbine exhaust systems. There is a solution in the form of a probe microphone. The sampling end consists of a small-diameter tube with the

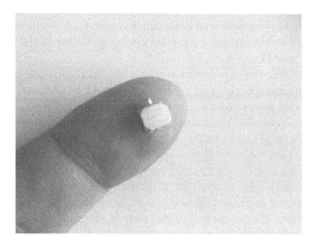

Figure 8.3 Example of a MEMS microphone.

microphone connected at a convenient distance back from the tip. If the tube is terminated by a larger-diameter coaxial microphone, it will reflect sound waves and the wave interference and associated acoustic resonances will cause the frequency response not to be flat. The problem is solved by connecting the microphone to the tube through a side aperture, and continuing the tube past the pressure measurement point either to a section of fibrous sound-absorbent material or to an exponential termination that mimics an infinite tube. Typically, probe microphones can be constructed to measure reliably up to 20 kHz. If possible, probe tube microphone systems should be calibrated at the temperature distributions associated with intended applications.

8.5.3 Artificial ear

There are many circumstances in the practice of acoustic modelling, testing and measurement where an ability to simulate the acoustical behaviour of the outer ear is extremely valuable. The outer ear comprises the external pinna and the ear canal terminated by the tympanic membrane (eardrum) (see Chapter 6). Two major advantages of such a simulation system are that it avoids the many problems associated with using human subjects, and the outputs are objective and quantifiable. Examples include the testing of ear defenders, telephonometry, recording of sound fields in concert halls, measurement of noise inside road vehicles, measurements on spatial sound reproduction systems and tests of the performance of headphones and hearing aids. As described in Chapter 6, the diaphragm of the human ear is not fully exposed to an incident sound field in the way that a microphone is. In cases such as insert headphones, hearing aids and telephone receivers, where the pinnae play no part in the auditory response, an artificial ear comprises a specialised coupler that simulates the impedance of the ear canal terminated by the tympanic membrane. Consequently, a sound source driving the occluded ear produces at the measurement microphone a close approximation to the pressure to which the tympanic membrane would be exposed by the same form of excitation. A famous example is the Zwislocki coupler.

A head and torso simulator, of which there are a number commercially available, places the artificial ear in a model head and upper body which diffracts the incident sound field in a way that is representative of a typical human. These devices are standardised (BSI 2005).

8.5.4 Directional microphones

Single sensor-element microphones, such as the half-inch condenser microphone described in Section 8.2, are closely omnidirectional over a substantial part of the audio-frequency range. In some areas of microphone use, such as sound reinforcement of an individual's speech or song, public address announcement or selective recording of an individual orchestral instrument in an orchestral performance, omnidirectionality is a disadvantage. This is because it does not 'discriminate' between the wanted sound and the extraneous sounds generated by background noise sources or other microphone users or instruments in the near vicinity. In such cases, it is important to be able to enhance the desired sound and suppress the unwanted sound. This can be done very effectively by the use of spatially extended arrays of pressure microphones, as described in Section 8.5.5; but these are too large for use in audio applications. Instead, various compact geometric combinations of pressure microphones, together with various forms of electrical combination of their output signals, such as addition, subtraction and delay, can be devised to produce various forms and degrees of directivity and frequency response (Kinsler et al. 2000). The simplest example is the first-order, bidirectional, gradient microphone in which the output is formed by the difference between the signals from a pair of closely spaced pressure microphones; this produces a figure-of-eight directivity pattern. This combination effectively forms an acoustic particle velocity sensor, as explained in Section 8.7. One of the most widely used compound microphones is the cardioid which combines the first-order gradient microphone described earlier with an omnidirectional pressure microphone. As its name indicates, it has a heart-shaped directivity pattern that strongly suppresses sound arriving from the rear and favours sound arriving from the forward arc. Such specialised compound microphones are not generally used for quantitative measurement because of frequency-dependent directivities and sensitivities.

A form of directional microphone system which can be used to locate a spatially discrete, concentrated source of sound is the 'gun' or 'shotgun' microphone; this takes various forms, but the principle is common. The simplest, and least effective, form consists of a rigid tube of half a metre or so in length in which there is either a series of uniformly spaced small holes or a thin slit coincident with one generator of the tube. One end of the tube contains a sound-absorbent insert, and a pressure microphone is set into the other end. The pressure of a plane sound wave arriving along the direction of the tube axis 'pumps' air in and out through each hole to act as a monopole sound source exciting the air within the tube. Each source is in phase with the external field and the set of holes contributes cumulatively to driving a wave along the tube that corresponds in wavelength and frequency to the external wave. The microphone senses impingement of this wave and the absorber at the other end suppresses wave reflection from the other end of the tube.

Now consider the case where a plane harmonic (single frequency) sound wave approaches from a direction orthogonal (perpendicular) to the tube axis. The oscillatory flows pumped in and out of all the holes (sources) are in phase. Each one sends a wave component along the tube; but the delay, and hence the phase, with which the wave component associated with each source arrives at the microphone increases with the distance of the hole from the microphone. If the tube is a wavelength in length, the range of the phases of arrival of wave components at the microphone will be 2π, and cumulative pressure on the microphone will be zero. At low frequencies, the phase range will be $<2\pi$, and the cancellation process will be incomplete; but the sound pressure on the microphone will be somewhat less than the external wave pressure. If the length of the tube is greater than a wavelength, but is not a whole number of wavelengths, a fraction of the sources will generate wave components that mutually cancel completely, but the remaining fraction will excite wave components that do not do so. Their cumulative contribution to the sensed pressure is less than the external pressure, this contribution decreasing with the increase of frequency.

The phase difference between the sound pressures on *adjacent* holes in a plane wave incident on the tube at angles between 0° (direction of tube axis) and 90° varies with the angle of incidence. This phase difference adds to, or subtracts from, the phase difference of arrival of the waves generated by the associated sources due to their relative distances from the microphone. The directional sensitivity is therefore a function of frequency, but maximum sensitivity always occurs along the tube axis. Hence, the device can, in principle, be used to locate spatially concentrated, discrete sources of sound. Of course, it will also detect sound waves that are strongly reflected from neighbouring surfaces, so it is best used in free-field conditions. A highly refined shotgun microphone is described in Schulman (1987).

It is interesting to note that if the terminal microphone of a shotgun microphone is replaced by a small loudspeaker, the radiation directivity matches that of the system as a receiver. Greater directivity can be achieved in free space at high frequencies by a line array of loudspeakers in which the input signal to any one loudspeaker is delayed relative to that of the first loudspeaker of the array by the time for a sound wave to travel from the first loudspeaker to the loudspeaker concerned. The directivity of the individual loudspeakers is superimposed on that of the equivalent line array of omnidirectional sources.

A simple, very low-cost directional sound radiator can be constructed from a length of straight tube that has either a line of uniformly spaced holes along its length, or a slit along a generator line, and which incorporates a loudspeaker at one end. A sound-absorbing plug is fitted in the other end to reduce multiple reflections within the tube and the associated acoustic resonances (Holland and Fahy 1991).

8.5.5 Microphone arrays

Noise sources are infinitely diverse in physical form, temporal character, frequency range and radiation directivity pattern. In the practice of noise control, it is imperative to be able to locate, identify, characterise and quantify the sources that are responsible for generating unacceptable noise levels. All these features can, in principle, be determined by sampling the sound field radiated by a source. However, in many cases there are considerable challenges to be overcome. Among these are the following: physical extension of large complex sources; complexity and degree of statistical dependence of source mechanisms; source motion (as with vehicles); temporal variation of operating conditions; interference from other noise sources; reverberant measurement conditions; air movement; harsh environments; and inaccessibility for reasons of source geometry or operational constraints. Some of these problems can be overcome by simultaneously sampling the radiated sound field at many different points using a spatial array of electret microphones, often, but not always, in close proximity to the source. This section presents a brief overview of acoustic array systems and signal processing procedures. Those readers who wish to learn more about the theoretical basis of the technology are referred to Nelson (2004).

There are two generic forms of array, source modelling assumptions and associated signal processing procedures; they are categorised as 'beamformer' and 'inverse' techniques. In the application of the most basic beamformer, which takes the form of a linear array of uniformly spaced microphones, it is assumed that there is either a single, spatially concentrated source, or two or more such well-separated sources that are mutually uncorrelated (statistically independent) and located at a distance much greater than the length of the array. The line beamformer is most effective for locating outdoor sources that cause environmental noise at considerable distances, such as the components of industrial plant (Boone and Berkhout 1984; Boone et al. 2000). When a plane sound wave approaches the array from a certain angle, the times of arrival of a wavefront at any neighbouring pair of microphones will differ by an amount depending on the speed of sound, the angle of incidence and the distance between adjacent microphones. The delay between arrivals at next-but-one neighbours will be twice the pair value, and so on. A simple summation of all the microphone outputs will favour plane waves arriving from the direction normal to the array line, irrespective of the time history of the incident wave. Hence, the presence of one or more well-separated and statistically independent (uncorrelated) sources located in the far field (so that arriving waves are effectively plane) will be detected by one or more maxima in the output of the array as its angular orientation is swept (or steered) physically through space. However, physical sweeping is generally not practicable for audio-frequency applications. Fortunately, a 'beam' can be virtually steered by delaying each microphone signal before signal summation by an amount of time proportional to its distance from one or other end of the array. In the frequency domain, delay implies phase shift. The maximum useful delay increment per microphone pair is equal to the pair separation distance divided by the speed of sound when the array 'looks' along its own length in the so-called end-fire mode, in which it behaves rather like a shotgun microphone (Section 8.5.4).

The simple line-array beamformer has limitations of angular ambiguity and directional sharpness. For acoustic wavelengths less than twice the microphone separation, spatial aliasing occurs in which more than one angular direction of maximum output occurs for a single physical source location. This ambiguity problem can be overcome by arranging the microphones in a sparse array with a range of different microphone spacings that can be combined to cover a large frequency range (Boone and Berkhout 1984; Boone et al. 2000). The sharpness of the principal directional lobe which is used to locate sources increases as the ratio of the array length to the acoustic wavelength increases. For example, when the ratio is 2, the main lobe lies within $\pm 30°$ from the normal; and when it is 5, it lies within $\pm 12°$. There is effectively no directional discrimination if the array is shorter than a wavelength. The simple 'delay and sum' beamformer does not discriminate between sound arriving from the forward and rear azimuthal arcs. This limits its effectiveness in noisy and reverberant environments.

A cruciform array comprising two orthogonally oriented line arrays in a T-configuration has been used to locate distant discrete noise sources in two dimensions, as described in Boone and Berkhout (1984) and Boone et al. (2000). There are many geometric arrangements of microphones in two-dimensional beamforming arrays; examples include + shape (Brühl and Röder 2000), X shape (Barsikow 1996), star shape (Mellet et al. 2006) or spiral (Nordborg et al. 2000). A commonly used spiral array is shown in Figure 8.4; see also Hald (2004). Figure 8.5 shows the results of imaging a set of four loudspeakers 0.5 m apart in an anechoic environment with a spiral beamformer. The resolution clearly improves as the frequency is increased.

The simple delay and sum array is appropriate for detecting compact sources operating at such large distances that the acoustic wavefronts arriving at an array are effectively plane. For the detection of compact sources close to an array, the signal processing is a little more complicated than the basic delay and sum because the acoustic wavefronts generated by a 'target' source element are spherical and the wave amplitude varies as the inverse of the distance between the source element and each microphone. Therefore, relative gains appropriate to the distance of each microphone from a selected source element and relative time delays (or equivalent phase shifts) associated with wavefront curvature are applied to the microphone signals. This constitutes the most basic case of a 'focused beamformer'. In order to scan a source region, the focal point can be moved in three dimensions by means of appropriate signal processing based on simple geometric considerations (Elias 1995). Beamformers are useful for scanning a region to detect the presence of discrete compact sources but have a limited ability to characterise the spatial distribution of the strength and statistical properties of spatially extended sources.

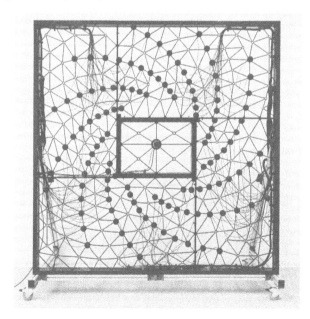

Figure 8.4 Example of a spiral microphone array. (Courtesy of gfaitech GmbH, Berlin, Germany; www.acoustic-camera.com.)

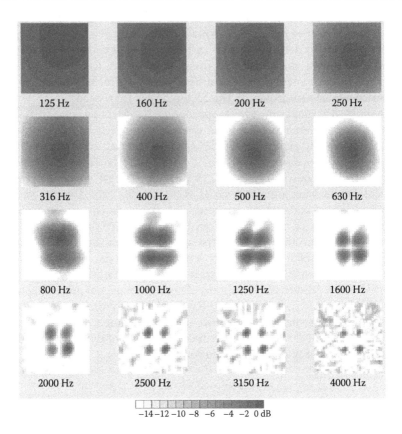

Figure 8.5 A set of four loudspeakers imaged by a beamforming spiral array at various frequencies. (Courtesy of K.R. Holland, University of Southampton.)

Acoustic 'images' (or 'sound maps') of source regions and source strength distributions derived from the signals of beamforming microphone arrays can be overlaid on video images of the physical system(s) generating the sound, together with any reflections of that sound from neighbouring passive objects that reach the microphone array. This combination provides assistance in the identification of active sources and with the selection of appropriate noise control measures. The technique is best suited to surveying sources operating under outdoor or anechoic conditions which minimise extraneous reflections and background sources.

The principal alternative technique to beamforming for surveying sound fields radiated by sources to characterise and quantify them is the 'inverse' technique (Holland and Nelson 2012). It has the advantage over beamforming that it can provide estimates of the spatial distributions of the source strength of spatially extended sound radiators, such as large vibrating machinery, and of systems of multiple correlated sources. But, it has the disadvantage that sophisticated signal processing techniques must be employed if the results are to be reliable. The inverse technique can be applied to cases in which the location(s) of individual sources are known, but their individual strengths are not. Alternatively, a spatially extended source region can be notionally divided (discretised) into an array of individual monopole sources. An array of microphones (which can take various geometric forms) is installed at some distance from the source region. In principle, the number of microphones should equal the number of sources (or assumed sources), although procedures exist to solve underdetermined cases in which the number of sources exceeds the number of microphones. If the source system is stationary in time, it is possible to reduce the equipment cost by selecting a master (or reference) microphone position and moving another microphone sequentially to form a virtual array.

The crucial element that most influences the effectiveness of the inverse method is the theoretical specification (or experimental determination) of the transfer function between each source (normally assumed to be a monopole of unit strength) and the sound pressure at each microphone position. These will depend on the relative locations of the sources and microphones, together with the geometrical and acoustical properties of the local environment (e.g. free field and reverberant). Thus, a matrix equation can be written, each line of which expresses the sound pressure at any one microphone, p_i, as the sum of the products of all the unknown source strengths q_j, and the transfer functions appropriate to that microphone G_{ij}:

$$\begin{bmatrix} G_{11} & & G_{1n} \\ & \ddots & \\ G_{n1} & & G_{nn} \end{bmatrix} \begin{Bmatrix} q_1 \\ \vdots \\ q_n \end{Bmatrix} = \begin{Bmatrix} p_1 \\ \vdots \\ p_n \end{Bmatrix} \qquad (8.1)$$

The source strengths can be estimated by *inverting* the transfer function matrix and multiplying it by the column matrix of microphone signals; hence the term *inverse* technique. The accuracy of the matrix inversion is especially sensitive to the presence of small values of some of the transfer functions, and of differences between some pairs, which become dominant through the process of inversion. There are many different ways of improving the reliability of the source strength estimates, which are too specialised for presentation here (Holland and Nelson 2012).

Most arrays are designed to sample only that component of the sound fields of sources that transport sound energy away from the sources. However, this means that they cannot 'image' features of the source structure that are less than an acoustic wavelength in spatial extent. There is another generic form of radiated sound measurement in which an array of microphones (sometimes in a geometric form that is conformal with the targeted radiator) is located so close to a source surface that the near field of the radiator, as well as its power-radiating component, is also sampled. This allows features of the radiating surface that are smaller than an acoustic wavelength to be imaged. This technique is known as *nearfield acoustic holography* (NAH). Its principle is too complex to be included in this chapter on fundamentals and the required instrumentation is rather expensive. Details of this and similar techniques are clearly presented in the book *Fourier Acoustics* (Williams 1999).

8.6 SOUND LEVEL METERS AND THEIR USE

8.6.1 Sound level meter

The SLM is the primary instrument for measuring acoustic noise in the environment. The simple handheld meter consists of a number of elements which together give a readout in decibels which can be interpreted and compared with recognised standards, regulations or criteria, as described in Chapter 6. The main elements of an archetypal SLM are shown in Figure 8.6. A physical example is shown in Figure 8.7. The standardised performances of SLMs are graded according to type numbers (IEC 2013).

Normally, a half-inch microphone would be mounted together with its preamplifier at the top of the meter. In some cases there is provision for an extension cable which enables the microphone to be placed remotely from the body of the meter, thereby minimising the scattering effect of the body. The analogue signal which comes from the preamplifier is usually digitised immediately at a sample rate that excludes signal aliasing (see Chapter 4) and all the subsequent

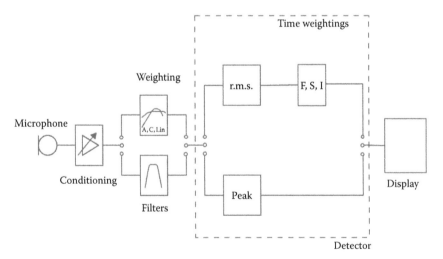

Figure 8.6 Elements of a sound level meter. F: fast; S: slow; I: impulse. (Courtesy of AcSoft Ltd, Aylesbury, UK; www.acsoft.co.uk.)

Measurement of audio-frequency sound in air 427

Figure 8.7 Example of a modern digital sound level meter. (Courtesy of Svantek UK Ltd., Aylesbury, UK; www.svantek.co.uk.)

functions of the meter are implemented in the digital signal processor (DSP). In order for the full dynamic range of the signal to be acquired, a true 24-bit analogue-to-digital (A/D) converter is necessary. A 24-bit conversion enables analysis over a 130 dB range, which in practice is sufficient for most purposes.

Once the signal is in digital form the whole of the rest of the processes take place in a DSP. However, it is convenient to divide the processes into blocks as shown in Figure 8.6 and discuss them individually. The signal time series output of the A/D converter can be recorded directly for further analysis. It is important that the recording medium uses lossless encoding; otherwise reanalysis of the signal is not possible. The live or recorded signal is then passed directly into the filter. The filter block implements a number of broadband or narrowband filters. Initially, the overall weighting filters are implemented as illustrated in Figure 6.6. The output of these filters is a time series and, in order to obtain r.m.s. levels, some form of squaring circuit and subsequent averaging* is required. For peak levels there is no squaring or averaging process; therefore, a separate detector is required.

Although the 'raw' decibel is the primary unit of measurement of noise, SLMs can compute an increasingly sophisticated set of derived metrics that can be used to express the level of a particular noise (see Chapter 6). In the memory and storage blocks the noise levels can be analysed to obtain statistical parameters. The output of the detector is sampled regularly and a statistical picture of the noise signal is formed in a cumulative distribution. The decibel scale is divided into bins, say every 2 dB, and the number of times the level occurs in each bin is counted as a percentage of the total. From this information a level distribution can be obtained which can indicate the level exceeded for a percentage of the time. Some noise sources, such as ventilation fans, produce noise at a steady level; others, such as vehicular noise, vary with time. It is therefore crucial to record the metric being used to describe a noise level.

In the filter block it is also possible to divide signals into frequency bands. One-third octave bands (see Section 5.2.2) are frequently used to analyse and describe acoustic data since they result in relatively few numbers and are related indirectly to the frequency discrimination of the human ear. Octave band filters are rarely used these days. The octave and one-third octave band series are constant percentage band filters in which the bandwidth is a fixed percentage of the centre frequency of the band: the octave bandwidth is 71% of the centre frequency and the one-third octave bandwidth is 23% of the centre frequency (see Table 5.1). The bandwidth of the filter increases as the frequency increases but the output of each filter is displayed in the same width on a graph using a logarithmic frequency scale. (Readers should carefully note that if a signal has a uniform distribution of 'energy' over a wide frequency range [e.g. pseudo white noise], the level of the one-third octave band centred on 1000 Hz will be 10 dB greater than that of the one-third octave band centred on 100 Hz.) It is necessary, therefore, to use a parallel process at this stage and, as indicated in Figure 8.6, it is possible to have many parallel processes. Since the output of the one-third octave filters is still a time series, it is possible to calculate any of the

---

* Typically, averaging using an exponential time weighting is used, given by

$$L_{p\tau}(t) = 10\log_{10}\left(\frac{1}{\tau}\int_0^\infty \frac{p^2(t-u)}{p_{ref}^2} e^{-u/\tau} du\right)$$

where:
$p(t)$ is the acoustic pressure at time t, after applying the A-weighting or similar as required
$L_{p\tau}(t)$ is the weighted sound pressure level at time t for a time constant τ

The time constant τ is set to 125 ms for fast (F) weighting and to 1 s for slow (S) weighting.

parameters mentioned earlier for each one-third octave band in real time. It is also possible to store the numbers and perform a Fast Fourier Transform (FFT). The FFT is a constant bandwidth filter, and the output is therefore usually displayed on a linear frequency scale. The FFT process means that results are only available for a block of numbers, and averaging has taken place within the transform. It is a very useful technique when it is necessary to try to identify a single constant frequency among other background noise. Figure 8.8 presents spectra of the same signal computed in very narrow, constant-width bands by FFT and in one-third octave bands. Note how the level of the one-third octave spectrum rises with frequency relative

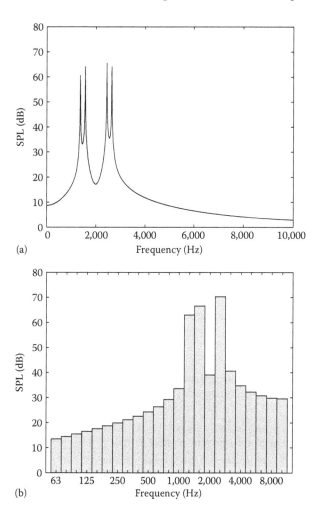

Figure 8.8 Narrowband (FFT) spectrum and one-third octave spectrum of the same signal.

to that of the narrowband filter. Moreover, the first two peaks appear in separate one-third octave bands but the two higher frequency peaks are too close together to be resolved in the one-third octave spectrum.

8.6.2 Sound level meter usage

One of the important aspects of SLMs is that their performance, calibration and use are highly regulated. This is vital for a number of reasons. First, that measurements made in a standardised fashion can be repeated by others. Second, that the conditions under which the measurements were made can be compared against the objective standards which depend on certain conditions being met. Third, a regime of regular calibration and performance checking ensures that accurate measurements are made at all times. This regulation takes the form of international standards for the construction of SLMs (IEC 2013) and also guides to the use of SLMs for a range of other activities. Measurements made in the environment are often used to justify expensive building works or other planning activities, and consistent and trustworthy sets of measurements, which can be objectively compared with sets made by other operators, are crucial to this process.

A number of procedures are necessary to conform to relevant codes of practice. The first which has been mentioned already is the regular use of a calibrator. Each and every series of measurements made in the field should commence with a calibration check and end with a calibration check, both of which should be recorded. Many modern SLMs store their calibration history, which can be used to validate measurements. In the discussion on microphones, it was clear that the microphone itself has an effect on the sound field in which it is placed. The same applies to the body of the SLM, and indeed to the body of the operator, both of which generate much larger reflections than the microphone itself. The influence of the presence of an operator can be removed by the use of a mounting tripod.

There are a number of environmental factors which can affect measurements made outdoors with SLMs. The principal problem is wind, which can generate unsteady (non-acoustic) pressure fluctuations on a microphone diaphragm in two ways. The oncoming wind may itself be unsteady and turbulent, and turbulence may be generated in the boundary layer flow over the surface of the microphone and housing. This problem is normally solved by the use of a windshield placed over the microphone (see Figure 8.7). The most common windshield consists of a ball of open-cell foam that is considerably larger than the microphone and which is easily penetrated by an incident sound wave but not by a turbulent pressure flow, which the windshield displaces from the immediate vicinity of the microphone. When measurements are made outdoors in adverse weather conditions, it is also necessary to use some form of acoustically transparent rain shield on the microphone; and, typically, a dehumidifier is mounted behind a microphone with a rear vent. This prevents condensation from forming in the cavity of the microphone itself.

Some microphones have to be left outside for considerable lengths of time, if not permanently. This is necessary in noise monitoring locations such as airports and outdoor entertainment venues. These microphones have built-in remote calibrators which can be actuated electronically at fixed intervals. Protection from the wind and rain is provided by superior microphone grids and windscreens. It is also usually necessary to fit a protective cage, or system of spikes, which prevents birds from perching on top with the inevitable precipitation.

8.6.3 Computer-based sound level meters

All the signal processing, spectral analysis, weighting, time averaging, peak detection and computation of metrics performed by current SLMs can, in principle, be performed by a portable computer, fed by the microphone signal digitised by an A/D converter. The latter equipment requirement will eventually be superseded by MEMS microphones that output pulse code modulation (PCM) data directly into the 'SLM' software. However, not all such systems meet the requirement of international standards for sound measurement.

8.7 SOUND INTENSITY

8.7.1 Introduction

Measurements made with an SLM are measurements of sound pressure. Pressure is a scalar quantity having no direction, and single sensor microphones are almost uniformly directional in their sensitivity over a wide frequency range. This is appropriate when measuring noise in relation to human response since the ear responds directly to local pressure. However, in the practice of noise control it is necessary to identify principal noise sources and their locations, and to determine their strengths, temporal and frequency characteristics, and physical mechanisms of noise generation. In many cases, the source(s) being investigated operate in the presence of other sources which may not be stopped (as in a factory), the sound of which is superimposed on that of the target source, as is the sound reflected from local obstacles and enclosure surfaces. These interfering sound fields can substantially 'corrupt' the sound pressure field directly radiated by a source and degrade the identification process. As explained in Section 8.5, there are methods of locating and quantifying sound sources by using multi-microphone arrays that sample the radiated pressure fields at an array of points. However, the implementation of these methods is quite expensive, complicated and computationally demanding.

Sound sources generate acoustic energy, which is transported by the sound waves that they radiate; the rate at which a source generates acoustic energy is termed its *sound power* (see Chapter 2). Sound pressures in the radiated field of a source operating in a largely unconfined space (free field) are highly dependent on the distance from the source and the directivity of its radiation field. Sound *pressures* generated by a source in an enclosed (or reflective) space depend on the volume of the space and the sound-reflective and sound-absorbent properties of its boundaries. In contrast, the *sound power* of many noise sources (except tonal varieties) is rather independent of their locations and environments. If the sound power of a source is known, the noise it generates can be fairly accurately estimated on the basis of rather simple specifications of its environment. Hence, source sound power is a very important quantity in noise control practice; the methods of its experimental determination are specified in a number of international standards (see Section 5.3).

When trying to identify the locations and mechanisms of sources of noise, and to quantify their outputs, it is extremely helpful to be able to examine the directions and quantities of energy flow within their radiated sound fields. Sound waves transport energy in two forms: kinetic energy of particle velocity and potential energy of density/pressure variation. The instantaneous flux (rate of flow) of sound energy passing through a unit area of space is defined as the *sound intensity*. (Note: the term *intensity* is also widely used in audiological literature to indicate SPL.) Sound intensity is a vector quantity equal to the *instantaneous* product of the pressure and particle velocity in a sound wave; it has the direction of the particle velocity vector (Fahy 1995). Thus

$$\vec{I} = p.\vec{u} \tag{8.2}$$

where the overarrow indicates a vector quantity.

Except in cases of transient noise sources such as drop hammers (Fahy and Elliott 1980), most intensity measurements are made on time-stationary sound fields. Therefore, it is common to define sound intensity as the time average of the instantaneous sound intensity vector, which in a steady sound field is steady in magnitude and direction.

Sound intensity has units of watts per square metre. It is expressed in decibels as

$$L_I = 10\log_{10}\left(\frac{I}{I_0}\right)(\pm) \text{ dB} \tag{8.3}$$

where:
- L_I is the sound intensity level
- I is the measured intensity
- I_0 is the reference intensity: $I_0 = 10^{-12}$ W m^{-2}
- $\pm$ indicates the direction of energy flow in relation to a specified surface

8.7.2 Sound intensity measurement probes

8.7.2.1 Two-microphone sound intensity probe

Measuring the instantaneous pressure at a 'point' in a sound field is straightforward with a pressure microphone. The difficult problem with intensity measurement is how to measure the associated instantaneous particle velocity at the same 'point'. A number of different transducer types and arrangements (probes) have been tried over the past 60 years. Since 1980, when the first reliable commercial intensity measurement system became available, a probe based on the linearisation of Euler's equation (see Section 2.2.4), which relates the particle velocity at a point in a sound field to the associated spatial gradient of pressure, has been the most widely accepted and applied form. The particle velocity is thus

$$\vec{u} = -\frac{1}{\rho_0} \int (\vec{\nabla} p) \mathrm{d}t \tag{8.4}$$

where ρ_0 denotes the mean fluid density and ∇ denotes the spatial gradient.

An approximation to the spatial gradient of sound pressure can be made by measuring the difference between the pressures at two closely adjacent points and dividing by the separation distance. (Note carefully that the sound intensity cannot strictly be computed from either the r.m.s. or the mean-square values of pressure or particle velocity; although in free field, far away from a source, the sound intensity is closely proportional to the mean-square sound pressure. But in most practical cases this approximation is invalid.) If two microphones are placed close together at positions 1 and 2, a distance Δr apart, the particle velocity can be approximated by

$$u \approx \frac{1}{\rho_0 \Delta r} \int_{-\infty}^{t} (p_1(\tau) - p_2(\tau)) \mathrm{d}\tau \tag{8.5}$$

The distance between the points must be small compared with a wavelength, but not so small that the pressure difference is too small to be accurately measured. The pressure is approximated by the average of the measured pressures. Hence

$$p \approx \frac{p_1 + p_2}{2} \quad (8.6)$$

The instantaneous intensity can therefore be expressed as

$$I(t) \approx \frac{p_1(t) + p_2(t)}{2\rho_0 \Delta r} \int_{-\infty}^{t} (p_1(\tau) - p_2(\tau)) d\tau \quad (8.7)$$

In the frequency domain, the corresponding spectral equation for the time-averaged intensity of time-stationary sound fields is

$$I(\omega) \approx \frac{-1}{\rho_0 \omega \Delta r} \operatorname{Im}\{G_{p_1 p_2}(\omega)\} \quad (8.8)$$

where G is the single-sided cross-spectral density function (see Section 4.4.5.9). This form is universally implemented in computer software. The international standard for instruments measuring sound intensity is IEC (1993). It is also possible to evaluate the sound intensity fields of transient noise sources by applying Equation 8.7 (Fahy and Elliott 1980).

A typical two-microphone intensity probe consists of two, nominally identical, pressure microphones placed face to face, with a solid coaxial spacer in between them. This arrangement is termed a *p–p probe*. The grids remain on the microphones, thereby allowing the sound to access the microphone diaphragms, and the spacer defines precisely the spacing Δr, as shown in Figure 8.9. An example of a two-microphone intensity probe is shown in Figure 8.10.

The frequency range is limited by the microphone sensitivity and the electrical noise floor of the microphones and preamplifiers, but more fundamentally by the distance between the microphone diaphragms. At high

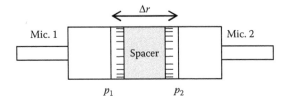

Figure 8.9 Arrangement of face-to-face condenser microphones in a p–p sound intensity probe.

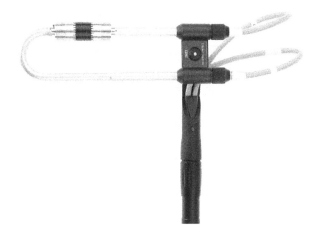

Figure 8.10 Example of a commercial p–p sound intensity probe. (Courtesy of Brüel & Kjær, Nærum, Denmark. Copyright © 2012.)

frequency, the wavelength becomes too short for the approximate evaluation of the particle velocity from the difference between the two pressure signals to be valid. At low frequency, the difference between the two pressure signals becomes too small and the derived particle velocity signal is affected by the system noise. A range of spacers of different sizes can be used to set the frequency range. A modern half-inch p-p probe has a useful range of 100 Hz to 10 kHz.

The dynamic range of a probe can be established by placing the sensing elements of the microphones close together in a small cavity and exposing them to broadband acoustic excitation which is very closely spatially uniform within the operational frequency range of the probe. Because the pressure field is uniform, the air undergoes zero particle acceleration and velocity and the actual intensity is zero. Because no microphones are perfectly phase matched, they register a finite pressure difference, and hence indicate a false particle velocity. The apparent intensity measured by the instrument is called the *residual intensity*. The difference between the measured SPL and the indicated residual sound intensity level is called the *Pressure-Residual Intensity Index* (PRI). It varies with frequency and is expressed in decibels. This indicates the dynamic range of the instrument. Typically, the PRI is smallest at low frequencies and rarely more than 20 dB.

In the practical application of sound intensity measurement it is necessary to ensure that the difference between the indicated SPL and the indicated sound intensity level, L_p–L_I, defined as the *pressure–intensity index* (PI), does not approach the PRI of the measurement system in the measurement band. The difference (PRI − PI) should exceed 7 dB for the phase mismatch measurement error to be less than ±1 dB. It should be clearly understood that the PI is a function not only of the form of the sound field being measured but also a function of the orientation of the probe in the sound field. If the direction of the axis of a probe is varied in a steady sound field, the SPL will change very little, but the indicated sound intensity will vary with axial direction; therefore, the PI and the associated measurement accuracy will also vary.

Conditions producing high PI levels include highly reverberant measurement environments, high levels of background noise, measurements of intensity normal to high impedance surfaces and attempts to measure the sound energy flow into surfaces having a low sound absorption coefficient.

Although a conventional p-p intensity probe is, in principle, capable of providing both pressure and particle velocity information, modern systems, unlike earlier analogue equipment, do not provide direct access to the pressure sum and pressure difference signals. This is a pity, since otherwise a p-p intensity probe could be used to estimate directly the specific acoustic impedances of surfaces in the laboratory and in the field.

8.7.2.2 Microflown

Other intensity probes that employ direct methods of detecting acoustic particle velocity have been constructed. A system based on the ultrasonic measurement of particle velocity did not achieve commercial success. The only other widely used form of sound intensity measurement system was invented in the Netherlands in the mid-1990s. It is known as the 'Microflown' (available at www.microflown.com). It incorporates two very thin resistance wires that lie close and parallel to each other and are heated electrically to over 200°C (Figure 8.11). The wires are suspended across a gap in the

Figure 8.11 Heated sensing wire pair in a Microflown. (Courtesy of Microflown Technologies, Arnhem, the Netherlands; www.microflown.com.)

supporting structure through which sound can travel transversely across them. They are mounted in a gap between two parallel circular section bodies that amplifies the particle velocity of a sound wave passing across them.

For a single heated wire in quiescent air, the surrounding air has a symmetric temperature distribution normal to the wire axis in the surrounding air which peaks sharply at the location of the wire. When the wire is exposed to airflow, the temperature distribution changes in the air normal to the wire, and convection causes it to fall on the upstream side and to rise on the downstream side. However, the temperature of the wire remains unchanged and the electrical resistance of the wire does not change. When two closely spaced wires are heated, the temperature distribution across the plane occupied by the wires is the sum of the temperature distributions around the individual wires, and the temperature peak on the downstream wire exceeds that on the upstream wire. The resulting difference between the electrical resistances of the two wires is proportional to the particle velocity across the wires, the direction being indicated by the sign of the difference. The sensor is linear over a wide range of acoustic particle velocities from zero up to a level of 135 dB (re 5×10^{-8} m s^{-1}). The particle velocity sensor does not have a flat frequency response over the audio-frequency range. The magnitude and phase responses are compensated through signal conditioning by digital filters.

In its intensity measurement configuration the Microflown incorporates a small pressure microphone which shares the same air channel as the heated wires. The product of the two signals yields the component of sound intensity in the direction of measurement axis.

The capacity of the Microflown to sense particle velocity, together with its small size, allows it to be deployed very close to vibrating surfaces over which it can be scanned. Multiple units may be arranged in various forms of array for the purpose of identifying the location and strengths of sound sources. Because particle velocity is a vector quantity it is sensitive to the wave direction, and the p-u configuration provides a powerful combination for suppressing extraneous noise and reflections in the application of array technology to source location and quantification. The complex acoustic impedance of material surfaces may be determined by evaluating the frequency-dependent transfer function between the particle velocity normal to the surface and the local sound pressure. The Microflown particle velocity signal is, by the nature of the sensor, vulnerable to corruption by unsteady airflow such as wind turbulence and mechanically driven airflow. Windscreens are commercially available and research is continuing to optimise their design. Microflowns that can sense three vector components of acoustic particle velocity and of sound intensity are also commercially available. A comparison of intensity measurements using the two-microphone and the Microflown systems is presented by Jacobsen and de Bree (2005).

8.8 APPLICATIONS OF SOUND INTENSITY MEASUREMENT

8.8.1 Determination of source sound power

The principal advantage afforded by the ability to measure sound intensity directly is that the sound powers of individual sound sources (or individual components of extended sound sources) operating under conditions of high background noise generated by other sources (or components) may be determined. Theory demonstrates that if a virtual *closed* surface is described around any region of a *time-stationary* sound field, the spatial integral over the surface of the component of sound intensity directed outward and *normal* to the surface is determined solely by the sound power generated within enclosed volume (Fahy 1995). This equality is true only if both the enclosed sources and the sources outside the circumscribed volume that substantially affect the sound intensity field on the measurement surface are themselves time stationary. It is also necessary that there are no large areas of sound-absorbent materials within the measurement surface because some of the source power would be absorbed before it reaches the surface and some of the radiation of sources outside the surface will be absorbed.

In practice, it is impossible to effect a complete surface integral, but it may be substituted by either an array of discrete point measurements or continuous sweeps over a set of subareas. Methods of determination of source sound power that ensure various grades of quality are specified in a series of international standards (ISO 1993, 1996, 2002b). Sound intensity measurement may be used as a general diagnostic method to evaluate the distribution of sound powers of discrete components of extended sound sources such as industrial machinery as an aid to the specification of noise control procedures (Fahy 1995).

8.8.2 Determination of sound transmission through partitions

Sound intensity measurement is used extensively for *in situ* evaluation of the sound powers radiated by the various surfaces of rooms in buildings in response to noise sources operating elsewhere within the same building (flanking transmission). It may also be used for the evaluation of vehicular noise transmitted through building facades. It is possible to determine the sound transmission loss of partitions using only one reverberation chamber. The test sample is installed in one wall of the chamber. The sound power incident on one side of a partition is estimated indirectly from the space-average mean-square pressure in the source room, as in the conventional test procedure. The transmitted sound power is estimated directly by intensity scanning over the other surface of the partition using the standardised procedure for source sound power (ISO 2000, 2002a, 2003). This general procedure can be extended to the *in situ* estimation of sound transmission through many forms of structure including the windows and roofs of buildings and traffic noise barriers.

8.8.3 Leak detection

One of the most commonly encountered noise transmission problems with doors and windows of buildings and passenger vehicles is that of sound leaks. Sound intensity scans over suspect regions can be used to estimate the relative amounts of sound power transmitted through the partition structure and associated leak paths in order to establish whether the latter are significant. Leaks through thin gaps tend to be significant in the frequency range above about 2 kHz in which the transmission loss of the associated structures would normally be high, thereby short-circuiting the structure and requiring remedial treatment.

An effective, but only qualitative, alternative leak detection technique involves a flexible plastic tube of about 8 mm in diameter, one end of which is inserted and sealed into a hole drilled in one cup of a pair of ear defenders which are then worn by the investigator. The investigator then moves the end of the tube around the suspect regions and easily picks up local leaks. Ultrasonic sources and detectors are among the tools used in the automobile industry to detect faulty seals.

8.8.4 Measurement of sound absorption

Sound intensity probes can also be used to make *in situ* measurements of the rate of sound energy absorption (absorbed sound power) by material surfaces when exposed to incident sound, provided the random incidence sound absorption coefficient exceeds about 0.3; below that value, the error of the estimate can be unacceptable. Applications include measurements of sound absorption by room surfaces, by furnishings such as curtains and concert hall seating, and by room contents such as members of the audience.

REFERENCES

Barsikow, B. (1996) Experiences with various configurations of microphone arrays used to locate sound sources on railway trains operated by the DB AG, *Journal of Sound and Vibration* **193**, 283–293.

Boone, M. M. and Berkhout, A. J. (1984) Theory and applications of a high resolution synthetic acoustic antenna for industrial noise measurements, *Noise Control Engineering Journal* **23**, 60–68.

Boone, M. M., Kinneging, N., and van der Dool, T. (2000) Two-dimensional noise source imaging with a T-shaped microphone cross array, *Journal of the Acoustical Society of America* **108**(6), 2884–2890.

Brühl, S. and Röder, A. (2000) Acoustic noise source modelling based on microphone array measurements, *Journal of Sound and Vibration* **231**, 611–617.

BSI (British Standards Institution) (2005) *PD 6545:2005: Intended Use of Human Head and Ear Simulators: Guide*. BSI Standards Institution, London.

Elias, G. (1995) Source localization with a two-dimensional focused array: Optimum signal processing for a cross-shaped array, *Proceedings of Inter-Noise 95, International Congress on Noise Control Engineering*, Newport Beach, CA, 10–12 July, pp. 1175–1178.

Fahy, F. J. (1995) *Sound Intensity*, Spon Press, London.

Fahy, F. J. and Elliott, S. J. (1980) Acoustic intensity measurement of transient noise sources, *Journal of the Acoustical Society of America* **67**, 569.

Hald, J. (2004) Combined NAH and beamforming using the same microphone array, *Sound and Vibration*, 18–27.

Holland, K. and Fahy, F. (1991) A low-cost end-fire acoustic radiator, *Journal of the Audio Engineering Society* **39**(7/8), 540–550.

Holland, K. R. and Nelson, P. A. (2012) An experimental comparison of the focused beamformer and the inverse method for the characterisation of acoustic sources in ideal and non-ideal acoustic environments, *Journal of Sound and Vibration* **331**, 4425–4437.

IEC (International Electrotechnical Committee) (1993) *IEC 1043: Electroacoustics: Instruments for the Measurement of Sound Intensity: Measurements with Pairs of Pressure Sensing Microphones*, International Electrotechnical Committee, Geneva, Switzerland.

ISO (International Organization for Standardization) (1993) *ISO 9614-1:1993: Acoustics: Determination of the Sound Power Levels of Noise Sources Using Sound Intensity. Part 1: Measurement at Discrete Points*, International Organization for Standardization, Geneva, Switzerland.

ISO (International Organization for Standardization) (1996) *ISO 9614-2:1993: Acoustics: Determination of the Sound Power Levels of Noise Sources Using Sound Intensity. Part 2: Measurement by Scanning*, International Organization for Standardization, Geneva, Switzerland.

ISO (International Organization for Standardization) (2000) *ISO 15186-1:2000: Acoustics: Measurement of Sound Insulation in Buildings and of Building Elements Using Sound Intensity. Part 1: Laboratory Measurements*, International Organization for Standardization, Geneva, Switzerland.

ISO (International Organization for Standardization) (2002a) *ISO 15186-3:2002: Acoustics: Measurement of Sound Insulation in Buildings and of Building Elements using Sound Intensity. Part 3: Laboratory Measurements at Low Frequencies*, International Organization for Standardization, Geneva, Switzerland.

ISO (International Organization for Standardization) (2002b) *ISO 9614-3:2002: Acoustics: Determination of the Sound Power Levels of Noise Sources Using Sound Intensity. Part 3: Precision Method for Measurement by Scanning*, International Organization for Standardization, Geneva, Switzerland.

ISO (International Organization for Standardization) (2003) *ISO 15186-2:2003: Acoustics: Measurement of Sound Insulation in Buildings and of Building Elements Using Sound Intensity. Part 2: Field Measurements*, International Organization for Standardization, Geneva, Switzerland.

Jacobsen, F. and de Bree, H-E. (2005) A comparison of two different measurements of sound intensity, *Journal of the Acoustical Society of America* **118**(3), 1510–1517.

Kinsler, L. E., Frey, A. R., Coppens, A. B. and Sanders, J. V. (2000) *Fundamentals of Acoustics*, 4th edn., John Wiley & Sons, New York.

Mellet, C., Létourneaux, F., Poisson, F. and Talotte, C. (2006) High speed train noise emission: latest investigation of the aerodynamic/rolling noise contribution, *Journal of Sound and Vibration* **293**, 535–546.

Nelson, P. A. (2004) Source identification and location, in Fahy, F. and Walker, J. (eds), *Advanced Applications in Acoustics, Noise and Vibration*, chapter 3, pp. 100–153, Spon Press, London.

Nordborg, A., Wedemann, J. and Willenbrink, L. (2000) Optimum array microphone configuration. *Proceedings Inter-Noise 2000*, Nice, France.

Schulman, Y. (1987) Reducing off-axis comb-filter effects in highly directional microphones, *Journal of Audio Engineering Society* **35**(6), 463–469.

Williams, E. G. (1999) *Fourier Acoustics: Sound Radiation and Nearfield Acoustical Holography*, Academic Press, London.

Chapter 9

Vibration testing

Tim Waters

9.1 INTRODUCTION

The commercial pressure to reduce design costs and cycle times has driven dramatic progress in the virtual world of simulation. The field of structural dynamics is no exception and it is common nowadays, through computer-aided design (CAD) and subsequent finite element (FE) analysis, to obtain predictions of the forced response of complex, linear structures with relative ease. These models necessarily incorporate assumptions regarding boundary conditions and element types, as well as parameter estimates such as material properties, joint stiffnesses and damping factors. At various stages in a product's design and development, measurements may become necessary to provide partial experimental validation for a structural dynamic model such that it can be exercised with some degree of confidence to predict that which is unmeasured. For this purpose, frequency response functions (FRFs), which are transfer functions pertaining to the steady-state vibration of linear systems, are most commonly sought by experimental means. The FRF, which is more generally referred to as a transfer function by the experimental community, can be defined and named in a number of ways depending on the input and output quantities to which it relates, as listed in Table 9.1. Natural frequencies, mode shapes and modal damping values may be subsequently inferred through the application of experimental modal analysis techniques. An experimental modal model of a structure is also an invaluable aid *per se* in diagnosing noise and vibration problems of existing systems for which palliative vibration control solutions are required.

Vibration isolation is a key consideration in many engineering systems for reasons of structural integrity (satellite electronics), precision (optical devices) or comfort (road vehicles). Steady-state vibration isolation performance can be measured *in situ* via transmissibility functions, as defined in Table 9.1, which relate the responses of the source and receiving structure. The overall performance is dependent on the dynamic behaviour of the source, isolator and receiver, all of which can be characterised independently

Table 9.1 Definitions of frequency response functions and their relationships via differentiation in the frequency domain

Standard type of FRF	Inverse type of FRF
Receptance = $\dfrac{\text{displacement}}{\text{force}}$	Dynamic stiffness = $\dfrac{\text{force}}{\text{displacement}}$
$\times i\omega$ ↓	$\div i\omega$ ↓
Mobility = $\dfrac{\text{velocity}}{\text{force}}$	Mechanical impedance = $\dfrac{\text{force}}{\text{velocity}}$
$\times i\omega$ ↓	$\div i\omega$ ↓
Accelerance = $\dfrac{\text{acceleration}}{\text{force}}$	Apparent mass = $\dfrac{\text{force}}{\text{acceleration}}$
Transmissibility = $\dfrac{\text{displacement}}{\text{displacement}} = \dfrac{\text{velocity}}{\text{velocity}} = \dfrac{\text{acceleration}}{\text{acceleration}}$ or $\dfrac{\text{force}}{\text{force}}$	

Note: For multiple degrees of freedom systems, the inverse FRF is not simply the reciprocal value of the corresponding standard FRF.

by measured transfer functions. Isolators are characterised by standardised measurements of their dynamic stiffness (ISO 1999), whereas source and receiver mobilities can often be obtained from instrumented hammer testing, as described later in this chapter. In the specific case where the receiver is a human subject it is standard practice to measure the apparent mass (see Chapter 7).

Besides system characterisation, it is necessary to gain knowledge of the dynamic loads that are experienced during operation. Where these are unknown, one may pursue transfer path analysis (TPA) to infer them indirectly (Verheij 1997). On a frequency by frequency basis, TPA estimates a linear combination of forces such that their combined effect most closely replicates a set of measured operational vibration (or acoustic) responses. A matrix of FRFs, which may be obtained by experimental means, is required for this purpose.

There are, of course, many reasons for conducting vibration tests besides measuring FRFs. Examples include order analysis and condition monitoring of rotating machinery, output-only modal analysis, qualification testing of components for rugged environments, simulation of seismic disturbances, fatigue testing and satisfying health and safety legislation. Such applications are not explicitly addressed in this chapter which is limited in scope to FRF measurement. However, some discussion will be pertinent to vibration measurements in general.

This chapter begins in Section 9.2 by outlining the general set-up adopted for FRF vibration testing with particular reference to electrodynamic shaker and instrumented hammer techniques. Section 9.3 discusses force and acceleration sensors, their principles of operation, limitations and practical use. Section 9.4 considers vibration testing using an electrodynamic shaker. Particular emphasis is placed on the necessity and precautionary measures for preventing the shaker's dynamics from altering the measured frequency response of the structure under test, and the correct use of commonly available excitation signals. Section 9.5 covers the essential know-how to conduct an instrumented hammer test including frequency analyser settings such as triggering, windowing and averaging. Sections 9.4 and 9.5 both conclude with a case study to illustrate some aspects of good practice as well as common pitfalls in shaker and hammer testing.

9.2 GENERAL TEST SET-UP

The equipment set-up adopted for a vibration test depends most notably on the chosen means of actuation, of which there are many options, depending on the application (Ewins 2000). The measurement of FRFs in particular requires a measurable input over which some control can be exercised regarding excitation frequency and amplitude. The most commonly employed excitation devices are the electrodynamic shaker and the instrumented impact hammer, the general set-ups for which are outlined here and discussed in some detail in later sections.

Electrodynamic shaker testing is arguably the most versatile option wherever practically possible and, if the shaker size and excitation signal are appropriately chosen, enables a unidirectional force of almost arbitrary spectrum to be delivered at a single point. In the case of large built-up structures, such as aircraft, a number of electrodynamic shakers can be deployed simultaneously to ensure adequate excitation of the modes of vibration, in which case multi-input multi-output (MIMO) methods are required for FRF estimation rather than the single-input methods cited in this chapter; see, for example, Brandt (2011).

Alternatively, electrohydraulic shakers are widely used in applications where very large dynamic loads are required, such as testing civil engineering structures and automotive vehicles. Electrohydraulic shakers can apply static and dynamic loads simultaneously, which is useful where static loading alters the dynamic response of the structure. However, the useable frequency range is typically limited to 1 kHz at best due to oil compressibility effects (Huntley 1979).

Figure 9.1a shows a schematic of a typical configuration for an FRF test using an electrodynamic shaker. Key to proceedings is the *frequency analyser* or *spectrum analyser*. Most modern analysers take the form of

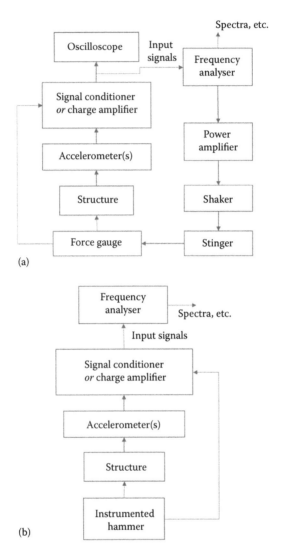

Figure 9.1 Schematic of the set-up for a single-input FRF vibration test using (a) an electrodynamic shaker and (b) an instrumented hammer.

hardware to acquire (i.e. digitise and store) the analogue signals from the sensors and software hosted by a personal computer (PC) to compute their discrete Fourier transforms (DFT) using the fast Fourier transform (FFT) algorithm (see Section 4.4.5) and to display derived quantities in real time in various formats. The analyser also normally provides the source signal which defines the frequency content of the excitation; alternatively, an external waveform generator is required. The source signal is amplified by a power amplifier in order to drive the shaker. The shaker is connected to the structure by a mechanical coupling device referred to here as a stinger, and the transmitted force at the surface of the structure is measured using a force gauge. The response is most commonly measured using one or more accelerometers which, depending on their type, require either a charge amplifier or a signal conditioner to function. Although expensive, laser velocimeters are becoming a more widely available alternative to accelerometers for densely spaced measurements or where noncontact sensing is paramount, such as for very light structures (Halliwell and Petzning 2004). Strain gauges can also be employed, principally for low-frequency applications. An oscilloscope is indispensable for viewing analogue signals, especially while setting up and troubleshooting.

The set-up for vibration testing using an instrumented hammer, shown in Figure 9.1b, is appealingly simple and eminently portable. The excitation is provided by the impulse resulting from a tap to the structure and the input force is measured by a force gauge immediately behind the hammer tip. Some degree of control can be exercised over the amplitude and bandwidth of the excitation, although the measurement duration and hence the frequency resolution is inherently limited by the decay time of the response. A defining feature of hammer testing is that, contrary to shaker testing, the excitation position can be varied even more readily than the response position. It becomes expedient to obtain a set of single-input multi-output (SIMO) transfer functions reciprocally by roving the hammer and siting the sensor at a fixed reference position. The principle of reciprocity holds only for linear systems and is stated formally in Section 3.6.3.

9.3 SENSORS

Vibration sensors are required to measure the mechanical input and output quantities related by the FRF of interest, as listed in Table 9.1. These quantities are only meaningful for the steady-state behaviour of linear systems, under which assumptions it is straightforward to integrate or differentiate between displacement, velocity and acceleration in the frequency domain. Consequently, it is normally sufficient to have instrumentation that is capable of measuring one of these three quantities and, of course, force. This section briefly examines the use of piezoelectric accelerometers and force gauges in respect of their principle of operation, some limiting characteristics, mounting issues, associated electronics and calibration.

9.3.1 Principle of force gauges

Gauges for measuring dynamic forces almost invariably operate on the piezoelectric effect. This is the ability of a material to generate an electric

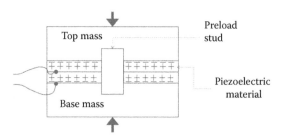

Figure 9.2 Schematic of a typical piezoelectric force gauge.

field in response to a mechanical strain. (By the inverse piezoelectric effect such materials can also generate mechanical strain by the application of a voltage and so be employed as actuators.) Quartz is a naturally occurring crystal that exhibits the piezoelectric effect and is widely used as the sensing element in force gauges owing to its stability with respect to time and temperature variations. Man-made ceramics such as lead zirconate titanate (PZT) are also widely used.

A simple schematic of a typical piezoelectric force gauge is shown in Figure 9.2 in which a pair of back-to-back quartz discs are held in compression between a top and a bottom mass by a preload stud. When a dynamic force is applied to the ends of the gauge the force is shared between the sensing element, the preload stud and the casing (not shown) in accordance with their relative stiffnesses. The component of force taken up by the sensing element causes it to experience a modulation in compressive strain. With each disc acting as a capacitor, a small electrical charge (in the order of a picocoulomb) accumulates on the electrodes in proportion to the applied strain. Knowledge of the sensitivity of the force gauge (in e.g. pC/N) enables calibrated measurements to be performed.

An impedance head is a sensor that incorporates a force gauge and an accelerometer arranged in series. The device is so called because the apparent mass transfer function between the two output signals can be readily estimated by a frequency analyser and integrated in the frequency domain to obtain the driving point mechanical impedance (see Table 9.1).

9.3.2 Principle of accelerometers

Accelerometers are available that operate on a number of different sensing principles such as piezoelectric, piezoresistive and variable capacitive effects. The most appropriate type depends on the particular application. For general transfer function measurements, which are the focus of this chapter, piezoelectric accelerometers are ideally suited on account of their wide bandwidth (typically a few hertz to a few tens of kilohertz).

There are various geometrical configurations in which the sensing element is subjected to compressive, shear or flexural deformation. They are also available in single axis or triaxis forms in order to measure one or three orthogonal components of motion. Figure 9.3a shows a schematic of a single axis compression-type accelerometer which is conceptually the simplest and also most similar in principle to a force gauge. A seismic mass m is mounted on a sensing element comprising a pair of back-to-back piezoelectric discs inside a casing, and is maintained in compression by a preload spring. When the base of the casing is subjected to dynamic motion, the sensing element applies a force to the seismic mass causing the latter to accelerate, the piezoelectric discs to deform and a proportional electrical output to be generated. The mechanical sensitivity of the accelerometer is governed by the relative displacement across the element per unit acceleration of the base. This FRF may be obtained by considering a time-harmonic acceleration of the base $Ae^{i\omega t}$ and the resulting relative displacement $X_r e^{i\omega t}$ of the equivalent single degree of freedom system shown in Figure 9.3b. Here, k is an effective stiffness attributable to the sensing element and the preload spring.

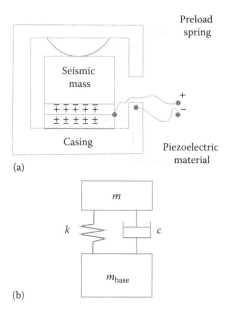

Figure 9.3 (a) Schematic and (b) dynamic model of a typical compression-type piezoelectric accelerometer.

An analysis similar to that outlined in Section 3.3.3 results in

$$\frac{X_r}{A} = \frac{1}{\omega_n^2} \cdot \frac{-1}{1-\left(\dfrac{\omega}{\omega_n}\right)^2 + 2i\zeta\dfrac{\omega}{\omega_n}} \qquad (9.1)$$

where:
- ω is the excitation frequency in radians per second
- ω_n is similarly the natural frequency of the seismic mass on the supporting sensing element when the accelerometer base is grounded (as is usually specified), that is

$$\omega_n = \sqrt{k/m} \qquad (9.2)$$

- ζ is the damping ratio

The FRF given in Equation 9.1 is shown for illustrative purposes in Figure 9.4 for two accelerometers with different natural frequencies. When the frequency of the base motion is much less than the natural frequency of the accelerometer, that is, $\omega/\omega_n \ll 1$, then the mechanical sensitivity is practically independent of frequency and

$$\left|\frac{X_r}{A}\right| \approx \frac{1}{\omega_n^2} \qquad (9.3)$$

Larger accelerometers tend to have a lower natural frequency and so are more sensitive in general but at the expense of a lower useable bandwidth and, of course, greater modification or 'mass loading' of the structure. As a rule of thumb, the bandwidth of use should not exceed 20% or so of the accelerometer's natural frequency in order to ensure an adequately flat frequency response. Of course, the overall sensitivity of an accelerometer depends also on its electrical properties, as discussed in Section 9.3.5.

9.3.3 Accelerometer mounting considerations

The previous section has illustrated that the frequency response of an accelerometer is flat only up to a point, beyond which the sensitivity is affected by the resonance of the seismic mass on the stiffness of the sensing element. For typical measurements at audio frequencies, this amounts to choosing an accelerometer with sufficient bandwidth. However, the analysis assumed

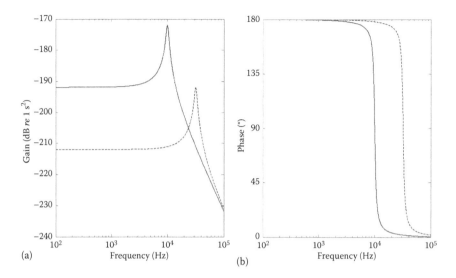

Figure 9.4 Frequency response function between relative displacement and base acceleration as given by Equation 9.1 for two different values of natural frequency: (a) magnitude and (b) phase.

that the accelerometer was rigidly connected to the structure under test. Figure 9.5 shows the most commonly used methods for attaching an accelerometer to a structure. Each inevitably introduces compliance at the interface, giving rise to rigid body mounting resonances of the accelerometer heaving and rocking on finite contact stiffnesses. If one considers the accelerometer as a single degree of freedom system moving rigidly in its sensing direction then its sensitivity will be modified in accordance with the transmissibility across the interface, as shown in Figure 3.23. The useable bandwidth is therefore further restricted to about 20% of the mounting resonance frequency.

The stiffest contact and hence the widest measurement bandwidth (typically more than 10 kHz) is obtained using a mechanical connection such as a mounting stud screwed into a blind tapped hole in the structure, as shown in Figure 9.5a. The mounting stiffness can be maximised by ensuring that the surfaces are flat, smooth and clean, and by lubricating the contact with a coupling fluid such as silicone grease to ensure intimate and uniform contact. The stud should not bottom out in the mounting hole of either the accelerometer or the structure, so as to ensure that transmission occurs across the accelerometer–structure interface. The disadvantage of the stud mount is that the structure needs to be drilled and tapped at each accelerometer location to accommodate the stud. Alternatively, one can avoid permanent modifications to the structure by bonding an adapter to the surface as shown in Figure 9.5b. Cyanoacrylate or similar adhesive is recommended since only a very thin (stiff) layer is required for a strong

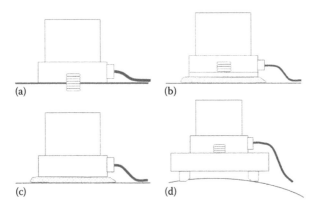

Figure 9.5 Various methods of attaching an accelerometer to a structure. (a) Stud mount; (b) adapter bonded with adhesive; (c) direct attachment using wax, double-sided tape or adhesive; and (d) dual-rail magnetic mount.

bond and the surfaces can be readily cleaned with solvent. Accelerometers without a mounting thread can be bonded directly onto the structure, as shown in Figure 9.5c. Where high bandwidth is not required, wax or double-sided tape may be used with caution. Wax is not suitable for high-temperature environments. Magnetic mounts are convenient for measurements on ferromagnetic surfaces at low frequencies, typically well below 1 kHz. Figure 9.5d shows a dual-rail magnetic mount which is designed for curved surfaces such as pipes.

Some accelerometer manufacturers provide experimental data to illustrate the effects of the mounting technique on an accelerometer's frequency response. These should be interpreted qualitatively since the mounting resonances are dependent on, for example, the type and thickness of the adhesive or wax used.

9.3.4 Signal conditioning

Piezoelectric force gauges and accelerometers share common signal conditioning problems. The sensing element acts as a current source and so has a high electrical impedance. The voltage observed by an oscilloscope or data acquisition system can be significantly reduced by the capacitance of the interconnecting cable resulting in very poor signal to noise. A charge amplifier, which has a low output impedance, must therefore be employed and connected to the sensor via a special low-noise cable, as illustrated in Figure 9.6a. The cable should also be taped down to minimise so-called triboelectric noise caused by electrostatic charges arising from the flexing of the cable. Standard coaxial cable can then be used to transmit the signal from the charge amplifier to the acquisition system.

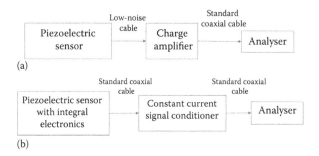

Figure 9.6 Connecting an accelerometer to a frequency analyser. (a) A charge mode sensor with remote electronics and (b) a voltage mode sensor.

Increasingly common, except in high-temperature environments, is the use of sensors with built-in electronics which eliminates the need for relatively costly charge amplifiers and any low-noise cabling. The integral amplifier requires a constant current supply provided by a signal conditioner. The output signal from the sensor is in the form of a modulation on the power supply line voltage from which the direct current (dc) component is removed. This configuration is shown in Figure 9.6b. Note that some frequency analysers have internal signal conditioning to allow sensors to be connected directly without the need for a separate signal conditioning unit. Sensors with integral electronics are known under various trademarked names but are sometimes termed *voltage mode sensors*, in contrast to charge mode sensors that require an external charge amplifier.

Piezoelectric transducers are intrinsically dynamic devices and cannot measure dc. The reason for this limitation is that the charge on the electrodes resulting from an applied force leaks away exponentially at a rate that is dependent on the *discharge time constant* of both the transducer and of the coupling circuit, if alternating current (ac) coupled. Due consideration may be required when measuring below a few hertz.

9.3.5 Sensitivity and calibration

Assuming again a focus on measuring FRFs, there are then many purposes for which calibration is unnecessary. FRFs, in units of volts per volt, are perfectly adequate for estimating natural frequencies and modal damping parameters, for example. However, for other applications such as driving point mobility measurements or dynamic stiffness measurements of viscoelastic materials (see Table 9.1), then measurements are required in engineering units. Calibration values are then required that specify the electrical output per unit response quantity. In the case of voltage mode transducers with integral electronics, calibration values are specified in terms of volts per engineering unit (force or acceleration). For charge mode transducers the calibration values are specified in terms of picocoulombs per engineering unit and the overall sensitivity is obtained by multiplying by the charge amplifier gain in volts per picocoulomb.

By way of an example, a measured accelerance FRF in volts per volt can be converted to engineering units as follows:

$$\text{Accelerance in } \frac{\text{ms}^{-2}}{\text{N}} = \frac{S_f}{S_a} \times \text{accelerance in } \frac{\text{V}}{\text{V}} \quad (9.4)$$

or equivalently

$$\text{Accelerance in dB } re \ 1\frac{\text{ms}^{-2}}{\text{N}} = \text{accelerance in dB } re \ 1\frac{\text{V}}{\text{V}}$$

$$+ 20\log_{10}\left(\frac{S_f}{1\,\text{V/N}}\right) \quad (9.5)$$

$$- 20\log_{10}\left(\frac{S_a}{1\,\text{V/ms}^{-2}}\right)$$

where S_f and S_a are the sensitivities of the force gauge and the accelerometer in units of volts per newton and volts per metre per second squared, respectively.

The sensitivity of a particular transducer is stated on its calibration certificate and this will deviate from the nominal value for that model of transducer within a tolerance band. It should be noted, however, that sensitivities can change with age and damage and so periodic recalibration is advisable, especially to check that it is functioning properly and exhibiting a flat frequency response.

One calibration method particularly worthy of mention in the context of this chapter involves a shaker test, as discussed in Section 9.4, on a calibrated steel mass m_{cal} suspended as a pendulum. Far above the resonance frequencies of the swinging pendulum, the response is governed simply by Newton's second law and the magnitude of the acceleration is known to be the reciprocal of the mass. It follows that

$$\text{Accelerance of structure in dB } re \ 1\frac{\text{ms}^{-2}}{\text{N}}$$

$$= \text{accelerance of structure in dB } re \ 1\frac{\text{V}}{\text{V}} \quad (9.6)$$

$$- \text{accelerance of calibration mass in dB } re \ 1\frac{\text{V}}{\text{V}}$$

$$- 20\log_{10}\frac{m_{\text{cal}}}{1\,\text{kg}}$$

A transducer will, due to manufacturing imperfections, inevitably produce a small electrical output due to motion or force in directions that are orthogonal to its primary sensing direction. The *cross-axis sensitivity* quantifies the maximum sensitivity in the orthogonal directions as a percentage of the sensitivity in its primary sensing direction. While typically only by a few per cent at most, significant contamination of the measurement can occur when, for example, motion in an orthogonal direction is much larger than in the primary sensing direction.

9.4 USING AN ELECTRODYNAMIC SHAKER

The electrodynamic shaker is by far the most versatile and widely used type of shaker for FRF measurements. This section outlines its working principle including factors that limit its range of operation, its interaction with the structure under test and the selection and usage of commonly available excitation signals.

9.4.1 Working principle

An electrodynamic shaker produces magnetically induced forces by passing a current through a coil suspended in a magnetic field, as also exploited in loudspeakers. The current may be alternating, to generate a harmonic force, or any other waveform with a desired spectrum (although, as discussed later in the section, the spectrum of the actual input force will also depend on the dynamics of the shaker and other factors). A schematic of a typical small shaker is shown in Figure 9.7. The moving part of the shaker, or armature, comprises a coil wound around a thin-walled tube or *coil form* and connected to the protruding part, the table. The armature is constrained to move only in the vertical direction, as depicted, by flexures which are flexible in bending but stiff in-plane. A permanent magnet is configured to produce a radial magnetic field in a small air gap through which the coil can pass orthogonally to the flux lines. The induced magnetic force is given by

$$f_s(t) = BLI(t) \tag{9.7}$$

from which it is clear that large forces require a proportionately long coil wire L, a high current I or a large magnetic field strength B. Commercially available shakers typically range from 10 N to more than 100 kN peak force and, unsurprisingly, their weight and cost also vary by orders of magnitude. Permanent magnets are commonly used for small shakers whereas electromagnets are used for large shakers.

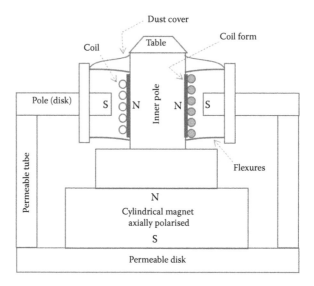

Figure 9.7 Schematic of a typical small electrodynamic shaker.

When selecting a shaker for a particular application there is the additional requirement of bandwidth. At low frequencies, a shaker's armature vibrates as a rigid body on its flexures and is effective at transmitting the electromagnetically induced force to the structure under test (with notable exceptions that are discussed later in Section 9.4.2.2). However, at a frequency much higher than its rigid body resonance the armature exhibits an elastic resonance of its own, above which transmission of force is highly attenuated. This resonance tends to be lower for larger shakers and so small shakers have a higher bandwidth of operation.

The peak sinusoidal acceleration A_{peak} deliverable by a shaker is limited by its maximum travel X_{peak} in accordance with

$$A_{peak} = \omega^2 X_{peak} \tag{9.8}$$

Consequently, a large shaker with a long stroke is required to deliver high acceleration at low frequencies.

9.4.2 Structure–shaker dynamics

An electrodynamic shaker is often considered to be the most controllable and repeatable means of exciting a structure to obtain reliable and accurate measured FRFs. The waveform, duration and magnitude of the excitation force can be almost arbitrarily chosen to achieve an input spectrum of the desired shape, frequency resolution and mean-square amplitude. It should

Figure 9.8 Electromechanical system comprising structure, shaker and associated components.

be borne in mind, though, that the electromechanical system formed by connecting the structure, the shaker and its associated components is complex. The equipment chain, shown in Figure 9.8, can affect not only the fidelity of the measurements but also, more profoundly, what it is that is being measured. This section discusses the dynamics of the various links in the excitation chain from which some vital principles emerge on how to configure the test equipment.

9.4.2.1 Force generation

The desired waveform for the excitation force is prescribed by a low-power control voltage from a signal generator. This signal needs to be amplified by a power amplifier of which there are two types. A voltage mode amplifier has a low output impedance and provides an output signal for which the voltage is proportional to the input voltage. A current mode amplifier has a high output impedance and provides an output signal for which the current is proportional to the input voltage. The advantage of the latter is in ensuring that the spectral content of the force that is generated exactly matches that of the control signal. This is not the case when a voltage mode amplifier is used, since the velocity of the shaker's coil through the magnetic field generates a back electromotive force that reduces the flow of current. The effect is most marked at resonances causing a 'drop-out' in the generated force. In the case of a current mode amplifier, the output voltage increases to maintain a constant current.

9.4.2.2 Force transmission

It is an occasionally held misconception that the force generated by the shaker is the same as that transmitted to the structure. This is not true, and it has repercussions on the accuracy of measurements, particularly at resonance, and how we choose to measure force, as discussed in Section 9.4.2.3.

Consider a single degree of freedom representation of a shaker with suspension stiffness k_s, moving mass m_s and viscous damping coefficient c_s. In the vicinity of one of its modes of vibration the structure can be similarly represented by an equivalent stiffness k, mass m and viscous damper c. Figure 9.9a shows the two systems coupled, that is to say that they share

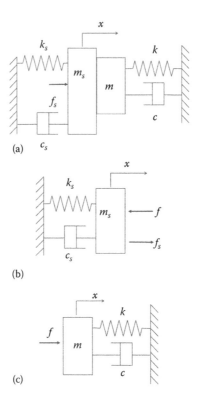

Figure 9.9 (a) Single degree of freedom representations of a shaker and a structure coupled together; (b) free-body diagram for shaker; and (c) free-body diagram for structure.

the same displacement $x(t)$. The electromagnetic force that is generated, $f_s(t)$, is given by Equation 9.7. The free-body diagrams for the shaker's mass and the structure's mass are shown in Figure 9.9b and c which depict an equal and opposite internal force $f(t)$ at the interface. What is required is the receptance of the structure when uncoupled from the shaker which is given by the transfer function between its displacement and the interfacial force:

$$\frac{X}{F} = \frac{1}{-\omega^2 m + i\omega c + k} \tag{9.9}$$

and exhibits a natural frequency, in angular units, of $\sqrt{k/m}$. The transfer function with respect to the *generated* force on the other hand is given by

$$\frac{X}{F_s} = \frac{1}{-\omega^2(m + m_s) + i\omega(c + c_s) + (k + k_s)} \tag{9.10}$$

which exhibits a natural frequency of $\sqrt{(k+k_s)/(m+m_s)}$ corresponding to that of the coupled structure–shaker system.

It is further informative to look at the force transmissibility from the shaker to the structure, which is obtained by dividing Equation 9.10 by Equation 9.9:

$$\frac{F}{F_s} = \frac{-\omega^2 m + i\omega c + k}{-\omega^2 (m+m_s) + i\omega(c+c_s) + (k+k_s)} \qquad (9.11)$$

If it were not for the dynamics of the shaker (i.e. if $m_s = k_s = c_s = 0$) then the force transmissibility would of course be unity at all frequencies. Given these dynamics, though, there is a peak in the force transmissibility close to the natural frequency of the coupled structure–shaker system, $\sqrt{(k+k_s)/(m+m_s)}$, as illustrated in Figure 9.10. The response of the coupled system is largest at this frequency and so too is the interfacial force between the structure and shaker.

When the coupled structure–shaker system is driven at the natural frequency of the uncoupled structure, that is, $\sqrt{k/m}$, then the coupled system is being driven off-resonance. Much of the force supplied by the shaker is consumed by accelerating the shaker's own moving mass and little is transmitted to the structure, especially if lightly damped. As a result, the measured force signal may be noisy, resulting in a poor FRF estimate in the narrow frequency band often of most interest – the resonance of the structure under test.

A measured force transmissibility from a practical case study is presented in Section 9.4.4, and the reader is referred to Wright and Skingle (1997)

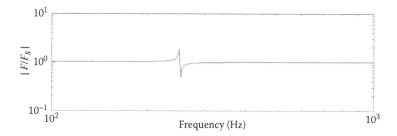

Figure 9.10 Magnitude of force transmissibility from shaker to structure for the single degree of freedom system shown in Figure 9.9a. The peak corresponds to the natural frequency of the coupled structure–shaker system whereas the dip corresponds to the natural frequency of the structure when uncoupled from the shaker.

and McConnell and Varoto (2008) for further detailed discussions on structure–shaker interaction effects.

9.4.2.3 Force measurement

For the purposes of system identification, it is required to measure the response to a *known* input, from which FRFs and subsequently modal properties can be identified. Equation 9.7 might misleadingly suggest that the input force can be obtained from the measurement of the current passing through the shaker's coil and, if calibrated FRFs are required, knowledge of coefficient BL. However, this results in the FRF of the coupled structure–shaker mechanical system, as illustrated in Section 9.4.2.2. Even more enticing is the prospect of adopting the source voltage signal as representative of the input force. The FRFs would then also incorporate the frequency response of the power amplifier. In the case of voltage mode amplifiers, the back electromotive force mentioned in Section 9.4.2.1 would lead to an overestimate of the force near resonance.

It is imperative for most purposes of FRF measurement to measure the input force directly with a force gauge. The gauge should be mounted directly to the structure so as to quantify the force transmitted to the structure under test. Even then, some modification to the structure is inevitable due to the mass of the gauge itself. The top mass shown in Figure 9.2 should be attached to the structure, since it is generally lighter than the base mass. The FRF between the response and interface force is thus that of the structure coupled to the top mass of the gauge. It should be noted that so-called mass loading of the structure by the force gauge also occurs orthogonally to the principal sensing direction, and inertia loading arises from the rotation of the structure at the forcing point.

9.4.2.4 Supporting the structure

The frequency response of a structure depends on its boundary conditions, and so a structure will sometimes need to be tested *in situ* in its operating environment to obtain fully representative results. However, this is not always practical or necessary. Where the boundary conditions in service are variable and uncertain then *in situ* testing may even be undesirable, at least in the first instance. Characterising the dynamic behaviour of the structure or component under known and repeatable boundary conditions and without extraneous sources of vibration can provide valuable insight. Where the purpose of the test is to validate a theoretical model, it is advantageous to choose boundary conditions that are comparable with those readily achievable in the model. In this respect, fixed boundary conditions can be problematic because a perfect constraint in translation or rotation is impossible to achieve in practice. Furthermore, an imperfect constraint

Vibration testing 461

is inherently prone to experimental unrepeatability. Consequently, structures are most commonly resiliently mounted, for example suspended on bungees or placed on isolation mounts. Rigid body modes are introduced which are dependent on the rigid body inertia of the structure and the stiffness of the supports. Provided these modes occur at low frequencies (less than 20% of the first elastic mode is a commonly used rule of thumb), then the natural frequencies of the elastic modes will be as if the structure is entirely free of supports.

9.4.2.5 Supporting the shaker

Over the intended frequency range of operation, a shaker can be represented as a two degrees of freedom system, as shown in Figure 9.11, in which the armature mass and base mass (typically 100 times larger) are coupled by the stiffness of the flexures. The shaker may be supposedly fixed (commonly mounted in a trunnion) or freely suspended, in which case the shaker support stiffness, shown in Figure 9.11, is either very large or very small. The support stiffness affects the shaker dynamics but this does not alter the true FRF between the structure's response and the force measured at the structure–shaker interface, as discussed in Section 9.4.2.2. In this respect, it does not matter whether the shaker is fixed or free. A potential disadvantage of fixing the shaker is that the reaction force on the supporting bench or floor can be inadvertently transmitted to the structure through flanking paths. Subsequent measured responses are then due to multiple inputs, only one of which is measured and assumed causal to the response. Resiliently mounting the shaker serves to isolate

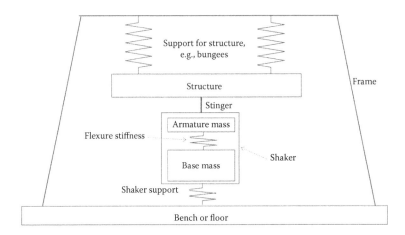

Figure 9.11 Schematic of a freely suspended structure connected to a shaker. The shaker can be fixed or freely suspended as represented by either a stiff or a flexible shaker support spring.

unintended transmission paths from the shaker to the structure. The need to isolate flanking paths is usually more acute when the structure under test is not resiliently mounted.

9.4.2.6 Connecting the shaker to the structure

The purpose of the shaker is to provide excitation in a single translational degree of freedom so as to facilitate the measurement of FRFs. A mechanical coupling is required between the structure and the shaker to transmit force effectively in the direction of motion of the armature. The mechanical system becomes that of the structure, coupling mechanism and shaker assembly, that is to say, the measurement instrumentation has modified the structure. In the direction of excitation, the structure under test is uncoupled from the shaker by virtue of measuring the interface force, as discussed previously. However, coupling still exists in the other five degrees of freedom. For example, should the method of connection enable rotation of the structure at the attachment point to be transmitted to the shaker, then the structure will experience the shaker's inertia. This problem is alleviated by selecting a coupling device, commonly referred to as a stinger, which is stiff axially but flexible in all other degrees of freedom. An effective stinger often takes the form of a straight steel rod of about 1 mm in diameter and 3–30 cm in length. Alternatively pretensioned piano wire can be used, although this prohibits free suspension of the structure or shaker.

An additional complication is the existence of resonances of the stinger itself. Axial resonances are problematic in that they cause a drop-off in the force spectrum with a corresponding loss of signal to noise. Bending resonances of the stinger, which occur at much lower frequencies than axial resonances, can also be excited, especially if the stinger is damaged or misaligned. Shifted and additional resonances in the measured FRFs can result. An additional reason for uncoupling the motions of the shaker and structure through the use of a stinger is to reduce moments on the force gauge which can cause damage to the sensing element. Studies on the effects of stingers on measured FRFs are reported in, for example, Cloutier and Avitabile (2009) and McConnell and Cappa (2000).

9.4.3 Excitation signals

The principal advantage of an electrodynamic shaker is that one can choose an arbitrary and repeatable excitation signal, thereby prescribing the spectrum of the input force, albeit filtered by the dynamics of the shaker and amplifier. The signal is usually broadband, requiring FFT-based spectral analysis to be performed on the excitation and response signals to determine the FRF. Given that the FFT implicitly assumes signal periodicity, it is perhaps counterintuitive that excitation signals chosen for vibration testing are not necessarily *periodic* but can be *transient* or even *random*. The distinctions between these three classes of signal are important to appreciate since they require different signal processing decisions of the user regarding averaging and windowing. Provided the structure under test is excited so as to behave linearly, its true frequency response is actually independent of the choice of excitation signal. The latter is usually chosen for reasons such as signal-to-noise ratio, or simply prevailing habit. Where a structure is behaving nonlinearly the measured frequency response *will* depend on the type and level of excitation signal, some being ideal for investigating nonlinear behaviour while others conveniently average out (weak) nonlinearities. This section discusses the use of harmonic and more generally periodic, transient and random signals for the purposes of FRF vibration measurement.

9.4.3.1 Harmonic

The simplest form of excitation signal is a harmonic (single frequency) waveform. It is the author's opinion that preliminary shaker tests should always be carried out harmonically and the signals monitored on an oscilloscope to check that the instrumentation is functioning correctly. For the purpose of measuring FRFs over a range of frequencies, however, manual setting of the excitation frequency is not practical and must be automated by the analyser. The so-called stepped-sine signal dwells at each of many user-specified frequencies in turn. Before acquiring data at a particular frequency, the analyser must first wait for steady state to be established, as illustrated in Figure 9.12 for the simple case of a single degree of freedom system responding to the switching on of a time-harmonic force. The acquired signals are effectively periodic (harmonic in this case) provided

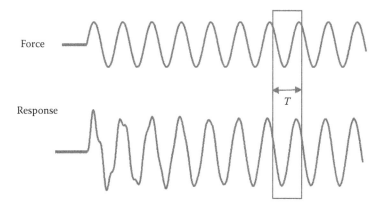

Figure 9.12 Transient response of single degree of freedom system to the switching on of a time-harmonic force of period *T*.

that the initial cycles are discarded. The relative amplitudes and phases of the output and input signals are then obtained without recourse to the FFT before automatically proceeding to the next frequency. Despite automation this process is time consuming, especially for lightly damped structures where the steady state can take many cycles to reach. The principal reason for selecting the stepped-sine technique is that it is usually the only appropriate way of investigating nonlinearities, as opposed to broadband signals that tend to average out nonlinear behaviour. An additional benefit worth noting is that non-uniform frequency spacing can be specified to achieve fine resolution near resonances and coarse resolution off-resonance.

9.4.3.2 Periodic

Broadband signals are the excitations of choice for most FRF measurements, for ease and speed, and they require application of the FFT to obtain the spectra of the excitation and response signals. The premise of the FFT is that all signals concerned are periodic. Periodicity is easily achieved in forced vibration by exciting a structure with any signal of finite duration T which is then 'looped' *ad infinitum*. Initially, the input and response signals are contaminated by start-up transients as previously described for harmonic excitation. These decay due to damping of the structure and shaker, culminating in steady-state periodic signals that can be acquired and fast Fourier transformed. Of fundamental importance is that the duration of data acquisition is exactly the same as (or a multiple of) the intrinsic period of the signals, otherwise the acquired data may not be assumed to be periodic. This condition is assured when using the source signal generated by the frequency analyser, and a rectangular window should be used on all channels. The FRFs between response and input channels can be estimated simply from the ratio of the Fourier transforms of one segment of data. Alternatively, cross-spectral FRF estimation can be used, such as the H_1 or H_2 estimator (see Section 4.6.3.1), to obtain an averaged estimate over a number of data segments. Strictly speaking, averaging is not required for periodic signals but it is very often beneficial to reduce the influence of extraneous measurement noise or persisting transients. Note that both the H_1 and H_2 estimators are identical to the ratio of the Fourier transforms when only one average is acquired.

A popular choice of periodic signal is a rapid sine sweep between specified lower and upper frequencies which is then looped. This excitation signal is referred to here as a *chirp* (although this term is sometimes used instead to describe the *burst chirp* signal discussed in Section 9.4.3.3). A chirp signal of 2 s duration sweeping from 1 to 400 Hz is illustrated in Figure 9.13a. The actual input force to the structure is, as always, modified by the dynamics of the shaker system, as is apparent from Figure 9.13b. A corresponding acceleration response is shown in Figure 9.13c. Note that the

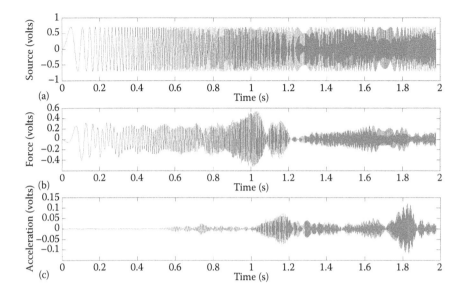

Figure 9.13 Example of a single chirp excitation which can be looped to obtain steady-state periodic response. (a) Source signal, (b) force signal and (c) response signal.

instantaneous response is large as the excitation frequency passes through resonances. The crest factor (ratio of peak to root-mean-square amplitude) of the response can be large for lightly damped structures and care may need to be exercised to avoid clipping the response signal or inducing non-linear behaviour.

Another widely used periodic signal for vibration testing is a finite random sequence which, when looped, results in what is commonly referred to as a *pseudorandom* signal. The random sequence can, for example, be generated by taking the inverse FFT of a uniform or shaped spectrum with an appropriate bandwidth and randomly generated phase. As is the case for other periodic excitation signals, the steady-state response should be reached before data acquisition is initiated.

9.4.3.3 Transient

A transient excitation signal that is zero except for an initial burst will now be considered. This is illustrated in Figure 9.14a for the particular case of a *burst random* signal lasting 30% of the segment. Of interest is the transfer function between the measured force and the response signals shown in Figure 9.14b and c. (Note that the force signal again differs from the source signal due to the dynamics of the system.) These signals represent a single transient event as opposed to a steady-state periodic behaviour and so the

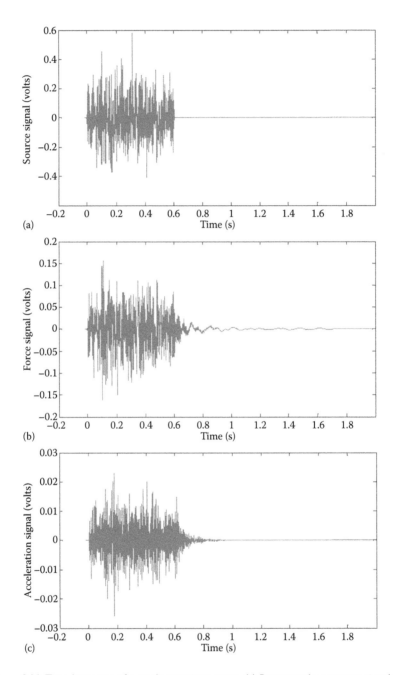

Figure 9.14 Time histories of a single transient event. (a) Burst random source signal, (b) burst random force signal, (c) burst random response signal, (d) burst chirp source signal, (e) burst chirp force signal and (f) burst chirp response signal.

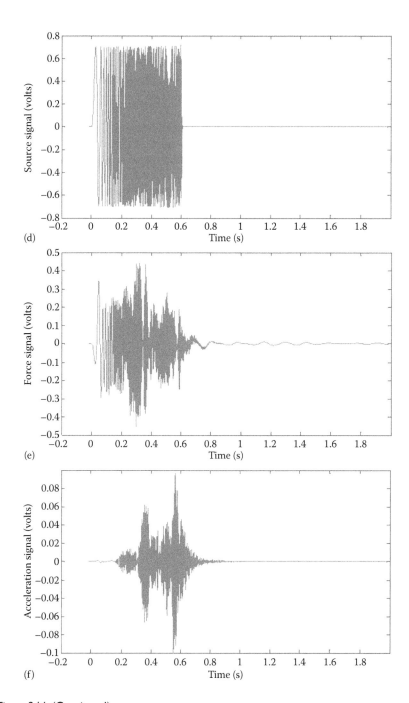

Figure 9.14 (Continued)

periodicity requirement of the FFT would appear not to be met. In fact, it is fully met provided that the whole transient event is acquired such that the input and response signals are zero (and have zero gradient) at the start and end of the segment. The data can then be legitimately taken as representative of the periodic response, were the event actually to be looped. A rectangular window should be used provided that the segment length T is sufficient for the decaying response, or *ring-down*, to be captured fully. Conversely, an excessively long segment length compared with the duration of the transient event is detrimental to the signal-to-noise ratio.

Unlike the pseudorandom signal, the spectrum of a burst random signal is not sufficiently uniform to ensure adequate excitation at all frequencies and so averaging over multiple events is required using a different burst random signal each time. This is actually a defining benefit of the signal when wishing to average out (weakly) nonlinear behaviour, in that the input spectrum encapsulates different random magnitudes and phases for each event and should result in a better linearised FRF than the pseudorandom technique of Section 9.4.3.2.

In a manner analogous to burst random excitation, a transient version of the chirp signal described in Section 9.4.3.2 can also be formed by arranging for the chirp excitation to occupy only a proportion of the segment, with the remainder at zero, to allow for the response to ring down to practically zero. A typical excitation signal is shown in Figure 9.14d and the corresponding measured force and response signals are shown in Figure 9.14e and f. In contrast to the periodic chirp signal, the burst chirp does not require steady state to be attained prior to data acquisition, and unlike the burst random signal it does not, in principle, demand averaging.

Hammer testing is discussed in detail in Section 9.5. However, it is appropriate at this stage to note that the impulse due to a hammer tap is also a transient signal and the preceding discussion relating to periodicity, averaging and windowing applies. Commercial analysers may also offer an impulse-type excitation signal for shaker testing, although it should be noted that the waveform of the actual input force will differ significantly from the source signal due to the dynamic characteristics of the shaker and the test item.

9.4.3.4 Random

The premise of the DFT permits spectral analysis of any signal that is genuinely periodic or can be thought of as such because the inputs and outputs are entirely contained within the data segment. It is, therefore, counterintuitive that one might choose a continuous random excitation signal which contravenes this basic condition. However, a random excitation is sometimes chosen for the same purpose as a burst random excitation, that is, to average out weakly nonlinear behaviour over tens or hundreds of averages.

The problem of aperiodicity is circumvented by resorting to a non-rectangular window, as described more fully in Section 4.3.3. A Hanning window is commonly chosen so as to remove discontinuities between the ends of the segment that arise from the periodicity assumption of the FFT process, which would otherwise result in spectral leakage.

By way of a simple example to illustrate windowing and averaging of a random signal, Figure 9.15a shows a random waveform of length 300 samples which might represent an excitation signal or a corresponding response-time history. Then one may choose to divide it into three segments of 100 samples each. By applying a Hanning window to each segment, resulting in Figure 9.15b, an FFT can be performed on each segment in turn without discontinuities arising across the ends. One can see that by assigning a low weighting to the data near the start and end of each segment the information therein is largely discarded. Alternatively, one could divide the time history into overlapping segments, as shown in Figure 9.15c through f, so as to 'recycle' the available data to produce more (but no longer independent) estimates for averaging. Four averages are chosen here for illustrative purposes, corresponding to an overlap of one-third of a segment. An overlap factor of two-thirds is commonly used since it weights the available data uniformly (McConnell and Varoto 2008).

The coherence function, equal to the ratio of H_1 to H_2 (see Sections 4.6.2 and 4.6.3.1), is widely regarded to assess the extent of the measurement noise on the transfer functions based on periodic, transient and random excitation signals alike. However, its puritanical use is in the context of random signals as a statistic to quantify the causality between the output and input of a supposedly linear system, and hence an indicator of nonlinear behaviour. Note, though, that for random excitation poor coherence may arise in the vicinity of resonances for a reason unrelated to nonlinearity. When using short data segments, the acquired response in one segment due to excitation in preceding segments can be nonnegligible, particularly for lightly damped structures. Selecting a finer frequency resolution will improve coherence at resonances if this is the cause (see Section 4.7.5.2).

9.4.4 Case study

In this section, measured results from a shaker test are presented to illustrate some of the issues discussed previously. The structure under test is an aluminium aerofoil section of chord 1 m with a span of 0.5 m. The structure was suspended on elastic bungees from a rigid frame and the instrumentation set-up was as depicted in Figure 9.1a. A small electrodynamic shaker was attached to the aerofoil via a short stinger and the input force was measured with a force gauge. *A priori* knowledge of the mode shapes from a preliminary hammer test is useful in placing the shaker so as to avoid nodal lines of modes in the frequency range of interest. The

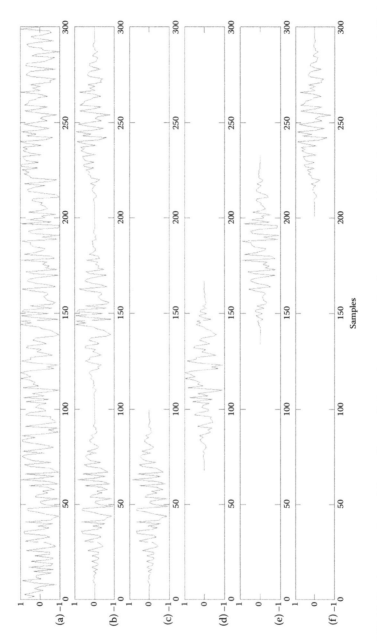

Figure 9.15 (a) Random waveform, (b) waveform split into three nonoverlapping segments with Hanning window applied and (c) through (f) waveform divided into four overlapping averages (33% overlap) with Hanning window applied.

shaker was driven by a voltage mode power amplifier and the source signal was provided by a frequency analyser. The response of the structure was measured at a position remote from the excitation point by a small voltage mode accelerometer. The signals from both transducers were connected to the analyser via a signal conditioning amplifier. Periodic, transient and random excitation signals were employed in turn.

Figure 9.16 shows the FRF magnitude (uncalibrated) and phase for periodic versions of chirp and random excitations. The results are presented for the first data segment, that is, before steady state is reached, in which

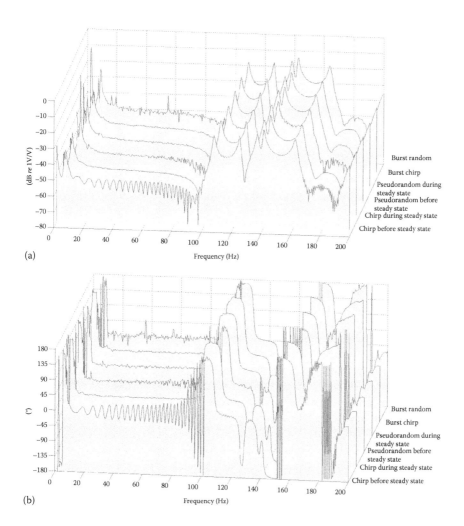

Figure 9.16 Comparison of measured FRFs using chirp, burst chirp, pseudorandom and burst random excitation signals (no averaging). (a) Magnitude and (b) phase.

the effects of initial transients are clearly visible, and for the last of many segments when steady state is assured. The latter appear to be satisfactory without averaging, as theory suggests. Also shown in Figure 9.16 are the results for a single burst chirp and a burst random transient event. The burst chirp result is satisfactory owing to its reliably flat input spectrum but the burst random result clearly requires averaging. Note that there are several resonance peaks below 10 Hz which correspond to rigid body modes of the structure vibrating on its supports. The first elastic mode of the structure itself occurs at about 100 Hz and so the rule of thumb quoted in Section 9.4.2.4 for emulating free–free boundary conditions is met in this case.

There appear to be some rapid fluctuations in phase at, for example, about 180 Hz. On closer inspection these are fluctuations of about 360° and so represent deviations of only a few degrees due to noise. This is a consequence of the *wrapped* phase representation that constrains the phase to lie between ±180°. The FRF presented can be clearly identified as a transfer FRF. Were it a driving point FRF then antiresonances would occur between every pair of neighbouring resonance peaks (see Section 3.6.3).

Figure 9.17 shows the FRF magnitude for a random excitation where 1, 10 and 100 averages have been used. One average (i.e. no averaging) is clearly inadequate, as expected from the discussion in Section 9.4.3.4. The distinction between the results for 1, 10 and 100 averages is better seen as a polar plot in the complex plane (see Section 3.6.3), as shown in Figure 9.18 for the narrow frequency range around the first elastic mode of the structure. Clear lobes are apparent after 10 averages but they are more 'circular', notwithstanding the limited frequency resolution, when 100 averages are used. The H_1 and H_2 estimators produce identical phases at all frequencies but different magnitudes ($H_1 = \gamma^2 H_2$) and so yield slightly differently shaped lobes. The H_1 estimator has underestimated the response at resonance (which leads to an overestimate in damping) due to its bias with respect to noise on the input signal (see Section 4.6.3.1). With the notable exception of resonances, the response signals are usually more susceptible to noise, caused by extraneous excitations, and so the H_1 estimator is more often the default choice.

The coherence is also shown in Figure 9.17b. By definition the coherence is unity without averaging and is shown here only to draw attention to its lack of meaning. The results for 10 and 100 averages show dips not only at resonance but also at antiresonance frequencies where the H_2 estimator is biased due to noise on the response signal.

In Section 9.4.2.2 it was argued, through analysis of a single degree of freedom system, that the force generated by the shaker differs from that transmitted to the structure and that the latter should be measured in order to obtain the correct FRF for the structure alone. To illustrate the

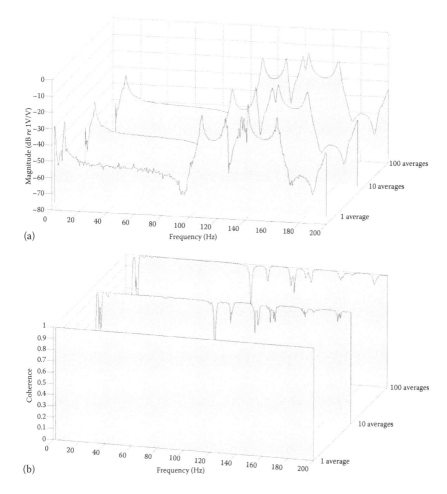

Figure 9.17 Random shaker test using 1, 10 and 100 averages (no overlap). (a) FRF magnitude and (b) coherence.

effect in practice a driving point accelerance of the structure was measured incorrectly by taking the voltage across a resistor in series with the shaker coil (proportional to coil current) as a measure of the input force. The magnitudes of the correct and incorrect FRFs are shown in Figure 9.19a. The peaks in the incorrect FRF correspond to the natural frequencies of the coupled structure–shaker system and are generally lower than those for the correct FRF, since the shaker is behaving in its mass-controlled region above 50 Hz or so. The effect is typically slight for the fundamental modes but more marked at higher frequencies, where the response is generally more sensitive to localised structural

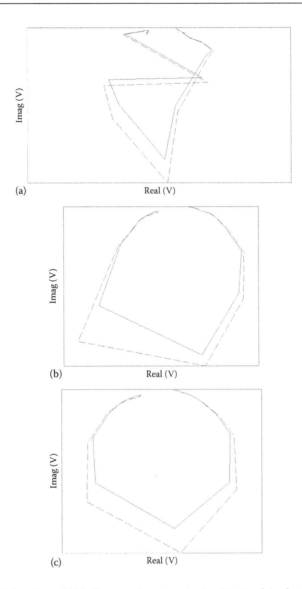

Figure 9.18 Polar plots of FRFs from random tests in the vicinity of the first elastic body resonance. H_1 estimator (solid line), H_2 estimator (dashed line). (a) 1 average, (b) 10 averages and (c) 100 averages.

modification. Figure 9.19b shows the ratio of the magnitudes of the directly measured (transmitted) and indirectly measured (generated) forces. Respective dips and peaks occur close to the natural frequencies of the uncoupled structure under test and the coupled structure–shaker system.

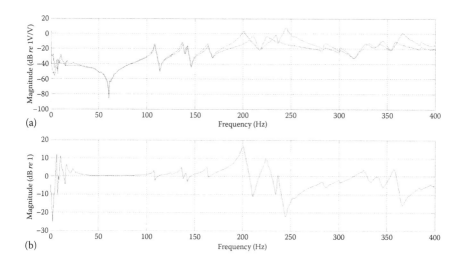

Figure 9.19 Comparison between directly measured and indirectly measured force. (a) Magnitude of point accelerance: —, force measured directly with force gauge; ·······, force measured indirectly. (b) Magnitude of force transmissibility (the transfer function between directly and indirectly measured force signals).

9.5 USING AN INSTRUMENTED HAMMER

Electrodynamic shaker testing is often the method of choice for measuring FRFs where control over the spectrum and duration of the input is desirable. However, the equipment required is not readily portable and requires appreciable electrical power which may not be convenient or possible for field testing. Significant time and expertise is also required to set up the instrumentation correctly and check for anomalies. By comparison, a hammer test requires minimal equipment (see Figure 9.1), is normally quick to set up and perform, affords greater access to awkward locations and can be reasonably trouble-free. This section discusses the principle of operation of impact hammer testing, highlights some of its limitations and intends to convey the fundamental know-how to avert flawed measurements.

9.5.1 Force spectrum

A single tap to a structure imparts a force similar to that of a half-sine pulse, as shown in Figure 9.20a, where, generally, the duration of the contact T_c will be very short compared with the resulting transient response. The amplitude spectrum of a half-sine pulse is shown schematically in

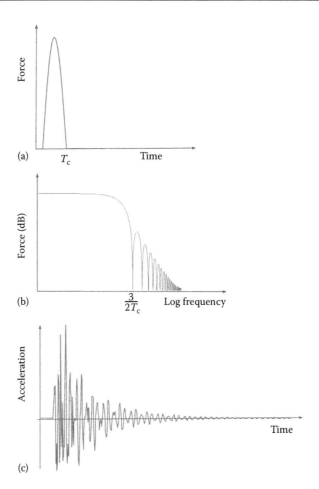

Figure 9.20 Schematic showing idealised force and corresponding structural response due to a hammer tap. (a) Time history of input force, (b) amplitude spectrum of input force and (c) time history of response.

Figure 9.20b and is uniform at low frequencies, that is, where the time period is much longer than the impulse. The level at 0 Hz is defined by the time integral of the force. Notches in the spectrum occur at a series of harmonically related frequencies that are inversely related to the contact time T_c. The first notch occurs at $3/2\, T_c$ for the idealised case of a half-sine pulse.

Ideally, one would arrange for T_c to be sufficiently short that the bandwidth of interest lies within the flat region of the spectrum in order to ensure adequate excitation at all frequencies. Conversely, too short an impact will spread the input force thinly over an unnecessarily wide bandwidth to the detriment of the signal-to-noise ratio. The contact duration depends on the impedance of both the structure and the hammer tip, the latter being more readily controllable. Commercially available hammers are usually supplied with interchangeable tips, ranging from a soft rubberlike tip for a low bandwidth to a stiff metal tip for a wide bandwidth.

The size of hammer chosen should obviously reflect the amplitude of force required, bearing in mind the structure under test and its susceptibility to nonlinear response. Small hammers have delicate force gauges with a lower peak force capability. They are lighter, however, which allows for shorter impulse durations for higher-frequency applications.

Figure 9.20c shows schematically the response of a structure to an impulsive force. High frequencies decay rapidly due to the passing of many cycles of vibration, whereas lower frequencies are seen to ring on for longer. Provided that the complete transient event is captured then the force and response signals can be considered periodic and their spectra obtained via the FFT without fear of leakage, as discussed in Section 9.4.3.3.

9.5.2 Triggering

In contrast to most shaker tests, the excitation due to a hammer tap is not initiated by the analyser, which therefore needs to know when the event is about to start in order to begin acquiring data. Starting too early may cause truncation of the transient response and introduce leakage; starting too late will most likely result in complete failure to capture the impact on the force channel. It is commonplace for analysers to trigger the start of the acquisition by monitoring when either the force or one of the response channels exceeds a threshold value, or *trigger level*. A typical force signal is shown in Figure 9.21 for which the polarity of the force gauge is positive for compression. The trigger level can usually be set in absolute terms in either volts or engineering units, or as a percentage of full scale. It is also possible to impose the additional condition that the slope of the

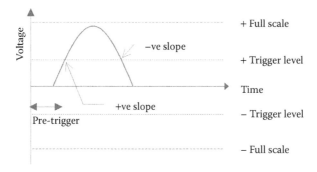

Figure 9.21 Schematic of waveform of force due to hammer tap showing trigger parameters.

signal is either positive or negative. As illustrated in Figure 9.21, a positive slope would trigger more closely to the start of the event. The trigger level should be low enough to trigger from a typical hammer tap but not too low so as to trigger accidentally. The optimal trigger level can be expected to change during a test, for example when the hammer tip is interchanged or any amplifier gains altered. When triggering instead from a response channel, the trigger level may also need to be varied according to the position of the sensor.

By the time data acquisition has been triggered, part of the event has already been missed and must be retrieved from a buffer. A pre-trigger, typically in milliseconds or a percentage of the segment length, should be specified so as to recover the signals from all channels to just prior to the start of the event. Note that some analysers instead request the user to prescribe the delay between triggering and starting to acquire data, which should of course be negative. A visual inspection of the measured time histories is vital to check the correct setting of the pre-trigger.

9.5.3 Windowing

Another compelling reason for previewing the acquired time histories is to check visually that the responses have decayed practically to zero within the segment length. Rectangular windows can then be used on all channels without incurring leakage problems. The required ring-down time is case dependent and is easily established by trial and error. The maximum duration that can be measured depends on the maximum FFT size permitted by the analyser and the sampling frequency which is dictated by the measurement bandwidth. Where these restrictions prevent the complete event from being acquired, then one may resort to using an *exponential window* on the response channels, which in its simplest form is given by

$$w(t) = e^{-t/\tau} \tag{9.12}$$

where τ is a time constant chosen by the user to be sufficiently short that the windowed response signals are practically zero by the end of the data segment.

The effect of the window is to add artificial damping, thereby increasing the modal damping ratios by an amount of $1/\tau\omega$. The true structural damping can be estimated by subtracting this frequency-dependent quantity from each measured modal damping value. Note that the subtraction of significant artificial damping from noise-prone modal damping estimates of lightly damped structures can prove nonrobust. This compensation is also only strictly valid when an exponential window is applied to the response and force signals. Instead, analysers often advocate a *transient rectangular* or *force window* for the force channel in conjunction with an exponential window on the response channels. The force window is unity until after the impulse is deemed to have finished and zero thereafter. The intention is to zero-weight the signal, which is inevitably contaminated by measurement noise to some small extent, when the true force is known to be zero.

While some structures are problematic for hammer testing on account of their long transient decay, too brief a ring-down can pose a different problem. A fine frequency resolution requires a long segment length (since $\Delta f = 1/T$) and this may tempt one to measure far beyond the end of the transient event. However, the resulting transfer function will (i) be estimated from response signals which are near zero and therefore noisy for the most part; and (ii) the frequency resolution appears fine, but does not genuinely resolve close frequency components. This is an implicit limitation of hammer testing when compared with shaker testing, although multiple impacts can be used with caution to lengthen the transient event, as explained in the following section.

9.5.4 Averaging

The response and force signals due to a single hammer tap are transient, but can legitimately be considered as periodic for the purposes of FFT analysis as explained in Section 9.4.3.3. Averaging is not strictly required and the transfer functions can be reliably estimated simply by taking the ratio of their Fourier transforms rather than via their averaged spectra. The observed signals obtained by the transducers, however, may be contaminated by random noise, in which case a small number of averages may be beneficial. Coherence, which is meaningless (always unity) in the absence of averaging, may then be helpful in quantifying statistical similarity between averages.

One particular application of averaging in hammer testing is where multiple taps occur unintentionally due to the structure rebounding onto the hammer tip. Figure 9.22a shows an idealised representation of (the initial part of) a segment featuring two taps a duration T_{gap} apart, where $T_{gap} \ll T$. The corresponding spectrum is shown in Figure 9.22b. A main lobe is caused by the finite duration of the impulse, as discussed in Section 9.5.1, onto which are superimposed many smaller lobes of width $1/T_{gap}$. The numerous notches in the force spectrum can cause poor signal-to-noise ratio at these discrete frequencies and it is usual practice to discard such a measurement. Although not ideal, where multiple taps are persistent or unavoidable then averaging can be employed on the reasonably safe assumption that the time separation between multiple taps, and hence the frequencies of the notches in the input spectrum, will differ from one average to another. Multiple taps can also be employed intentionally to obtain arbitrarily long transient

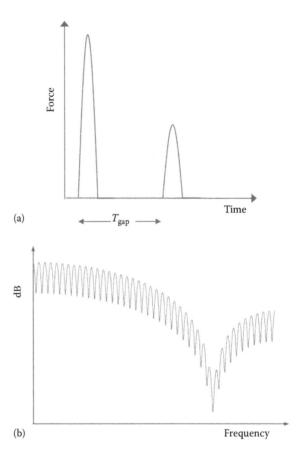

Figure 9.22 A double tap. (a) Schematic of time history and (b) resulting spectrum.

events, and hence finer frequency resolution, at the expense of necessitating averaging.

9.5.5 Case study

The following case study concerns an instrumented hammer test of the same freely suspended aerofoil structure as presented in Section 9.4. The instrumentation set-up was as depicted in Figure 9.1b. A segment length of 2 s was chosen to capture completely the decaying transient response, resulting in a frequency resolution of 0.5 Hz. Acquisition was triggered on the force channel with a trigger delay of −1% of the segment length. FRFs were measured for three different hammer tips (hard, soft and very soft) from a single transient event, that is, no averaging. The auto power spectra of the impulses are shown in Figure 9.23a and the magnitudes of

Vibration testing 481

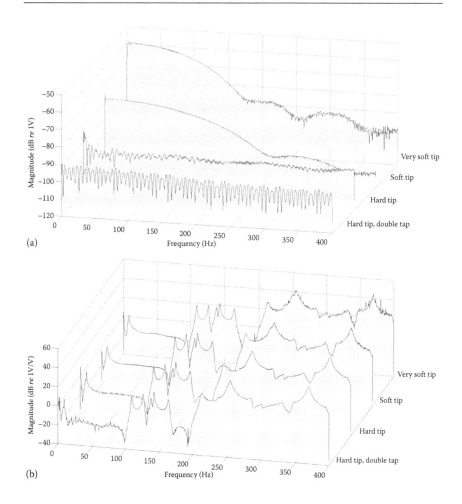

Figure 9.23 Effects of hammer tip and multiple taps on (a) the auto power spectrum of the input force and (b) the magnitude of the resulting FRF estimate.

the FRFs are shown in Figure 9.23b. The hard tip results in a flat input of relatively low level over the whole frequency range of interest: that is, except where a double tap is experienced, in which case there are numerous uniformly spaced notches in the input spectrum causing modest deterioration in the FRF in frequency regions of low response. The soft tip delivered a higher (20 dB or so) input at low frequencies at the expense of roll-off above about 200 Hz due to the longer contact duration. Some noise is apparent on the FRF at these frequencies. This problem is exacerbated when a very soft tip is used although excitation at low frequencies is boosted by a further 20 dB.

REFERENCES

Brandt, A. (2011) *Noise and Vibration Analysis*. John Wiley & Sons, Chichester.
Cloutier, D. and Avitabile, P. (2009) Shaker/stinger effects on measured frequency response functions, Proceedings of International Modal Analysis Conference XXVII, Orlando.
Ewins, D.J. (2000) *Modal Testing: Theory, Practice and Application*, 2nd edn. Research Studies Press, Baldock, England.
Halliwell, N.A. and Petzning, J.N. (2004) Vibration measurement techniques using laser technology: Laser vibrometry and TV holography, in Fahy, F. and Walker, J. (eds), *Advanced Applications in Acoustics, Noise and Vibration*, chapter 13, Spon Press, London.
Huntley, B.L. (1979) Electrohydraulic—The most versatile shaker, *Journal of Environmental Sciences*, 22(2), 32–35.
ISO (International Organization for Standardization) (1999) *ISO 10846-1 Acoustics and Vibration: Laboratory Measurement of Vibro-Acoustic Transfer Properties of Resilient Elements. Part 1: Principles and Guidelines*, International Organization for Standardization, Geneva.
McConnell, K.G. and Cappa, P. (2000) Transducer inertia and stinger stiffness effects on FRF measurements, *Mechanical Systems and Signal Processing*, 14(4), 625–636.
McConnell, K.G. and Varoto, P.S. (2008) *Vibration Testing Theory and Practice*, 2nd edn. John Wiley & Sons, New York.
Verheij, J.W. (1997) Inverse and reciprocity methods for machinery noise source characterization and sound path quantification. Part 1: Sources, *International Journal of Acoustics and Vibration*, 2(1), 11–20.
Wright, J.R. and Skingle, G.W. (1997) On the direct or indirect measurement of force in vibration testing, *Proceedings of the Society of Photo-Optical Instrumentation Engineers*, 3089(2), 1360–1370.

Index

Absorber, *see* Sound absorber; Vibration absorber
Absorption, *see* Sound absorption
Accelerometer, 3
   frequency range, 450–451
   mounting, 450
   principle, 448–450
Acceptability criteria, 314, 363
Acoustic short circuit, 245
Adiabatic process, 22, 30
Aeroacoustics sources, 221
Aerodynamic noise, 221
Aliasing, 162
Analogue-to-digital conversion (ADC), 160, 169–170
Annoyance, 355–360
Anti-resonance, 128–130
Aperiodic signals, 167
Apertures, 265, 268
Articulation index (AI), 347
Attenuation, geometric, 225
Attenuators, duct, 282
Auditory system
   anatomy, 312
   basilar membrane, 317
   capabilities, 322
   cochlea, 313, 316–317
   detectability, 334
   external ear, 314
   frequency response, 328
   function, 312
   hearing range, 322
   inner ear, 317–321
   just noticeable difference, 324
   masking, 318
   middle ear, 315
   nerves, 321
   perception, 314
   temporal response, 320
   thresholds, 322
Autocorrelation function, 179–183, 198–199
A-weighting, 216–217, 330, 426

Bandwidth, half-power, 102
Barrier diffraction, 266–269; *see also* Noise barriers
Base excitation, *see* Oscillator
Beam bending vibration, 134–139
   characteristic equation, 135
   modes, 136
   natural frequencies, 136
   wave dispersion, 139, 241
   wavenumber, 138, 241
   wave phase speed, 139
Bessel function, 43, 55
Biodynamic
   models, 367
   response, 367, 378, 389, 402
Bivariate processes, 176
Boundary
   conditions, 29, 37, 57
   element, 55
Building vibration, acceptability criteria, 363–384

Cancellation phenomenon, 246
Central limit theorem, 174
Cepstrum, 152
Characteristic
   frequency, *see* Natural frequency
   function, *see* Mode
   specific acoustic impedance, 27
Chirp, 159, 464

Chi-squared distribution, 200–201
Coherence function, 191, 205–208
Coincidence
  condition, 244
  diagram, 255
  phenomenon, 255
  transmission, 256–257
Comfort, *see* Equivalent comfort contours
Complex exponential representation (CER), 16, 82
Compressibility, 17
Confidence intervals, 197–198, 202–206
Conservation
  energy, 31
  mass, 29
  momentum, 21
Control volume, 19
Convection, 21
Convolution, 111, 152–154, 158, 200
Covariance (correlation) function, 176, 179–181, 184–189
Cross spectrum, 188–189

Damped oscillator, *see* Oscillator
Damped spring element, *see* Oscillator
Damper, viscous, *see* Oscillator
Damping, 107
  coefficient, 95
  control, 101, 257, 260, 286
  critical, 98
  hysteretic, 107
  mechanical loss factor, 107, 294–297
  radiation loss factor, 280
  ratio, 96, 287
  treatment, 298–299
  viscous, 95, 107, 273
Data analysis
  categorisation, 146
  decimation, 167–169
  generation models, 146
  methods, 147
  quantisation, 169
  reduction, 146, 167
  truncation, 154, 164, 166
  windowing, 154–157
dB(A), *see* A-weighting
Decibel (dB), 13
  arithmetic, 16
Deconvolution, 152

Degrees of freedom, 87, 116; *see also* Multimode vibration
Delay effect, 151, 159
Diffraction, *see* Barriers
Digital signal processing (DSP), 145
Dipole
  compact, 44
  directivity, 47
  flow speed dependence, 221
  force on fluid, 47, 218–219
  sound field, 46
  sound power, 42
  source geometry, 45
  source strength, 46
Dirac delta function, 147
Directivity, *see* Sound field
Discrete Fourier transform (DFT), 164–168
Dispersion, wave, 242
Dissipation loss factor, *see* Damping
Divergence theorem, 36
Duct sound attenuators, *see* Attenuators
Duhamel integral, *see* Convolution
Dummy head, 419
Dynamic, *see* Vibration absorber

Ear, *see* Auditory system
  artificial, 418
Effective Perceived Noise Level (EPNLdB), 332
Efficiency
  acoustic, 218, 220, 222
  radiation, 43
Enclosure
  absorption, 68–71
  acoustic natural frequencies, 61–62
  close-fitting, 266
  diffuse field model, 63, 66
  direct field, 70, 71
  modal density, 71
  mode count, 62
  noise control, 265
  reverberant field, 70
  reverberation time, 70
  sound power balance, 68
Energy
  acoustic, 30–31
  conservation, 31
  density, 30
  vibrational, 85, 92

Ensemble average, 177, 181
Equal loudness contours, see Loudness
Ergodic process, 181
Error
   least square, 193
   mean square, 197
   quantisation, 169
Estimation errors
   bias, 196
   coherence bias, 206
   FRF, 206
   random, 201–202, 206
Euler's equation, 21
Exciters, see Vibration exciters
Experimental design and methods, 3
Exposure duration, see Sound
      Exposure level (SEL)

Far field, see Sound field
Fast Fourier transform (FFT), 167–169
Feedback 185
Field incidence, see Sound reduction
      index
Filters
   analogue, 170
   anti-alias, 163
   digital, 152–153
   frequency, 215–217
Flanking transmission, 438
Flat plates, see Plates
Flow resistance, see Sound absorption
Fluid
   density, 19
   pressure, 10–11
   properties, 19
   temperature, 17, 22, 26
Fourier
   analysis, 147, 165–166, 207
   integral transform (FIT), 149, 165–166
   series analysis, 150
   transform, see Fourier analysis
Free vibration, see Oscillator
Frequency
   angular, 13
   coincidence, 244
   critical, 242
   damped oscillator resonance, 98, 236
   folding (Nyquist), 162
   instantaneous, 159
   mass-air-mass resonance, 260
   resolution, 158, 202

   response of enclosure, 60
   Schroeder (large-room), 71
   standard bands, 216
Frequency response function (FRF), 83, 190, 206
   characteristics, 127
   methods of measurement, 482
Fresnel number, 444

Gaps, see Apertures
Gas, equation of state, 22
Gaussian (normal) distribution, 174–176
Gauss integral, see Sound intensity
      measurement
Glass
   critical frequency, 243
   sound transmission, 243
Glass-fibre (wool), see Porous materials
Green's function
   free-space (harmonic), 41

Hair cells, 317–321
Hammer, instrumented, 447
Hand-arm vibration, see Human
      vibration
Hand-transmitted vibration, 363–365, 389–392, 395–399
   preventative methods, 401
Harmonic
   motion and frequency, 11
   wave representation, 24
Hearing
   damage risk 343, 427
   permanent threshold shift, 344
   spatial, 342
   temporary threshold shift, 344
   threshold curve, 14
Helmholtz equation, 26
Helmholtz resonator, 280
   impedance, 26
   radiation loss factor, 280
   resonance frequency, 280
Hilbert transform, 158
Human response to sound
   behavioural aspects, 311, 353
   loudness, 322–328
   objectives and standards, 311
   uncertainties, 311
Human vibration
   biodynamics, 366
   body impedance, 366

body transmissibility, 366
buildings, 384
comfort, 369
discomfort, 369–371
dose value (VDV), 380, 386
effects, 369–379
hand-arm, 390, 399
hand-transmitted, 389, 396
health, 365–372, 379–383, 392–395, 399
human perception, 369, 384, 390
measurement, 364
muscular effects, 395
preventative methods, 401
Raynaud's syndrome, 373
ride comfort, 385
seat dynamics, 384
seat transmissibility, 367, 386
vascular disorders, 392–395
weightings, 370–373
white finger, 392–396
whole body, 363–365

Impact
  control, 236
  frequency spectrum, 236
  reduction, 234
  time history, 237
Impedance
  beam, 84
  characteristic acoustic, 26
  characteristic specific acoustic, 27
  concept, 288
  Helmholtz resonator, 26
  layer, 276–277, 283
  mechanical, 84, 299
  mismatch, 290
  specific acoustic, 16
  surface, 278
  tube acoustic, 284
Impulse response, 152–153
  damped SDOF oscillator, 11
Insertion loss, 266–267, 293
  barrier, 236, 266–269
  close cover, 265
  noise control enclosure, 265
Isolation, vibration, *see* Vibration isolation

Large room (Schroeder) frequency, *see* Frequency
Laser Doppler velocimeter (LDV), 447

Leak detection, 439
Linearity and superposition, 27
Local reaction, *see* Sound absorption
Longitudinal wave in solid, *see* Rod
Loss factor, *see* Damping
Loudness, 322, 326–328

Machinery
  safety directive, 213, 382
Masking, 319, 347–351
Mass conservation, 20
Mass, effective, 92
Mass law, *see* Partitions
Mass-spring system (2DOF), *see* Oscillator
Material properties, 243
Microflown, 436–437
Microphone
  array, 421–425
  condenser, 412
  directional, 419
  MEMS, 417
  probe, 417–418
Mineral wool, *see* Sound absorption
Mobility
  definition, 84, 288
  examples, 289
  plate, 290
Modal
  analysis, 443–444, 482
  density of enclosure, 61, 71
  overlap factor, 296, 299
Mode
  beam, 136
  count, 62
  plate, 246
  rod, 133
  tube, 57–60
Modified wave equation, 273
Momentum equation, *see* Euler
Monopole
  intensity field, 42
  point, 40–41
  sound power, 41
  source strength, 41
Motion sickness, 387
Multimode vibration
  damped forced response, 124
  matrix notation, 119–123
  multiple degrees of freedom (MDOF), 116
  undamped forced response, 123–124

undamped free vibration,
  116–119
Multipole source model, 44
Music, 322, 341–345, 350–351

Near field *see* Field
Nearfield acoustic holography
  (NAH), 426
Noise
  aircraft, 325, 332
  annoyance descriptors, 355
  at-work regulations, 335, 343–344
  background, 339, 347–348, 351
  community, 337
  complaints, 356
  criteria (NC), 331
  effects on people, 343
  environmental, 337
  exposure, 323, 343
  impact, 234
  measurement, 191–195
  rating (NR) curves, 332
  road traffic, 325, 339, 350, 353
  white, 146, 184
Noise control, 4, 7
  at-source, 233–235
  barriers (screens), 236, 266–269
  constraints, 214
  damping, 234–237, 253–260
  enclosures, 265
  impacts, 234, 266, 288
  principles, 232–237
  strategies, 227
  targets, 227
  transmission paths, 236
Noise sources
  categories, 219
  localisation, 421–426
  quantification, 227
  sound power measurement, 227
Nonlinearity, 175
Normal surface derivative
Nyquist sampling criterion, *see* Aliasing

Orthogonality principle, 121
Oscillator (SDOF)
  free vibration, 88–92, 95–97
  frequency response to base
    excitation, 106
  frequency response to harmonic
    force, 98–101
  FRF, 101–103, 112

impulse response, 108–110
parameter estimation, 97–98, 103
pendulum, 92–95
periodic excitation, 112
polar plot, 103
random excitation, 138
rotating unbalance, 106
viscously damped, 95, 102
Oscillator (2DOF), 116–117

Panel absorber, *see* Sound absorbers
Parseval's theorem, 185
Particle
  acceleration, 411
  velocity, 19, 23, 27, 35, 39, 46, 47
Partitions, *see also* Sound reduction
    index
  double-leaf, 258–261
  single-leaf, 251–254, 254–258
  sound transmission measurement,
    262–265
  practical considerations, 261–262
Perforated sheets, 250, 279
Phase
  instantaneous, 159
  shifter, 158
  spectrum 188, 205
Phasor, *see* CER
Piston
  excitation of fluid in a tube, 56, 19,
    29, 33
  sound radiation, 54
Plane wave, 17, 25, 29, 31
  equation, 19
  harmonic form, 24
  impedance, 31, 50
  intensity, 32
  in three-dimensions, 35
  in tubes, 22
  in two dimensions, 43
Plastic foam, *see* Sound absorption
Plate, flat
  bending, 240
  bending wave modes, 246
  bending wavenumber, 242
  critical frequency, 243
  modal density
  radiation, 243
  radiation efficiency, 241
Porous materials, *see* Sound absorption
Potential energy, *see* Energy
Power, *see* Sound power

Power spectral density, 186, 190, 198–199
Preamplifier, 415
Pressure
  atmospheric, 10
  –density relation, 21–22
  gradient, 36
  sound, 10
Probability
  density function, 172
  distribution, 169, 174, 202
  theory, 170

Quadrupole source
  lateral, 48–51
  longitudinal, 48
Quantisation, see Data
Quasi-longitudinal wave, see Rod

Radiation
  efficiency (ratio), 237
  loss factor, 280
  index, 239, 250
  reduction, 247
Radiation from vibrating
  pistons, 54
  plates, 241–247, 249–250
  spheres, 239
Random
  burst, 465
  excitation of SDOF system, 138
  excitation (testing), 468
  processes, 195
  pseudo, 465
Rapid eye movement (REM), 353
Rapid speech transmission index (RASTI), see Speech intelligibility
Rayl (unit), 27
Rayleigh integral, 53–55, 239
Reciprocity, vibrational, 127
Reflecting plane, source near, 52
Reflex, acoustic, 317
Reverberation; see also Enclosures
  time, 70, 285
Ride value, 371, 373
Riemann form, 151
Rod
  axial (longitudinal) vibration, 131–134
  equation, 132
  mode shapes, 133

  natural frequencies, 133
  stiffness, 89
Room acoustics, see Enclosure

Sabine formula, 70
Sample space, 171
Schroeder (large room) frequency, 71
Sea sickness, see Motion sickness
Seat effective amplitude transmissibility (SEAT) 385–387
Seating dynamics, 364–370, 376–384
Sensory conflict theory, 388
Signal analysis; see Digital signal processing (DSP); data analysis
Signal processing, see Digital signal processing (DSP)
Signals, categorisation, 146
Single-degree-of-freedom system, see Oscillator (SDOF)
Single event level, see Sound exposure level
Single input–single output system, 152, 189
Sleep disturbance, 322–355
Smearing, 154, 156, 201
Software, 7
Sound absorbers
  Helmholtz resonator, 280
  microperforated sheets, 282
  panel, 280
  perforated cover sheet, 279
Sound absorption, 269
  air, 230
  air cavity, 278
  flow resistance, 273
  impedance surface, 270
  locally reactive surface, 269
  measurement methods, 283–286
  modified wave equation for porous materials, 273
  plane surface of uniform impedance, 270
  porous material parameters, 272–275
  random (diffuse) incidence, 68–69
Sound exposure level (SEL), 337
Sound field
  diffuse, 63, 66
  direct, 70
  enclosed, 60–63
  far, 226
  free, 225, 228
  near, 46

Index 489

reverberant, 70
spherically symmetric, 37
3D statistics, 37
Sound intensity, 30–34
  applications, 238
  definition, 32
  Gauss integral, 438
  instantaneous, 432
  in interference fields, 58
  level (SIL), 34
  measurement, 432–436
  monopole field, 41
  one-sided, 66
  in plane wave fields, 32
  source sound power determination, 42, 438
  transient, 432–434
Sound level, see Sound pressure level
Sound level meter (SLM), 426
  usage, 430
Sound measurement, 411
Sound power
  determination from intensity, 42, 438
  dipole in free field, 47
  level (PWL), 34
  monopole in free field, 41
  monopole near hard reflecting surface, 50–53
  radiated into tube, 33
  reasons for determination, 227
  reference, 34
  typical levels, 34
Sound pressure, 10
  –intensity index, 435
  mean square, 438
  reference value, 14, 412
Sound pressure level (SPL)
  definition, 14
  equivalent continuous (L_{eq}), 336, 344
  weightings, 328–331
Sound radiation
  bending waves, 240–247
  reduction, 247–250
  simple point sources, 239
Sound reduction index; see also Transmission loss
  double-leaf partitions, 258
  measurement, 262–265
  single-leaf partitions, 250
Sound sources
  aeroacoustic, 220–221

categorization, 214
dipole, 44–46, 218
directivity, 224
frequency spectra, 223–224
line, 226
mechanical, 221
monopole, 40–44, 218
near reflecting surface, 50–53
quadrupole, 48, 218
speed dependences, 222
Spatial Fourier analysis, see Wavenumber spectrum
Speech
  communication, 331, 346
  intelligibility, 347–349
  interference, 346
  transmission, 347–348
Sphere
  oscillating, 47, 240
  pulsating, 41, 240
Splitter attenuator, see Attenuators
Stationarity, 179
Statistical energy analysis (SEA), 127
Stevens loudness level, 332
Stiffness, 85–89
Stress, 345, 353
Structure-borne
  paths, 233, 236
  sound, transmission, 233
  sources, 237
Superposition, 28, 37
System identification, 189

Tectorial membrane, 314–318
Threshold of hearing, see Hearing threshold curve
Time-frequency analysis, 158–159
Time weightings, 426, 428
Tinnitus, 344
Trace wavenumber, see Wavenumber
Transfer function, 205
Transfer path analysis (TPA), 444
Transmissibility
  base excitation of SDOF system, 106
  human body, 366
  seat, 367, 367, 385
  vibration isolator, 457
Transmission loss (TL), see Sound reduction index (SRI)
Tube, 55
  impedance, 283

Two-degree-of-freedom system (2DOF), *see* 2DOF Oscillator

Uncertainty principle, 157

Variance, 169, 174–178, 197–201
Vascular disorders, 922
Vibration
　absorber (neutralizer), 129–130
　control, 101, 286
　damping control, 101
　energy, 77–78, 92–95
　exciter, 455
　force transmission, 457
　high frequency, 292
　impedance, 84
　mass addition, 299–301
　stiffness addition, 299–301
Vibration testing
　actuators, 448
　averaging, 480
　calibration, 453
　case study, 480
　force generation, 457
　hammer excitation, 475–477
　reasons, 444
　sensors, 445–447, 453
　shaker use, 455
　triggering, 477
　windowing, 478
Viscoelasticity, 298
Volume
　source strength density, 41
　velocity, 41

Wave, *see* under specific type entries
Wave
　dispersion, 139
　energy, 30
　evanescent, 139
　standing, 55–60
Wave equation
　1D acoustic, 17, 19, 22, 26–28, 35
　2D acoustic, 43
　3D acoustic, 35–39
　spherically symmetric acoustic, 37
Wavefront, 24, 39
Wavelength, 24
Wavenumber
　acoustic, 24
　vector, 61
Wave phenomena
　diffraction, 236, 266
　dispersion, 139
　reflection 138–139
　scattering, 55–57
Wave speed
　beam, 139
　sound, 9, 17
　sound in water, 27
White finger, *see* Human vibration
　constitutional, *see* Raynaud's disease, 373
　vibration-induced, 396
Whole body motion, perception, *see* Human vibration
Wiener–Khinchin theorem, 186
Windowed records, *see* Data windowing
Work–energy, 80

For Product Safety Concerns and Information please contact our EU representative GPSR@taylorandfrancis.com
Taylor & Francis Verlag GmbH, Kaufingerstraße 24, 80331 München, Germany

www.ingramcontent.com/pod-product-compliance
Ingram Content Group UK Ltd.
Pitfield, Milton Keynes, MK11 3LW, UK
UKHW021445080625
459435UK00011B/376